THE ADAPTIVE ROLE
OF LIPIDS IN
BIOLOGICAL SYSTEMS

THE ADAPTIVE ROLE
OF LIPIDS IN
BIOLOGICAL SYSTEMS

NEIL F. HADLEY

Department of Zoology
Arizona State University
Tempe, Arizona

A Wiley-Interscience Publication

JOHN WILEY & SONS

New York   Chichester   Brisbane   Toronto   Singapore

Library of Congress Cataloging in Publication Data:

Hadley, Neil F.
  The adaptive role of lipids in biological systems.

  "A Wiley-Interscience publication."
  Includes index.
  1. Lipids.   2. Adaptation (Physiology)   I. Title.

QP751.H24   1985   574.19'47   84-17298
ISBN 0-471-09049-2

Printed in the United States of America

10 9 8 7 6 5 4 3 2 1

In memory of
Dr. and Mrs. Gordon Alexander

PREFACE

The advent of modern analytical instrumentation, coupled with the development of innovative techniques for separating and identifying lipid compounds, has produced an explosive growth in our knowledge of lipid biochemistry and the role of lipids in biological systems. Using these modern methods, scientists have shown the complexity that exists in the structure and composition of lipids. Moreover, these studies have greatly expanded our awareness of the diverse and often critical functions that can be attributed to this important class of organic compounds.

This book describes the uniqueness of lipids as a biological material and how they are used by plants, animals, and microorganisms in adapting to and functioning in a variety of physical environments. In Chapter 1, basic information is presented on the nomenclature and structure of lipids, along with pertinent physical and chemical properties of lipids that are relevant to subjects developed in subsequent chapters. Chapter 2 examines the origins, synthesis, and processing of lipids in plants and animals. Like Chapter 1, the broadly comparative treatment is designed to provide the reader with the background necessary to fully appreciate the material that follows. Each of the remaining chapters deals with the role of lipids in a specific biological function presented within the framework of lipid biochemistry, physiology, adaptation, and evolutionary biology. Within each chapter the biological problem is first defined in general terms and those features of lipids ideally suited to the solution of the problem are discussed before documenting specific adaptations in plants and animals. I have not attempted to include some of the more "clinically oriented" roles of lipids, except in those cases where the function has widespread biological importance.

I thank all my colleagues who graciously provided original figures for inclusion in the book and/or commented on specific portions of the text. I am especially indebted to Drs. Jeff Hazel and George Somero, who read the draft in its entirety and offered many constructive suggestions. Their strong endorsement of the worthiness of this effort was greatly welcomed. Many of the electron micrographs and artwork are original; Greg Hendricks' superb technical help in their preparation is gratefully acknowledged. I also want to thank the editorial and production staff at John Wiley & Sons for their assistance and cooperation, and Trev Leger, Editor, for his counsel and support. Finally, I apologize to my family—Mary, Cindy, David—for the many hours taken from their company. Their unfailing understanding and encouragement was a source of inspiration.

NEIL F. HADLEY

Tempe, Arizona
September 1984

CONTENTS

THE ADAPTIVE ROLE
OF LIPIDS IN
BIOLOGICAL SYSTEMS

1

LIPIDS: CLASSIFICATION, STRUCTURE, AND PROPERTIES

DEFINITION

Lipids are a large and diverse group of important organic compounds in living organisms. Unlike the other major groups of biomolecules, lipids are better characterized by their physical properties than by their chemical structure or function. With few exceptions, lipids are generally insoluble in water but are soluble in organic solvents such as chloroform, ether, benzene, or ethanol. Lipids derive this distinctive property from long hydrocarbon chains that form a major portion of their structure. These hydrocarbon moieties, which may be unbranched or branched, may form carbocyclic rings, and may obtain unsaturated linkages, are also the basis for the "fattiness" or "oiliness" associated with many lipid compounds (Davenport and Johnson, 1971). The terms "fats" and "oils" (if the substance is a liquid at room temperature), however, should not be used interchangeably with the term "lipids." Fats refer to a specific class of lipids, the triacylglycerols, which are major components of depot or storage lipids in plant and animal cells.

CLASSIFICATION

Although many classification schemes have been proposed for lipids, none is universally accepted. The grouping of lipid classes selected for use in this book follows the system presented by White et al. (1973) and includes all compounds discussed in the subsequent chapters. The system is based

1

on chemical composition, with the principal distinction being the presence or absence of glycerol as a backbone structure. The system also incorporates lipid molecules that complex with proteins and carbohydrates. The structure(s), nomenclature, and general occurrence of these lipid classes are presented in the following sections of Chapter 1. In addition, those physical and chemical properties that are primarily responsible for the unique function of a specific lipid group in biological systems are briefly summarized. A more detailed discussion of these properties and their relationship to the adaptive role of lipids in membrane function, waterproofing, energy storage, buoyancy, and so on, is included in the introduction to the chapters treating these subjects. The remainder of this chapter is organized as follows:

I. Fatty acids
II. Lipids containing glycerol
    a. Neutral fats
        1. Mono-, di-, and triacylglycerols
        2. Glycerol ethers
        3. Glycosylglycerols
    b. Phosphoglycerides
        1. Phosphatidates
        2. Diphosphatidylglycerols and phosphoinositides
III. Lipids not containing glycerol
    a. Sphingolipids
        1. Ceramides
        2. Sphingomyelins
        3. Glycosphingolipids
    b. Aliphatic alcohols
    c. Wax esters
    d. Hydrocarbons
    e. Terpenes
    f. Steroids
    g. Prostaglandins
IV. Lipids combined with other classes of compounds
    a. Lipoproteins
    b. Proteolipids
    c. Phosphatidopeptides
    d. Lipoamino acids
    e. Lipopolysaccharides

FATTY ACIDS

Fatty acids consist of a hydrocarbon chain with a terminal acidic carboxyl group. Although small quantities of "free" or "unesterified" fatty acids are present in the cells and fluids of living organisms, the majority of fatty acids occur linked to alcohols, such as glycerol, or to long-chain bases (see sphingolipids). Most naturally occurring fatty acids are straight-chain compounds with an even number of carbon atoms. Odd-numbered carbon atom fatty acids are occasionally found, as are fatty acids with hydrocarbon chains that are branched or that contain a variety of functional groups including cyclopropane and cyclopropene rings. These more complex and unusual fatty acids are generally found in plants and microorganisms.

Fatty acids can be divided into two main classes: saturated and unsaturated. Saturated fatty acids have the general formula

$$CH_3—(CH_2)_n—COOH$$

and are usually named after the hydrocarbon from which they are derived. In naming the acid, the "e" in the hydrocarbon is changed to "oic" in the acid. Thus, the saturated fatty acid derived from the hydrocarbon decane, $CH_3—(CH_2)_8—CH_3$, is known as decanoic acid. Many fatty acids also have trivial or nonsystematic names that were established before a standardized nomenclature was adopted. Some of these trivial names (e.g., palmitic acid, stearic acid) have become so firmly entrenched in fatty acid chemistry that they will never likely be replaced by the more meaningful systematic names. Table 1.1 lists the trivial and systematic names of some saturated fatty acids of biological importance, along with notes on their occurrence.

TABLE 1.1
Some common saturated biological fatty acids

Molecular formula	Systematic name	Trivial name	Occurrence or source
$C_2H_4O_2$	Ethanoic	Acetic	Vinegar, nettle sting
$C_3H_6O_2$	Propanoic	Proprionic	Rumen bacteria
$C_4H_8O_2$	Butanoic	Butyric ⎫	Mainly in milk fats; minor components in coconut and palm oil, and some seed fats
$C_6H_{12}O_2$	Hexanoic	Caproic ⎪	
$C_8H_{16}O_2$	Octanoic	Caprylic ⎬	
$C_{10}H_{20}O_2$	Decanoic	Capric ⎭	
$C_{12}H_{24}O_2$	Dodecanoic	Lauric ⎫	Coconut and palm kernel oils; myristic minor component of animal lipids
$C_{14}H_{28}O_2$	Tetradecanoic	Myristic ⎭	
$C_{16}H_{32}O_2$	Hexadecanoic	Palmitic	Most widely occurring saturated acid in fats and oils
$C_{18}H_{36}O_2$	Octadecanoic	Stearic	Widespread, but less common than palmitic
$C_{20}H_{40}O_2$	Eicosanoic	Arachidic ⎫	Major components in some seed oils
$C_{22}H_{44}O_2$	Docosanoic	Behenic ⎬	
$C_{24}H_{48}O_2$	Tetracosanoic	Lignoceric ⎭	
$C_{26}H_{52}O_2$	Hexacosanoic	Cerotic ⎫	Major components of some plant waxes; beeswax; predominant unesterified fatty acids in scorpion cuticle
$C_{28}H_{56}O_2$	Octacosanoic	Montanic ⎬	
$C_{30}H_{60}O_2$	Tricontanoic	Mellistic ⎭	

Unsaturated fatty acids may contain from one up to six double bonds. The simplest unsaturated fatty acids, those containing only one double bond (monoenoic), have the general formula

$$CH_3(CH_2)_m CH{=}CH(CH_2)_n COOH$$

and are named by replacing the "e" of the corresponding unsaturated hydrocarbon with "oic" (i.e., decene becomes decenoic acid). Fatty acids with two or more double bonds (polyenoic or polyunsaturated) are named by taking the stem of the name of the corresponding hydrocarbon and adding "-dienoic" (two double bonds), "-trienoic" (three double bonds), and so forth. The double bond in the singly unsaturated fatty acids of most organisms is generally in the 9,10 position (the carboxyl carbon is C-1). However, because the double bond in a fatty acid of a given chain length may occur in several positions, its specific location must be designated in any description of the acid. Moreover, introduction of a double bond gives rise to the possibility of *cis-trans* isomerism. For example, there are two isomeric forms of 9-octadecenoic acid, each with a different trivial name—oleic (*cis*) and elaidic (*trans*):

$$H{-}\underset{\|}{C}{-}(CH_2)_7{-}CH_3 \qquad CH_3{-}(CH_2)_7{-}\underset{\|}{C}{-}H$$
$$H{-}C{-}(CH_2)_7{-}COOH \qquad\qquad H{-}C{-}(CH_2)_7{-}COOH$$

Oleic acid (*cis*)   Elaidic acid (*trans*)

Cis forms predominate in nature; *trans* acids are rare, but do occur naturally and may be nutritionally significant. Table 1.2 lists the trivial and systematic names, and general occurrence of several important unsaturated fatty acids.

To simplify discussion of fatty acids, a shorthand notation has been developed to designate the length of the carbon chain, and the number and position of the double bonds. Thus, palmitic acid, which contains 16 carbons and no double bonds, is symbolized by 16:0. Palmitoleic acid, which has 16 carbons and 1 double bond, can be symbolized as either $^{9-}16{:}1$ ($\Delta^9 16{:}1$) or 16:1 (*n*-7). The Δ^9 notation indicates that the double bond is located between carbons 9 and 10 counting from the carboxyl carbon. The expression "*n*-7" indicates that there are seven carbon atoms from the double bond to the terminal methyl group. The position of all double bonds in polyunsaturated fatty acids can also be determined from this notation if the fatty acid is unconjugated (i.e., the double bonds are separated by a single methylene group). For example, the shorthand designation for arachidonic acid is 20:4 (*n*-6). The notation indicates that the acid has 20 carbons and 4 double bonds, with the last double bond (furthest from the carboxyl carbon) located between carbons 14 and 15. The "*n*-*x*" nomenclature is preferred by biochemists with interests in fatty acid metabolism, as the acids that can be derived from one another by chain elongation (i.e., are metabolically related) can

TABLE 1.2

Some common unsaturated biological fatty acids

Molecular formula	Number of double bonds	Systematic name	Trivial name	Occurrence or source
$C_{14}H_{26}O_2$	1	cis-9-Tetradecenoic	Myristoleic	Marine animal oils
$C_{16}H_{30}O_2$	1	9-Hexadecenoic	Palmitoleic	Common in animal fats and some seed oils
$C_{18}H_{34}O_2$	1	cis-9-Octadecenoic	Oleic	Occurs in practically all fats; most common of all fatty acids
$C_{18}H_{32}O_2$	2	9,12-Octadecadienoic	Linoleic	Widespread in plant and animal tissues
$C_{18}H_{30}O_2$	3	9,12,15-Octadecatrienoic	Linolenic	Important component in plants; minor component in animals
$C_{20}H_{32}O_2$	4	5,8,11,14-Eicosatetraenoic	Arachidonic	Major components of animal lipids, especially phospholipids
$C_{22}H_{34}O_2$	5	7,10,13,16,19-Docasapentaenoic	Cluponodonic	
$C_{24}H_{34}O_2$	1	cis-15-Tetracosenoic	Nervonic	

be clearly identified (Gurr and James, 1975) (see Chapter 2). A lengthier description of structure must be used when double-bonded carbon atoms are adjacent to each other or when the double bonds are separated by more than one methylene group.

Properties

The physical and chemical properties of fatty acids are of particular interest because other lipid classes are either derived from or composed essentially of fatty acids. Thus, many of these properties are shared by other biological lipids. Coverage here is restricted primarily to properties relevant to function in organisms; for more extensive discussion one is referred to treatises by Deuel (1951) and Gunstone (1967).

(a) Melting Point. The melting point of fatty acids is one of several physical properties that vary according to the length of the carbon chain and the degree of unsaturation. Except for some of the very short-chain fatty acids, the melting point rises progressively as the chain length increases. Even- and odd-numbered chain saturated fatty acids with 10 or more carbon atoms are all solids at room temperature (ca. 25°C). Introduction of a double bond significantly reduces the melting point of a fatty acid. For example, palmitic acid (16:0) melts at approximately 63°C, whereas palmitoleic acid (16:1) melts between -0.5 and 0.5°C. The presence of additional double bonds further depresses the melting point. All biologically important unsaturated fatty acids are liquids at room temperature.

(b) Solubility. The solubility of fatty acids varies inversely with chain length. Acetic (2:0), propionic (3:0), and butyric (4:0) acids are completely miscible with water, but fatty acids with eight or more carbon atoms are virtually insoluble. Solubility in water increases markedly in fatty acid molecules that contain hydroxyl groups or that combine with metallic cations to form "soaps."

(c) Monomolecular Films. Although long-chain fatty acids dissolve only slightly in water, they can form a fine film over the surface of the water. The hydrophilic carboxyl groups apparently dissolve in water, while the hydrophobic hydrocarbon chains are oriented parallel to one another and perpendicular to the surface (Figure 1.1). Such films, which often are only one molecule thick, if comprised of saturated fatty acids can be compressed laterally until the area occupied by each molecule is 21×10^{-16} cm^2 (Gunstone, 1967). This value corresponds to the area occupied by a CH$_2$ group, and is used as evidence that the acyl chains are arranged perpendicularly. The thickness of the film depends on the different chain lengths of the constituent fatty acid molecules. This oriented arrangement is the foundation of the lipid component of all biological membranes (Needham, 1965), and

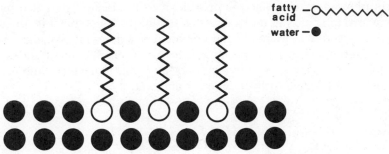

FIGURE 1.1. Diagrammatic representation of monomolecular film formed at the fatty acid/water interface.

also provides a basis for many of the observed enzymatic and protective functions associated with lipids at lipid-aqueous interfaces.

(d) Crystalline Structure and Polymorphism. Fatty acids crystallize in the solid state. Moreover, fatty acids with the same chemical composition may crystallize into more than one solid form (i.e., may be polymorphic). X-ray diffraction has shown that fatty acid molecules within a crystal unit are arranged in pairs, and that polymorphs differ in the way the paired molecules pack together (Deuel, 1951; Gunstone, 1967). Important biological properties related to variations in crystalline form include melting point, thermal transition points, heat and electrical conductivity, and compressibility. Crystalline polymorphs also occur in other long-chain aliphatic compounds including triacylglycerols, alcohols, and hydrocarbons.

(e) Density and Viscosity. The absolute density (i.e., g per cm^3) of fatty acids decreases as the length of the carbon chain increases. Except for acetic acid (2:0), all fatty acids are less dense than an equal volume of water; that is, they have specific gravities less than 1.0. The density of fatty acids further decreases as temperature is raised, but the change is nominal and may have little, if any, biological importance. Little difference is noted in the density of saturated versus unsaturated fatty acids of corresponding chain lengths when measured at the same temperature (Deuel, 1951).

Viscosity is the resistance to flow or to change in shape. Like density, it is frequently determined by comparison with water. The longer the chain length of the fatty acid, the greater its viscosity; at 70°C, the viscosities of palmitic acid and stearic acid are 7.8 and 9.9 times that of water (Deuel, 1951). The viscosity decreases as temperatures are raised, but the change is not linear.

(f) Chemical Reactivity. Fatty acids are representative of most biological lipids in that they are predominantly carbon and hydrogen compounds with oxygen occupying either one or very few sites on an otherwise large hydrocarbon molecule. The minimal degree of oxidation provided by such a composition results in molecules that exhibit both chemical inertness and chemical reactivity. Many of the chemical reactions involve the terminal carboxyl group. The most prominent and biologically important of these reactions is esterification, in which one fatty acid molecule combines with one alcohol molecule to form an ester and water:

$$CH_3-(CH_2)_n-C\overset{O}{\underset{OH}{}} + \overset{OH}{CH_2-(CH_2)_n-CH_3} \longrightarrow$$

Fatty acid   Alcohol

$$CH_3-(CH_2)_n-\overset{O}{\overset{\|}{C}}-O-CH_2-(CH_2)_n-CH_3 + H_2O$$

Ester

The above reaction is reversible and yields an acid and alcohol on hydrolysis. The carboxyl group of fatty acids, like other carboxylic acids, may be titrated with base to form a metallic salt or "soap." The commercial uses of the

$$CH_3-(CH_2)_n-C\overset{O}{\underset{OH}{}} + NaOH \longrightarrow CH_3-(CH_2)_n-C\overset{O}{\underset{O^-Na^+}{}} + H_2O$$

(Aqueous)   (Aqueous)

metallic soaps are well known; however, the extent to which they contribute to detergency and lubrication in biological systems is not clear.

The hydrocarbon chain of the fatty acid can also react chemically, especially if it contains unsaturated linkages or oxygen-containing functional groups. The addition of hydrogen to an unsaturated fatty acid can lead to partial or total saturation (e.g., linoleic acid, 18:2, to stearic acid, 18:0). Halogens such as chlorine, bromine, and iodine also combine readily with unsaturated acids to yield saturated compounds. Halogenation itself is of little direct biological importance, but it is the basis of the "iodine number method" for determining the number of double bonds present in a lipid molecule. Finally, the double bonds are often the site of cleavage during the metabolism of a fatty acid molecule.

LIPIDS CONTAINING GLYCEROL

Two biologically important groups of lipids, the neutral fats and phosphoglycerides, have glycerol as a structural backbone. Glycerol is a trihydroxy

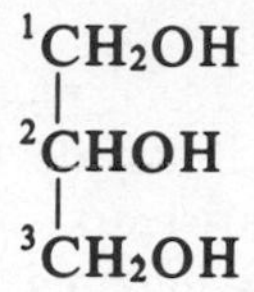

$$^1CH_2OH$$
$$^2CHOH$$
$$^3CH_2OH$$

alcohol that itself possesses some useful chemical properties. It is fluid at ordinary temperatures and is an excellent solvent. This small molecule is able to esterify three fatty acid molecules simultaneously. Glycerol also forms esters with inorganic acids (e.g., phosphoric acid), and thus serves as an important intermediate in both lipid and carbohydrate metabolism.

Neutral Fats

(a) Mono-, Di-, and Triacylglycerols. Fatty acid esters of glycerol are called acylglycerols or glycerides*. The most abundant of these are the triacylglycerols (triglycerides), in which all three of the hydroxyl groups of glycerol are esterified to fatty acids. The fatty acids are typically long chain and contain an even number of carbon atoms. Simple triacylglycerols contain the same fatty acid in each of the three positions; mixed triacylglycerols contain two or more different fatty acids. The primary function of triacyl-

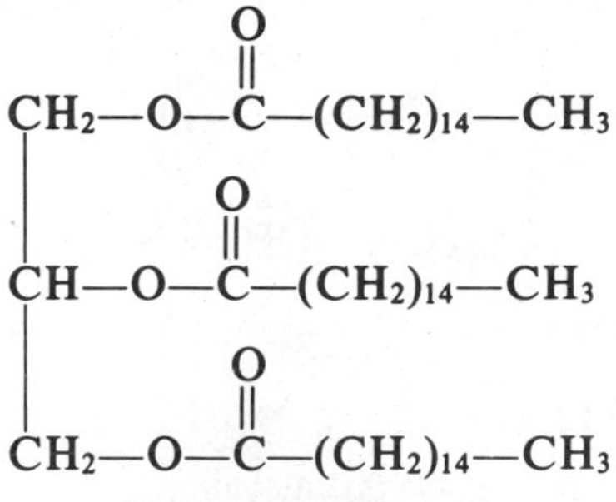

$$CH_2-O-\overset{\overset{\displaystyle O}{\|}}{C}-(CH_2)_{14}-CH_3 \qquad CH_2-O-\overset{\overset{\displaystyle O}{\|}}{C}-(CH_2)_{14}-CH_3$$

$$CH-O-\overset{\overset{\displaystyle O}{\|}}{C}-(CH_2)_{14}-CH_3 \qquad CH-O-\overset{\overset{\displaystyle O}{\|}}{C}-(CH_2)_{16}-CH_3$$

$$CH_2-O-\overset{\overset{\displaystyle O}{\|}}{C}-(CH_2)_{14}-CH_3 \qquad CH_2-O-\overset{\overset{\displaystyle O}{\|}}{C}-(CH_2)_{16}-CH_3$$

Simple triacylglycerol   Mixed triacylglycerol
(tripalmitin)   (1-palmitodistearin)

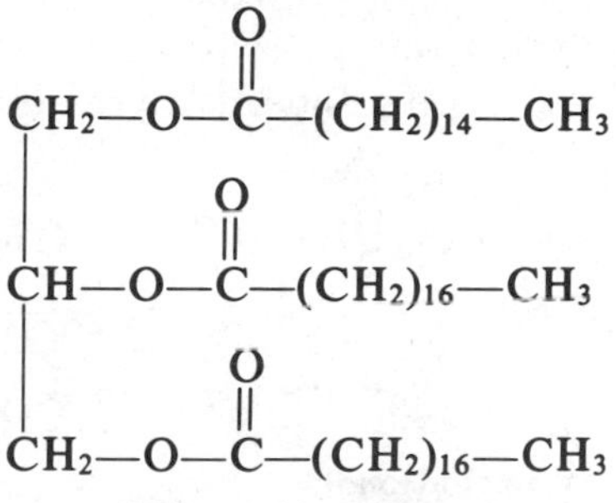

glycerols is to provide a store of chemical fuel that can be used by plant and animal cells for energy. Storage or depot fats typically contain a complex mixture of both types of triacylglycerols. Monoacylglycerols and diacylglycerols, in which only one or two of the hydroxyl groups of glycerol are esterified, respectively, are also found. They are less abundant, but are important intermediates in lipid metabolism.

The physical properties of triacylglycerols are basically similar to those exhibited by fatty acids. All triacylglycerols are insoluble in water, but are

*The terms mono-, di-, and triglycerides are chemically inaccurate and should be replaced by the corresponding -acylglycerols.

soluble in nonpolar solvents. Triacylglycerols are also less dense than water. The melting points of triacylglycerols, particularly simple types, are determined largely by the chain length and degree of unsaturation of their fatty acid components. For example, tristearin, which contains three stearic acid (18:0) residues, melts at 71°C, compared to 69.6°C for stearic acid alone. Unsaturation greatly decreases the melting point. Triolein, which is composed of three oleic acid (18:1) molecules esterified to glycerol, is a liquid at room temperature (melting point = −17°C). The melting points of mixed triacylglycerols, however, depend upon the position of the fatty acids, and may vary considerably from a simple mixture of similar fatty acids.

The most significant chemical reaction involving triacylglycerols is the hydrolysis of the ester linkage with the resultant formation of glycerol and three fatty acid molecules. This reaction can be accomplished by heating the triacylglycerol with a solution of sodium or potassium hydroxide at a boiling temperature. Hydrolysis by alkali is termed "saponification," and the fatty acids are ultimately converted into soaps (salts of the fatty acids). In living systems, the hydrolysis reaction is catalyzed by lipase enzymes such as those present in pancreatic juices. This process and its metabolic importance are discussed in greater detail in Chapter 2.

(b) Glycerol Ethers. Glycerol ethers are comprised of glycerol and an aliphatic alcohol (saturated or unsaturated) linked to the glycerol molecule in the one position. Two widely occurring glycerol ethers in animal tissues are batyl alcohol and selachyl alcohol, both of which contain alcohol chains

$$CH_2-O-CH_2(CH_2)_{16}CH_3 \qquad CH_2-O-CH_2(CH_2)_7CH=CH(CH_2)_7CH_3$$
$$CH-OH \qquad\qquad\qquad\qquad CH-OH$$
$$CH_2-OH \qquad\qquad\qquad\qquad CH_2-OH$$

Batyl alcohol   Selachyl alcohol

with 18 carbon atoms. Selachyl alcohol is liquid at room temperature, whereas batyl alcohol melts between 70 and 71°C (Deuel, 1951). These compounds are abundant in the liver of elasmobranchs, but have also been isolated from the tissues of higher vertebrates including humans. Closely related to glycerol ethers (and triacylglycerols) are alkyl ether acylglycerols in which the two remaining hydroxyl groups of glycerol are esterified to fatty acids.

(c) Glycosylglycerols. Glycosylglycerols contain a sugar (typically galactose) linked to the three position of either a mono- or diacylglycerol. These compounds are most commonly found in plant tissues, but are present in trace amounts in animals, where they are often overlooked (Christie,

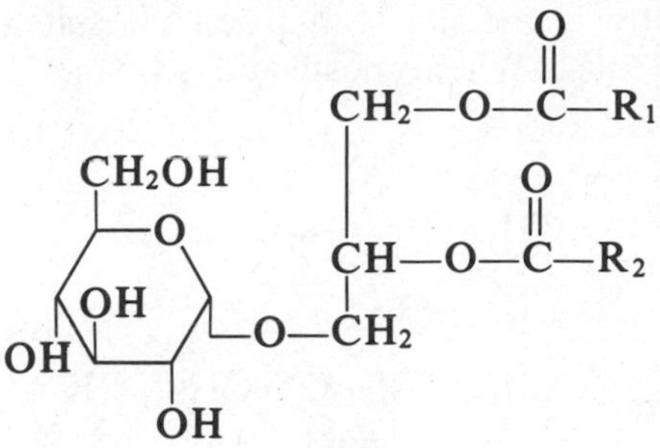

$$CH_2-O-\overset{\overset{\displaystyle O}{\|}}{C}-R_1$$

$$CH-O-\overset{\overset{\displaystyle O}{\|}}{C}-R_2$$

Galactosyl diacylglycerol

1973). Glycosyldiacylglycerols have also been isolated from the chromatophores of photosynthetic bacteria.

Phosphoglycerides

The second major class of lipids that have glycerol as a backbone structure is the phosphoglycerides. These compounds, upon hydrolysis, produce two fatty acids, glycerol, inorganic phosphate, and an organic base or polyhydroxyl compound. The phosphate group, which is linked to the third carbon of glycerol, has connected to it a small polar or ionized compound frequently containing nitrogen. The fatty acids of phosphoglycerides are both saturated and unsaturated, with the unsaturated fatty acid usually found on the two position of the glycerol. Included among the phosphoglycerides are the phosphatidates, diphosphatidylglycerols, and phosphoinositides.

(a) Phosphatidates. This group includes some of the most abundantly occurring and biologically important phosphoglycerides. Phosphatidates may be regarded as derivatives of phosphatidic acid in which the three position of glycerol is esterified with phosphoric acid instead of carboxylic acid.

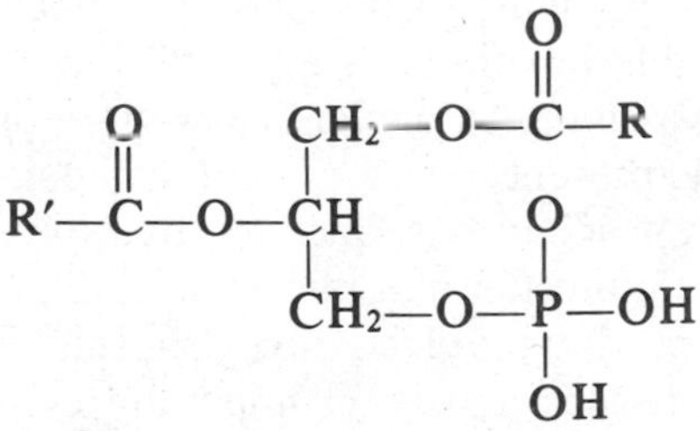

$$R'-\overset{\overset{\displaystyle O}{\|}}{C}-O-CH$$

$$CH_2-O-\overset{\overset{\displaystyle O}{\|}}{C}-R$$

$$CH_2-O-\overset{\overset{\displaystyle |}{P}}{\underset{\displaystyle OH}{}}-OH$$

L-α-Phosphatidic acid

Phosphatidic acid, although present in only small amounts in plant and animal tissues, is the parent compound for the following phosphoglycerides: phosphatidylcholine (lecithin), phosphatidylethanolamine, and phosphatidylserine. Phosphatidylcholine is an important constituent of plasma membranes, and typically is the most abundant phosphoglyceride present in animal and plant tissues. When phosphatidylcholine is present, it is often

$$
\begin{array}{l}
\quad\quad\quad\quad\quad\quad O \\
\quad\quad\quad\quad\quad\quad \parallel \\
\quad O \quad\quad CH_2{-}O{-}C{-}R \\
\quad \parallel \quad\quad\quad | \\
R'{-}C{-}O{-}CH \quad\quad O \\
\quad\quad\quad\quad | \quad\quad\quad | \\
\quad\quad\quad CH_2O{-}P{-}O{-}CH_2{-}CH_2{-}\overset{+}{N}(CH_3)_3 \\
\quad\quad\quad\quad\quad\quad | \\
\quad\quad\quad\quad\quad\quad O^-
\end{array}
$$

Phosphatidylcholine

$$
\begin{array}{l}
\quad\quad\quad\quad\quad\quad O \\
\quad\quad\quad\quad\quad\quad \parallel \\
\quad O \quad\quad CH_2{-}O{-}C{-}R \\
\quad \parallel \quad\quad\quad | \\
R'{-}C{-}O{-}CH \quad\quad O \\
\quad\quad\quad\quad | \quad\quad\quad \parallel \\
\quad\quad\quad CH_2O{-}P{-}O{-}CH_2{-}CH_2\overset{+}{N}H_3 \\
\quad\quad\quad\quad\quad\quad | \\
\quad\quad\quad\quad\quad\quad O^-
\end{array}
$$

Phosphatidylethanolamine

$$
\begin{array}{l}
\quad\quad\quad\quad\quad\quad O \\
\quad\quad\quad\quad\quad\quad \parallel \\
\quad O \quad\quad CH_2{-}O{-}C{-}R \\
\quad \parallel \quad\quad\quad | \\
R'{-}C{-}O{-}CH \quad\quad O \\
\quad\quad\quad\quad | \quad\quad\quad \parallel \\
\quad\quad\quad CH_2O{-}P{-}O{-}CH_2{-}CH{-}\overset{+}{N}H_3 \\
\quad\quad\quad\quad\quad\quad | \quad\quad\quad\quad | \\
\quad\quad\quad\quad\quad\quad O^- \quad\quad\quad COO^-
\end{array}
$$

Phosphatidylserine

accompanied by small amounts of lysophosphatidylcholine, in which only one of the two available positions of glycerol is esterified to a fatty acid (Christie, 1973). Phosphatidylethanolamine is often the second most abundant phosphoglyceride present in tissues, and may constitute the major lipid class in bacteria. In phosphatidylserine, the hydroxyl group of the amino acid L-serine is esterified to the phosphoric acid. Phosphatidylserine is not as abundant as either phosphatidylcholine or phosphatidylethanolamine, but it is present in most tissues and is an important component of brain and erythrocyte lipids.

Plasmalogens are also derivatives of glycerol phosphate and contain nitrogen as part of the polar group; however, they differ from the previously described phosphoglycerides in that they contain an α, β-unsaturated ether (rather than a fatty acid) esterified to the one position of glycerol. They appear to be more prevalent in animals, especially in heart and brain tissue.

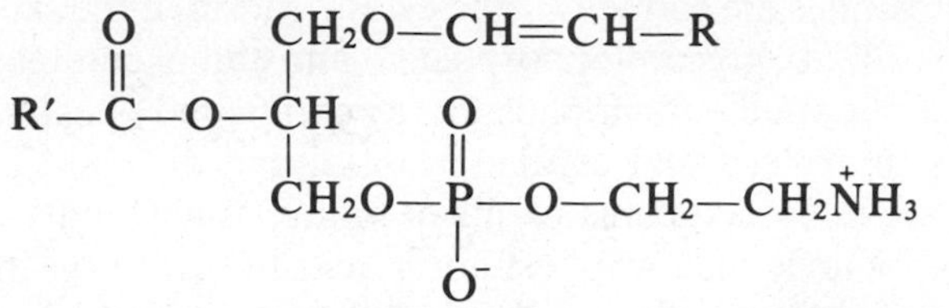

$$R'-\overset{\overset{\displaystyle O}{\|}}{C}-O-\underset{\underset{\displaystyle CH_2O-\underset{\underset{\displaystyle O^-}{|}}{\overset{\overset{\displaystyle O}{\|}}{P}}-O-CH_2-CH_2\overset{+}{N}H_3}{|}}{\overset{\overset{\displaystyle CH_2O-CH=CH-R}{|}}{CH}}$$

Phosphatidal ethanolamine (ethanolamine plasmalogen)

Ethanolamine plasmalogen is the most common form. Plasmalogens that lack a nitrogenous base (neutral plasmalogens) also occur in very small amounts in animal tissues.

(b) Diphosphatidylglycerols (DPG) and Phosphoinositides. These two lipid groups are nitrogen-free derivatives of glycerol phosphate. The most important example of the former in animal tissue is cardiolipin (DPG). Cardiolipin consists of three glycerol moieties joined by two phos-

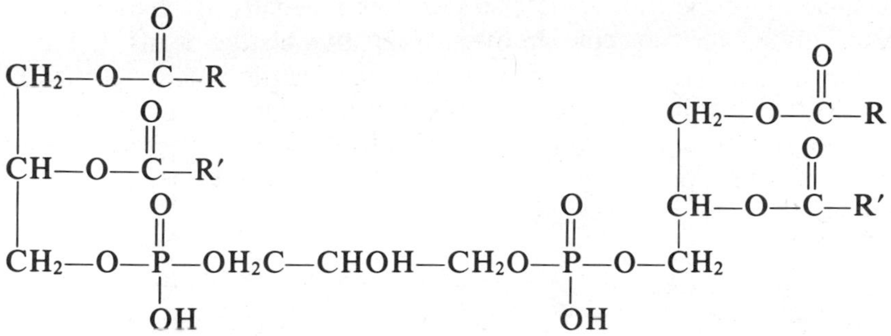

$$\begin{array}{l} CH_2-O-\overset{\overset{\displaystyle O}{\|}}{C}-R \\ CH-O-\overset{\overset{\displaystyle O}{\|}}{C}-R' \\ CH_2-O-\underset{\underset{\displaystyle OH}{|}}{\overset{\overset{\displaystyle O}{\|}}{P}}-OH_2C-CHOH-CH_2O-\underset{\underset{\displaystyle OH}{|}}{\overset{\overset{\displaystyle O}{\|}}{P}}-O-CH_2 \end{array} \qquad \begin{array}{l} CH_2-O-\overset{\overset{\displaystyle O}{\|}}{C}-R \\ CH-O-\overset{\overset{\displaystyle O}{\|}}{C}-R' \end{array}$$

Cardiolipin

phodiester bridges. The hydroxyl groups of the two external glycerol molecules are esterified to fatty acids, with linoleic acid being most common. Cardiolipin is primarily a mitochondrial lipid; it was first isolated from heart tissue. Phosphatidylglycerols (monophosphatidylglycerols) are similar compounds that are present in plants, but that contain only a single molecule of glycerol esterified to phosphatidic acid (White et al., 1973).

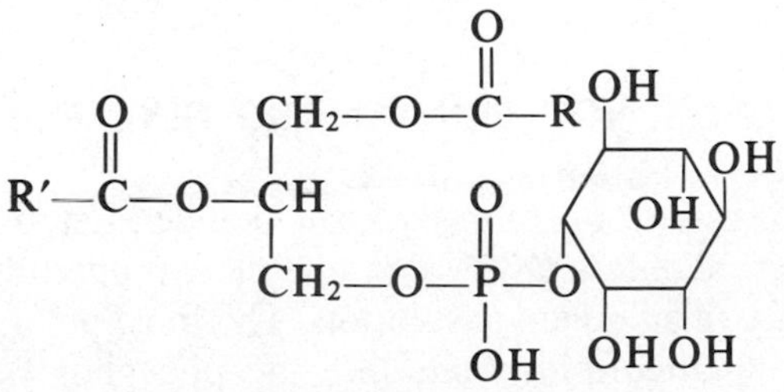

$$R'-\overset{\overset{\displaystyle O}{\|}}{C}-O-\underset{\underset{\displaystyle CH_2-O-\underset{\underset{\displaystyle OH}{|}}{\overset{\overset{\displaystyle O}{\|}}{P}}-O}{|}}{\overset{\overset{\displaystyle CH_2-O-\overset{\overset{\displaystyle O}{\|}}{C}-R}{|}}{CH}}$$

Phosphatidylinositol

Phosphoinositides are formed by the esterification of the six-carbon cyclic alcohol, inositol, to glycerol phosphate. The three commonly occurring phosphoinositides are the monophosphorus compound phosphatidylinositol, which occurs in skeletal and cardiac muscle, and di- and triphosphoinositides, which are important constituents of brain tissue (Johnston, 1971). The latter two compounds feature phosphoric acid linked to position four or to positions four and five of the cyclic alcohol, respectively.

(c) Properties of Phosphoglycerides. Although phosphoglycerides are chemically related to triacylglycerols, their physical properties are in many ways different, and they fulfill a quite different function in plant and animal cells. The key factor here is that the phosphate moiety and the nitrogenous base (when present) are ionized. The phosphate group of all phosphoglycerides has a negative charge at pH 7. In phosphatidylethanolamine and phosphatidylcholine, this negative charge is balanced by a positive charge on the nitrogenous base; hence at this pH these two phosphoglycerides are dipolar zwitterions with no net electrical charge. Phosphatidylserine, however, also contains a negatively charged carboxyl group, giving the molecule an overall negative charge at pH 7. Phosphoglycerides without a nitrogenous base (e.g., cardiolipin, phosphatidylinositol) are also quite polar because of their high content of hydroxyl groups. Thus, unlike the totally nonpolar triacylglycerol molecule, a phosphoglyceride contains a polar region at one end of the molecule and nonpolar fatty acid chains at the other end (Figure 1.2).

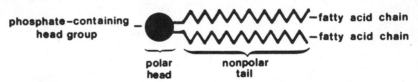

FIGURE 1.2.   Diagrammatic representation of phosphoglyceride molecule.

This amphipathic character (presence of hydrophilic and hydrophobic regions) allows phosphoglycerides to associate with polar and nonpolar molecules. This property is described in more detail when the structure of the cell membrane, which contains most of the phosphoglycerides present in biological tissues, is discussed (Chapter 3).

LIPIDS NOT CONTAINING GLYCEROL

Lipids that lack the glycerol carbon skeleton include a heterogeneous mixture of compounds that share few physical or chemical properties other than their general insolubility in an aqueous solution. The most polar members of this category are the sphingolipids that, like the phosphoglycerides, produce three or more products on hydrolysis. Sphingolipids are usually further classified into ceramides, sphingomyelins, and glycosphingolipids. The only other lipid class in this grouping that yields fatty acids following alkaline hydrolysis is the waxes. The remaining classes are nonsaponifiable and include the hydrocarbons, aliphatic alcohols, sterols, and terpenes.

Sphingolipids

Lipids that contain the amino alcohol sphingosine or a closely related compound are termed sphingolipids. All sphingolipids contain one molecule of

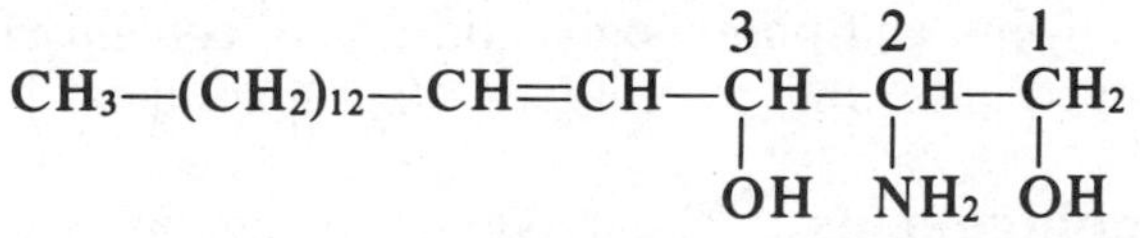

$$CH_3-(CH_2)_{12}-CH=CH-\overset{3}{\underset{|}{CH}}-\overset{2}{\underset{|}{CH}}-\overset{1}{\underset{|}{CH_2}}$$
$$\phantom{CH_3-(CH_2)_{12}-CH=CH-}OH\quad NH_2\ OH$$

Sphingosine

a fatty acid covalently bonded to the free amino group of sphingoside and a polar head group esterified to the number one position of the sphingoside base. The polar groups may include inorganic phosphate, carbohydrates, or other complex organic compounds.

(a) Ceramides.   The ceramides contain a saturated or unsaturated long-chain fatty acid linked to the amino group of the sphingosine (or related) base. The fatty acid residue may contain up to 26 carbon atoms, but most

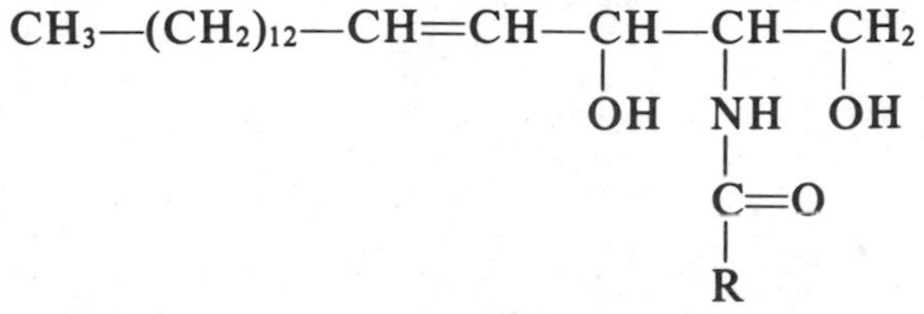

$$CH_3-(CH_2)_{12}-CH=CH-\underset{|}{CH}-\underset{|}{CH}-\underset{|}{CH_2}$$
$$OH\quad NH\quad OH$$
$$\underset{|}{C=O}$$
$$R$$

Ceramide

often is of the C_{16}, C_{18}, C_{22}, or C_{24} series. Although ceramides are widely distributed in both plant and animal tissues, they are usually minor components and primarily serve as the basic structural group for the more complex sphingolipids (Christie, 1973).

(b) Sphingomyelins.   Sphingomyelins are major components of plasma membranes and serum lipoproteins of higher animals, but are absent from plant tissues. The most common sphingomyelin contains the polar head group phosphoryl choline, which is linked to the primary alcohol of the ceramide. Sphingomyelin is structurally similar to phosphatidylcholine, which also contains the phosphoryl choline head group, and to phosphoglycerides in general because the molecules are geographically separated into polar

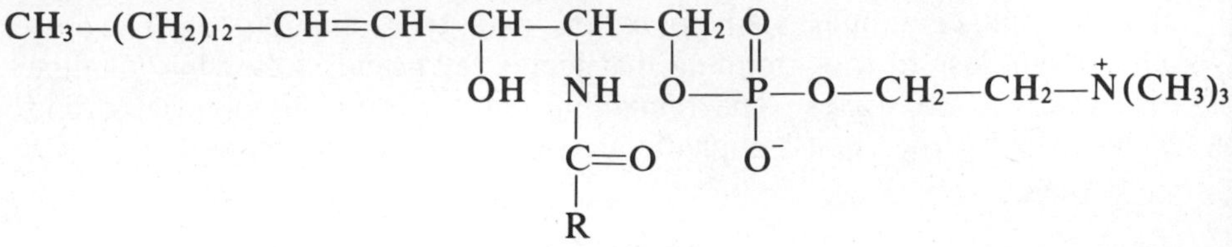

$$CH_3—(CH_2)_{12}—CH=CH—CH—CH—CH_2 \ O$$

$$\overset{|}{OH} \quad \overset{|}{NH} \quad O—\overset{||}{P}—O—CH_2—CH_2—\overset{+}{N}(CH_3)_3$$

$$\overset{|}{C=O} \qquad \overset{|}{O^-}$$

$$\overset{|}{R}$$

Sphingomyelin

(hydrophilic head) and nonpolar (hydrophobic tail) regions connected by a belt region of intermediate polarity (Barenholz and Thompson, 1980). Because sphingomyelin and phosphoglycerides both contain phosphate, they are often grouped under the general term ''phospholipids.''

(c) Glycosphingolipids.   This group of sphingolipids contains one or more moles of carbohydrate linked to the primary alcohol of the ceramide. The simplest of these are the cerebrosides, which contain a monosaccharide (usually galactose, sometimes glucose) present in a β-glycosidic linkage.

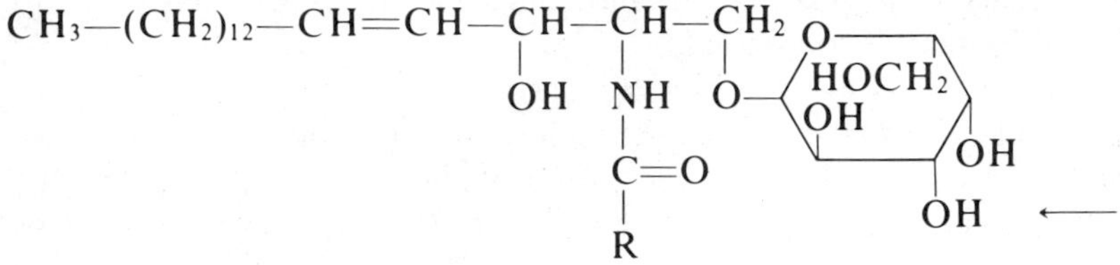

$$CH_3—(CH_2)_{12}—CH=CH—CH—CH—CH_2$$

Cerebroside

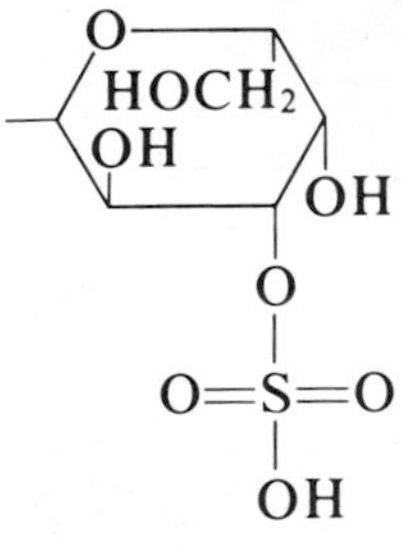

Sulfatide (when substituted)

Cerebrosides are important constituents of brain and nervous tissue, especially the myelin sheath, but also occur in small amounts in nonneural animal tissues as well as in plants. Closely related to the cerebrosides are sulfatides, in which a sulfate group is present in the three position of galactose (see above). Sulfatides are also abundant in nervous tissue. Ceramide polyhexosides are sphingolipids with a disaccharide, trisaccharide, tetrasaccharide, or similar compound as their polar head group; these compounds tend to be more common in the spleen, liver, plasma, and erythrocytes. Because cerebrosides and ceramide polyhexosides contain sugar residues as their polar head groups, and thus lack an electrical charge, they are sometimes referred to as "neutral glycosphingolipids" (Lehninger, 1975).

The most complex group of glycosphingolipids is the gangliosides. This class contains one or more molecules of sialic acid (*N*-acetylneuraminic acid, abbreviated NANA) linked to one or more of the sugar residues of a

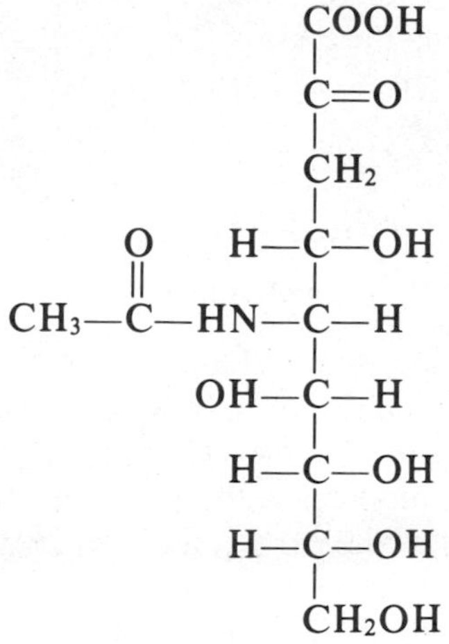

$$
\begin{array}{c}
\text{COOH} \\
| \\
\text{C}=\text{O} \\
| \\
\text{CH}_2 \\
| \\
\text{H}-\text{C}-\text{OH} \\
\end{array}
$$

$$
\text{CH}_3-\overset{\displaystyle \overset{\text{O}}{\|}}{\text{C}}-\text{HN}-\text{C}-\text{H}
$$

$$
\begin{array}{c}
\text{OH}-\text{C}-\text{H} \\
| \\
\text{H}-\text{C}-\text{OH} \\
| \\
\text{H}-\text{C}-\text{OH} \\
| \\
\text{CH}_2\text{OH}
\end{array}
$$

Sialic acid

ceramide polyhexoside. Gangliosides are named according to the number of sialic acid residues present in the molecule (i.e., mono-, di-, tri-, and tetrasialogangliosides). The structures of gangliosides are complex and variable. For example, a disialoganglioside may have each sialic acid linked to a separate sugar residue or both sialic acids linked to each other and one of them linked to a central sugar residue. Because of this complexity, several shorthand notations have been developed to express a specific structural arrangement (see Gurr and James, 1975). The major monosialoganglioside of the human brain has the following structure:

Cer (1 $\longleftarrow$ 1) Glu (4 $\longleftarrow$ 1) Gal (4 $\longleftarrow$ 1) Gal NAc (3 $\longleftarrow$ 1) Gal

$$
\left(\begin{array}{c} 3 \\ \uparrow \\ 2 \end{array}\right)
$$

NANA

In addition to their abundance in gray matter of the human brain, gangliosides also occur in smaller concentrations in other tissues. Gangliosides must be extracted from tissues with care, for unlike other lipids they are soluble in water as well as in polar organic solvents (Christie, 1973).

Aliphatic Alcohols

Alcohols are derived from saturated or unsaturated hydrocarbons by re-placing a hydrogen atom with a hydroxyl group. Aliphatic (fatty) alcohols include those molecules that contain more than eight carbons and, hence, are relatively insoluble in water but are soluble in organic solvents. Alcohols are named by taking the longest chain of carbon atoms that includes the hydroxyl group and adding the ending "-ol." Thus, the primary monohydric alcohol with a chain length of 18 carbons is called *n*-octadecanol; it is also

$$CH_3-(CH_2)_{16}-CH_2-OH$$

n-Octadecanol

known by the common name stearyl alcohol. Secondary alcohols, in which the hydroxyl group is located internally on the carbon chain, and polyhydroxyl alcohols (more than one hydroxyl group per molecule) are also present in biological tissues. In these cases the position of the hydroxyl group(s) is indicated by the number of the carbon atom(s) to which it is attached (e.g., pentacosan-8,9-diol).

Although most aliphatic alcohols in biological systems occur esterified to fatty acids (waxes) or connected to glycerol (see the section on glycerol ethers), small quantities of "free" or "unesterified" alcohols can often be recovered from various lipid-containing tissues. Potentially rich sources here include the head and blubber "oils" of marine mammals and the cuticle lipids of plants and arthropods. The majority of these are even-numbered primary alcohols having a chain length of 12 to over 30 carbons; however, branched and secondary alcohols have been reported in certain insect and plant species (Jackson and Blomquist, 1976; Tulloch, 1976).

Like fatty acids, many physical properties of aliphatic alcohols vary according to the length of the carbon chain. The melting point of alcohols rises progressively as chain length increases, but it is less than that of a corresponding acid (e.g., the melting point of *n*-octadecanol is 57.98°C versus 69.6°C for *n*-octadecanoic acid). Alcohols with 12 or less carbons are liquid at room temperature (ca. 25°C) (Deuel, 1951). The density of aliphatic alcohols is also less than water, but values of specific gravity are comparable to those of corresponding acids when determinations are made at similar temperatures. Finally, increased chain length results in a decrease in solubility in moderately polar solvents, whereas increased temperature increases solubility.

Wax Esters

Wax esters consist of fatty acids esterified to aliphatic alcohols. Wax esters can be separated into these two components by saponification in methanol

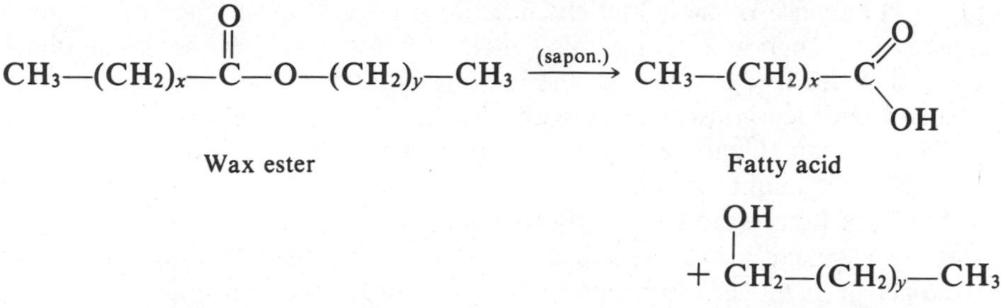

$$CH_3—(CH_2)_x—\overset{\overset{O}{\|}}{C}—O—(CH_2)_y—CH_3 \xrightarrow{\text{(sapon.)}} CH_3—(CH_2)_x—C\overset{\displaystyle O}{\underset{OH}{\diagup}}$$

Wax ester   Fatty acid

$$+ \overset{OH}{\overset{|}{CH_2}}—(CH_2)_y—CH_3$$

Primary alcohol

in the presence of a strong base such as KOH. The fatty acids are frequently those found as components of triacylglycerols (e.g., 14:0, 16:0, 18:0); however, acids containing up to 30 carbon atoms are not uncommon. The acids are typically saturated or may contain one double bond; branched or hydroxy fatty acids have also been reported. The aliphatic alcohols are typically saturated, primary alcohols, but small amounts of unsaturated, branched, or secondary alcohols are sometimes present in a mixture of wax esters.

Wax esters are relatively inert substances that, in most cases, are solid at biologically relevant temperatures. As they are warmed, however, wax esters tend to soften and become pliable. Because of their resistance to decomposition and their insolubility in water and most solvents, wax esters are often found as or contribute to the protective coatings (waxes) on the surfaces of plants and animals (Chapter 5). (The term "waxes" refers to wax esters and other lipids with similar physical properties, and is used in a broad functional sense). Wax esters are also abundant in marine animals, where they serve as an important energy source (Chapter 6) and also contribute to buoyancy control (Chapter 8).

Hydrocarbons

As the name implies, hydrocarbons contain only the elements carbon and hydrogen and, from this standpoint, represent the simplest of the lipid compounds. Naturally occurring hydrocarbons include saturated, straight-chain molecules (n-alkanes), methyl branched alkanes, and unsaturated mol-

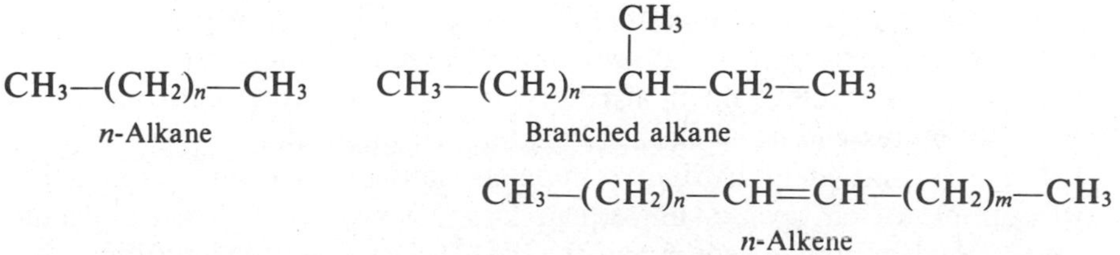

$$CH_3—(CH_2)_n—CH_3 \qquad CH_3—(CH_2)_n—\overset{\overset{CH_3}{|}}{CH}—CH_2—CH_3$$

n-Alkane   Branched alkane

$$CH_3—(CH_2)_n—CH{=}CH—(CH_2)_m—CH_3$$

n-Alkene

ecules (n-alkenes). The saturated hydrocarbons, which comprise the major portion of biological hydrocarbons, are also referred to as "paraffins" (= little affinity) because of their low chemical reactivity. The absence of oxygen-containing functional groups also makes them extremely nonpolar compounds. With few exceptions, the melting points, boiling points, and densities of hydrocarbons increase with increasing chain length. Hydrocarbons with more than 16 carbon atoms are waxlike solids at ordinary temperatures, and all are insoluble in water.

Hydrocarbons were previously thought to be compounds without biological significance; however, it has been shown that they are important constituents of plant and arthropod cuticular lipids (Jackson and Blomquist, 1976; Hadley, 1981). The principal hydrocarbon fractions in these groups are *n*-alkanes and branched hydrocarbons ranging in length from about 21 to over 40 carbon atoms. In contrast to fatty acids and aliphatic alcohols, which are invariably composed of an even number of carbon atoms, the predominant hydrocarbon molecules tend to have odd-numbered carbon chains. Hydrocarbons also occur in much smaller quantities in internal tissues (Kolattukudy, 1976), where their function is largely unknown. Additional information on the composition, distribution, and function of hydrocarbons associated with the integuments of plants and animals is presented in Chapter 5.

Terpenes

Terpenes are constructed of multiples of the five-carbon hydrocarbon unit called isoprene (2-methyl-1,3-butadiene). The isoprene unit and its derived

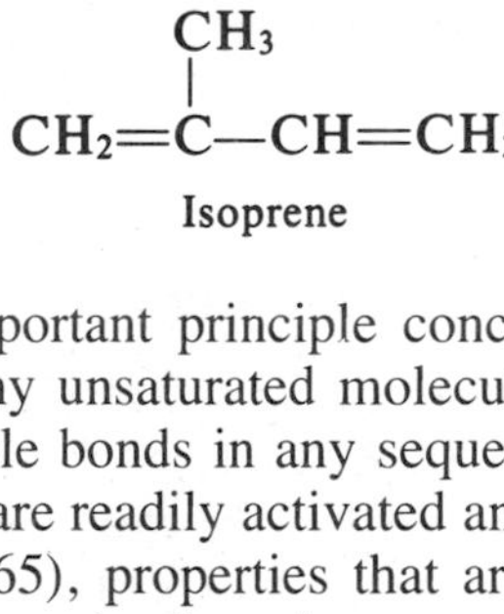

$$CH_3$$
$$|$$
$$CH_2{=}C{-}CH{=}CH_2$$

Isoprene

compounds reflect an important principle concerning the carbon atom in biological systems—in any unsaturated molecule, there is a regular alternation of single and double bonds in any sequence of carbon atoms. Molecules with this structure are readily activated and powerfully activate other molecules (Needham, 1965), properties that are certainly associated with terpenes. Most terpenes are simple polymers of the C_5 isoprene unit and, in fact, are the only lipid type that exhibits extensive polymerization. Naturally occurring terpenes may contain from one to eight isoprene units arranged in an open-chain or a ring structure; however, both linear and cyclic structures can occur in the same terpene molecule.

Terpenes are widely distributed in plants, animals, and microorganisms. An example of an open-chain terpene is squalene, a 30-carbon molecule that is found in human sebum and the oil of sharks; it is also an important intermediate in the synthesis of cholesterol. Terpene compounds also constitute the largest group of plant oils. Many of these terpenes are ternary

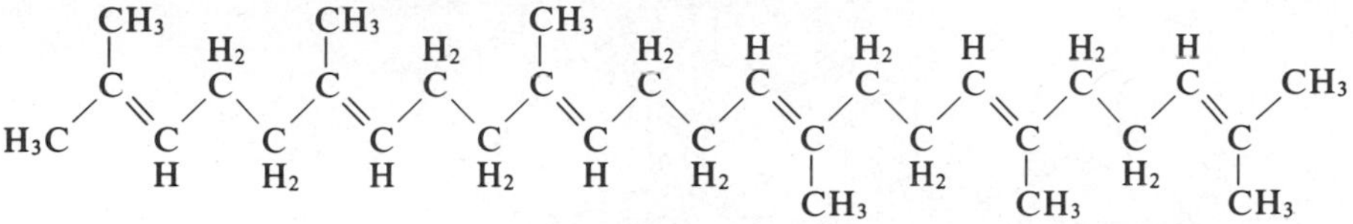

Squalene

compounds (i.e., contain oxygen along with carbon and hydrogen). The presence of oxygen greatly increases the scent, as well as increases the compound's solubility in water (Needham, 1965). Two examples of oxy-isoprenoid compounds with distinctive smells are citronellal (open chain) and camphor (cyclic). Another important class of terpenes are the carotenoid

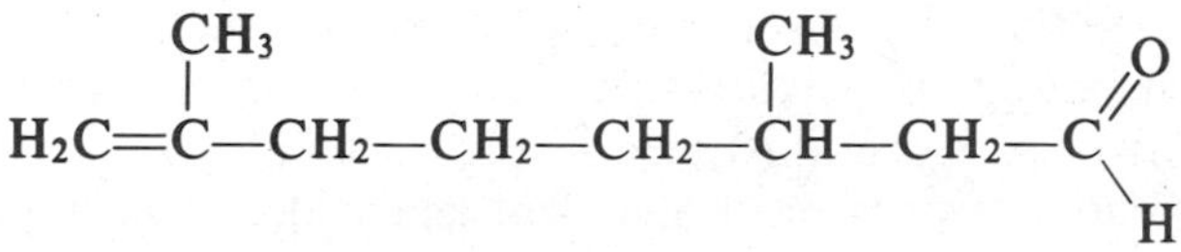

Citronellal 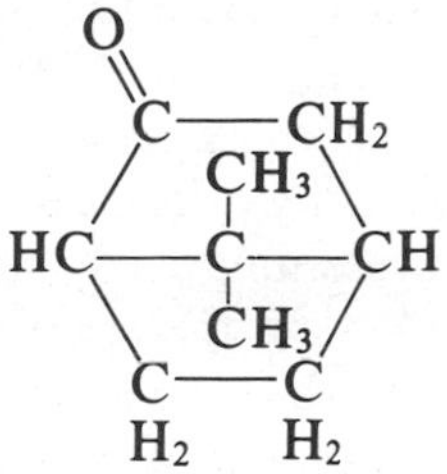   Camphor

pigments of plants, which consist of eight isoprene units with cyclic structures at both ends of the molecule. One important carotenoid, β-carotene, is the hydrocarbon precursor of vitamin A. The functions of terpenes are

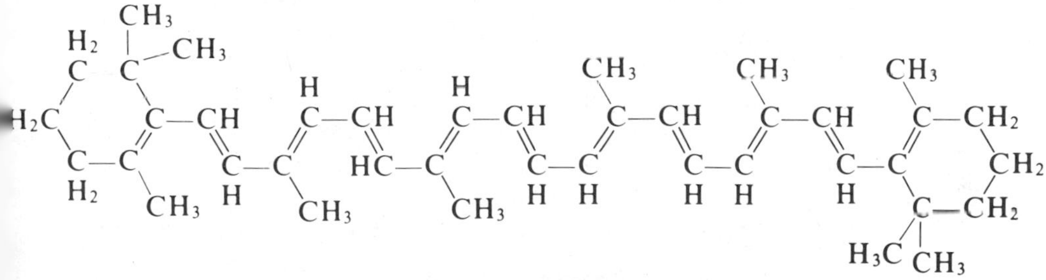

β-Carotene

as diverse as their structures and are discussed further in many of the chapters that follow.

Steroids

All steroids are derivatives of a complex ring system called perhydro-cyclopentanophenanthrene or, simply, a sterane ring. This complex consists of three fused cyclohexane rings (A, B, and C) and a terminal cyclopentane

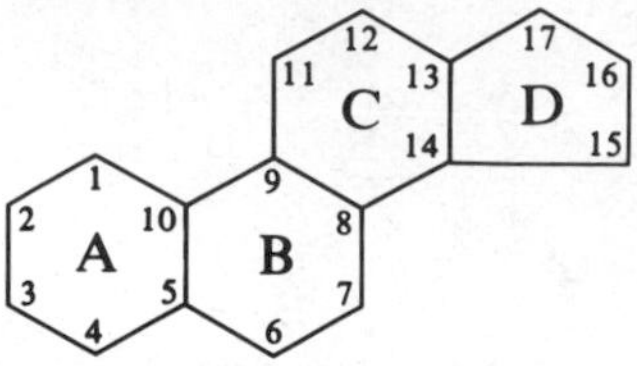

Perhydrocyclopentanophenanthrene

ring (D). A shorthand formula used to designate the rings and number the carbon atoms is shown in the diagram. In addition to the ring system, almost all biological steroids have an oxygenated substituent (either —OH or =O) on carbon atom 3 and methyl groups (CH_3) off of carbons 10 and 13. On carbon 17 there is often an aliphatic substituent that can be used to classify steroids (White et al., 1973). For example, sterols have a side chain that contains 8, 9, or 10 atoms, bile acids a chain with 5 carbon atoms, and progesterone and adrenocorticoid hormones a 2-carbon chain. Naturally occurring estrogens and androgens lack this side chain. Structures of steroids representing each of the classes are illustrated in Figure 1.3.

Among the sterols, cholesterol is most abundant and widely distributed in animal tissues, being an important constituent of plasma membranes and lipoproteins of blood plasma. Moreover, it acts as the precursor of many other steroids, including the bile acids and steroid hormones. Cholesterol can occur as a free alcohol or esterified to fatty acids (cholesteryl esters). The presence of an —OH group off of carbon three makes cholesterol one of the more hydrophilic steroids, and thus, able to function at lipid-aqueous interfaces. This property, along with its molecular size, shape, and structure, are discussed in greater detail in Chapter 3. Plants, which essentially lack cholesterol, have sterols with typically more than eight carbon atoms in the side chain. In higher plants these sterols are collectively termed phytosterols and include stigmasterol and sitosterol. Sterols from the lower plants are structurally different and are termed mycosterols. An important example is ergosterol of yeast, which, upon ultraviolet irradiation, is converted into vitamin D_2. Sterols are not found in bacteria.

Prostaglandins

Prostaglandins are a group of membrane-associated lipids composed of 20 carbon atoms of which 5 are joined to form a cyclopentane ring. Prostanoic acid, which is the parent compound (skeleton) for all prostaglandins, is shown, along with the system for numbering carbon atoms. Biologically active prostaglandins are all unsaturated; many are synthesized from arachidonic acid (20:4), which is also found in cell membranes. The prostaglandins are divided into four categories—E, F, A, B—depending on the nature of their ring substituents and the number of additional side-chain double bonds. They are further abbreviated PGE_1, PGE_2, PGA_1, and so on.

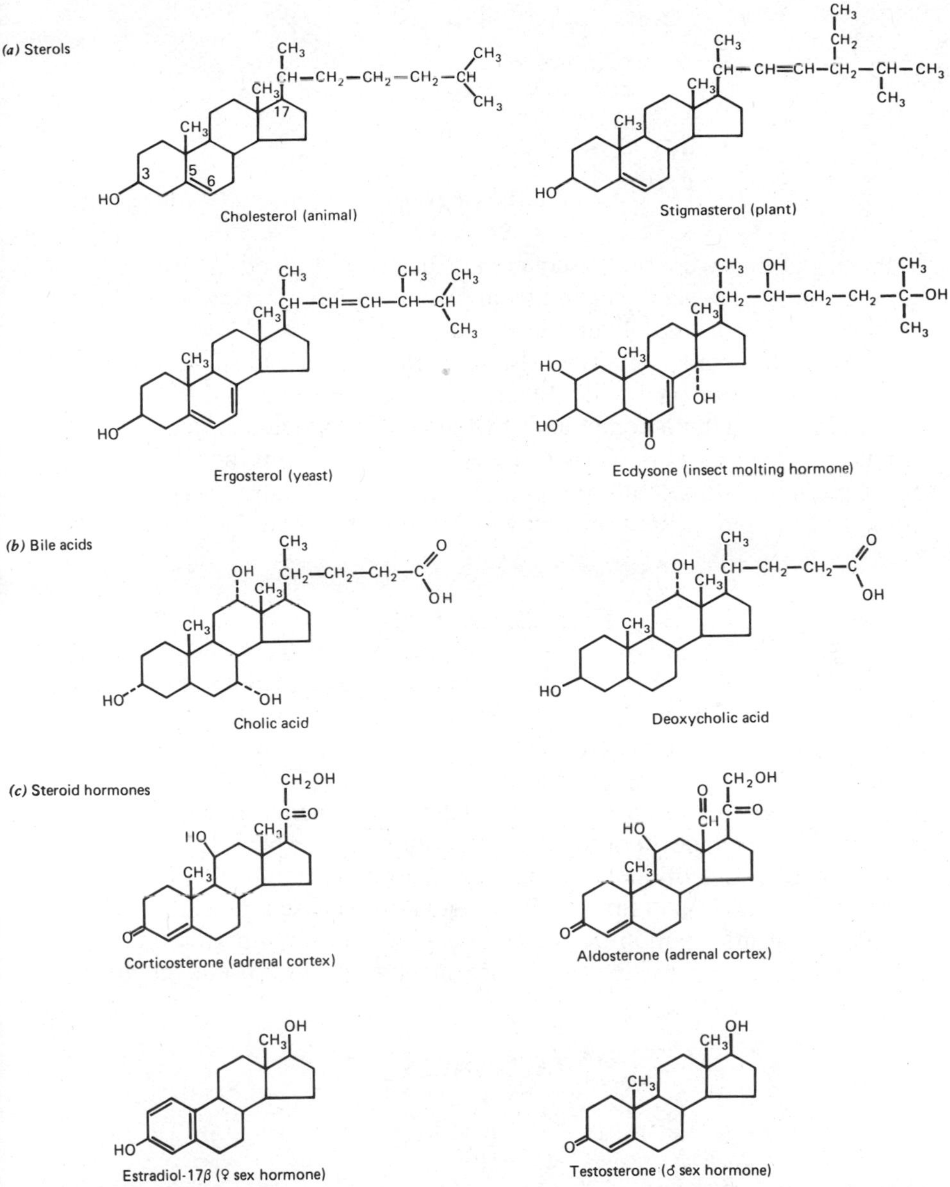

(a) Sterols

CH₃
CH—CH₂—CH₂—CH₂—CH
CH₃
17
CH₃
3 5 6
HO
Cholesterol (animal)

CH₃
CH₂
CH—CH=CH—CH₂—CH—CH₃
CH₃
CH₃
CH₃
HO
Stigmasterol (plant)

CH₃ CH₃
CH—CH=CH—CH—CH
CH₃
CH₃
CH₃
HO
Ergosterol (yeast)

CH₃ OH CH₃
CH₂—CH—CH₂—CH₂—C—OH
CH₃
CH₃
CH₃
HO
OH
HO
O
Ecdysone (insect molting hormone)

(b) Bile acids

CH₃ O
OH CH₂—CH₂—CH₂—C
CH₃
OH
CH₃
HO
OH
Cholic acid

CH₃ O
OH CH—CH₂—CH₂—C
CH₃
OH
CH₃
HO
Deoxycholic acid

(c) Steroid hormones

CH₂OH
C=O
CH₃
HO
CH₃
O
Corticosterone (adrenal cortex)

CH₂OH
O C=O
HO CH
CH₃
CH₃
O
Aldosterone (adrenal cortex)

OH
CH₃
HO
Estradiol-17β (♀ sex hormone)

OH
CH₃
CH₃
O
Testosterone (♂ sex hormone)

FIGURE 1.3. Some important biological steroids. Division into three classes based on number of carbon atoms in side chain off of carbon 17 (see text). Steroids also differ in number and position of double bonds in fused rings, and in the type, location, and number of substituent functional groups. Slight differences in chemical structure can result in significantly different physiological functions (e.g., corticosterone vs. aldosterone).

23

Prostanoic acid

Prostaglandins were first discovered in the prostate gland, hence the name, but they have now been isolated from a wide variety of mammalian tissues including the intestine, liver, heart, lung, and brain. Although they are synthesized in small quantities, they are very potent substances and have widespread effects on bodily functions. As in the case of terpenes and steroids, prostaglandin compounds that are closely related structurally often produce opposite actions. Known effects of prostaglandins include regulating blood pressure, stimulating uterine contractility, regulating stomach secretions, and inhibiting lipid breakdown (Tortora and Anagnostakos, 1981).

COMBINED LIPIDS

Discussion to this point has centered on the types and properties of lipids as individual molecules. Although they occur and function in a "free" state, many lipids in organisms are combined with proteins or carbohydrates to form macromolecular arrays whose physical and chemical properties may be quite different from those of the isolated components. Examples of such combinations include lipoproteins, proteolipids, phosphatidopeptides, lipoamino acids, and lipopolysaccharides. The most abundant of these, the lipoproteins, function in lipid transport and are discussed in detail in Chapter 2. Additional information on the composition and function of some of the other complexes is provided as they are encountered in subsequent chapters.

REFERENCES

Barenholz, Y. and T. E. Thompson (1980). Sphingomyelins in bilayers and biological membranes. *Biochim. Biophys. Acta* **604**:129–158.

Christie, W. W. (1973). *Lipid Analysis*. Pergamon Press, Oxford. 338 pp.

Davenport, J. B. and A. R. Johnson (1971). The nomenclature and classification of lipids. In: *Biochemistry and Methodology of Lipids* (Johnson, A. R., and Davenport, J. B., Eds.), pp. 1–28. Wiley-Interscience, New York. 578 pp.

Deuel, H. J., Jr. (1951). *The Lipids. I. Chemistry*. Interscience, New York. 982 pp.

Gunstone, F. D. (1967). *An Introduction to the Chemistry and Biochemistry of Fatty Acids and Their Glycerides*. Chapman and Hall, London. 209 pp.

Gurr, M. I. and A. T. James (1975). *Lipid Biochemistry: An Introduction,* 2nd ed. Chapman and Hall, London. 244 pp.

Hadley, N. F. (1981). Cuticular lipids of terrestrial plants and arthropods: A comparison of their structure, composition, and waterproofing function. *Biol. Rev.* **56**:23–47.

Jackson, L. L. and G. J. Blomquist (1976). Insect waxes. In: *Chemistry and Biochemistry of Natural Waxes* (Kolattukudy, P. E., Ed.), pp. 201–233. Elsevier, Amsterdam.

Johnston, P. V. (1971). *Basic Lipid Methodology.* Pierce Chemical Co., Rockford, Illinois. 100 pp.

Kolattukudy, P. E. (1976). Introduction to natural waxes. In: *Chemistry and Biochemistry of Natural Waxes* (Kolattukudy, P. E., Ed.), pp. 1–15. Elsevier, Amsterdam.

Lehninger, A. L. (1975). *Biochemistry.* Worth Publishers, New York. 1104 pp.

Needham, A. E. (1965). *The Uniqueness of Biological Materials.* Pergamon Press, London. 593 pp.

Tortora, G. J. and N. P. Anagnostakos (1981). *Principles of Anatomy and Physiology,* 3rd ed. Harper and Row, New York. 826 pp.

Tulloch, A. P. (1976). Chemistry of waxes of higher plants. In: *Chemistry and Biochemistry of Natural Waxes* (Kolattukudy, P. E., Ed.), pp. 235–287. Elsevier, Amsterdam.

White, A., P. Handler, and E. L. Smith (1973). *Principles of Biochemistry,* 5th ed. McGraw-Hill, New York. 1296 pp.

2

LIPID ORIGINS, SYNTHESIS, AND PROCESSING

The preceding chapter has provided information on the structure, occurrence, and properties of lipids relevant to their function in biological systems. To fully appreciate the adaptive role of lipids, it is also necessary to have a basic understanding of how lipids are acquired, digested, absorbed, transported, and stored. Because these processes are complex and have been intensively investigated, it is not possible to provide comprehensive coverage in a single chapter. I have chosen, therefore, to present the essential fundamentals of these processes by comparing and contrasting each as it occurs in plants and animals. Emphasis again is on those aspects of the physiology and biochemistry of lipids that pertain to specific functions described in subsequent chapters.

ORIGIN OF LIPIDS

Plant lipids are ultimately derived from substrates produced during photosynthesis. The initial steps involve the conversion of sugars into acetyl-CoA and its combination with CO_2 to form malonyl CoA. Subsequent reactions lead to the synthesis of palmitic acid (16:0), the most abundant fatty acid in plants and a major building block for triacylglycerols, which are the principal lipids of oil seeds and the fleshy parts of certain fruits, and phospholipids and glycolipids, which are present mainly in leaves, roots, and shoots. Animal lipids are directly or indirectly obtained from plants, and thus, from this standpoint, are also dependent on the photosynthetic process. Many lipids found in animal tissues, especially adipose triacylglycerols, are incorporated unchanged from the diet or subjected to minor restructuring following their intake and digestion. Other lipids are synthesized from non-lipoidal substrates utilizing basically the same metabolic scheme as that for the *de novo* synthesis of fatty acids by plants. Described below are the sites and major steps in the *de novo* and elongation systems in the two groups, followed by a brief examination of the synthesis of triacylglycerols and phosphoglycerides.

BIOSYNTHESIS OF FATTY ACIDS

Originally it was assumed that fatty acids were synthesized via the reverse of the pathway by which they are broken down. This concept led to difficulties, however, in reconciling the need for certain substrates and enzymes for fatty acid synthesis that were not required for their degradation. Research begun in the late 1950s eventually demonstrated that synthesis was a different process, requiring a multienzyme complex and two cofactors—Mn^{2+} and NADPH—plus ATP and CO_2 (Green, 1960). Although this sequence of metabolic events initially was established based on a process that occurs in pigeon liver, it has since been shown that the process is essentially the same in plants, animals, and aerobic microorganisms. A notable difference between plants and animals does exist in the specific site for the formation of fatty acids. In animals, the synthetic process occurs in the cytoplasm of fat bodies (insects) and the cytoplasm of liver, adipose, and mammary gland cells (vertebrates). Plants, in contrast, synthesize fatty acids in chloroplasts and nonphotosynthetic plastids in seeds (Stumpf, 1980).

The ultimate precursor of the carbon atoms in a fatty acid is acetyl-CoA, a two-carbon fragment (acetic acid) linked to coenzyme A. It is formed in mitochondria primarily from the oxidation of carbohydrates and proteins. Before acetyl-CoA can participate in the synthetic process, it must first be transported from the mitochondria to the cytoplasm. This requires conversion of acetyl-CoA to some other chemical form, for acetyl-CoA itself cannot cross the mitochondrial membrane. In animal cells, it is common for acetyl-CoA to combine with oxaloacetate to form citrate, which can be transported into the cytosol. The citrate is then regenerated to acetyl-CoA by a citrate lyase enzyme. A similar conversion of citrate to acetyl-CoA occurs in some plants (Nelson and Rinne, 1975, 1977a,b), but recent evidence suggests that other mechanisms are more important (Stumpf, 1980).

Whereas acetyl-CoA is the ultimate precursor in fatty acid synthesis, the principal donor of carbon atoms is not acetyl-CoA but the three-carbon compound linked to CoA, malonyl-CoA. Malonyl-CoA is made from acetyl-CoA and CO_2, the latter usually in the form of HCO_3^-:

$$CH_3-\overset{\overset{\displaystyle O}{\|}}{C}-CoA + ATP + CO_2 \longrightarrow$$

Acetyl-CoA

$$COOH-CH_2-\overset{\overset{\displaystyle O}{\|}}{C}-CoA + ADP + P_i$$

Malonyl-CoA

The reaction is catalyzed by the enzyme acetyl-CoA carboxylase, which requires the vitamin biotin and the divalent cation Mn^{2+}. In ensuing reactions, malonyl-CoA condenses first with acetyl-CoA and later with the fatty acyl-CoA compounds that result from prior condensations. Thus, the original acetyl portion of acetyl-CoA becomes the two terminal carbons of palmitic acid, while the seven remaining acetyl residues are derived from two of the three carbon atoms of malonyl-CoA (Lehninger, 1975). A previously incorporated CO_2 is removed with each condensation reaction. The seven condensation reactions required to produce palmitic acid are each catalyzed by six enzymes that are part of the fatty acid synthetase system. Also part of this system is a seventh protein (acyl carrier protein, ACP) that has no enzymatic activity, but instead serves as an anchor to which the growing fatty acid can attach. The process terminates with the formation of palmitoyl-ACP, which can be enzymatically converted to palmitic acid. The overall equation for palmitic acid biosynthesis starting from acetyl-CoA is as follows:

$$\text{acetyl-CoA} + 7 \text{ malonyl-CoA} + 14 \text{ NADPH} + 14 \text{ H}^+ \rightarrow$$
$$\text{palmitic acid} + 7 \text{ CO}_2 + 8 \text{ CoA} + 14 \text{ NADP}^+ + 6 \text{ H}_2\text{O}$$

The fatty acid synthetase system is also responsible for the formation of saturated fatty acids containing an odd number of carbon atoms. In this case, proprionyl-CoA replaces acetyl-CoA in the initial condensation reaction.

ELONGATION OF PALMITIC ACID

The end-product of the fatty acid synthetase system in most organisms is palmitic acid or palmitoyl-CoA, despite the fact that the second most abundant saturated fatty acid, stearic acid (18:0), would require only one more condensation reaction. One reason for the termination at 16 carbons is the fact that palmitoyl-CoA functions as a feedback inhibitor of the fatty acid synthetase system. Synthesis of stearic acid and longer chain saturated fatty acids requires one of two different types of enzyme systems. In animals, the first of these systems is associated with the mitochondria and employs acetyl-CoA as the C_2 donor. The second system is associated with the endoplasmic reticulum and utilizes malonyl-CoA. The reactions of the latter are similar to those of the fatty acid synthetase system except that ACP is not used as the acyl carrier. Two distinct elongation systems also function in plant tissues; however, the predominant mechanism employs malonyl-ACP as the C_2 donor rather than acetyl-CoA. Sites for the elongation reactions are the same as those for *de novo* synthesis of palmitic acid.

SYNTHESIS OF SHORT- AND MEDIUM-CHAIN FATTY ACIDS

An exception to the general rule that palmitoyl-CoA is the end-product of fatty acid synthetase occurs in the mammary tissue of lactating females. The milk fat of most mammalian species characteristically contains a much higher proportion of medium-chain (C_8–C_{12}) fatty acids than is found in the depot fat of the same species. In ruminants, these medium-chain fatty acids are accompanied by appreciable amounts of short-chain butyric acid (4:0) and hexanoic acid (6:0) (Davies et al., 1983). Both short- and medium-chain fatty acids are synthesized *de novo* in the mammary gland by a fatty acid synthetase complex similar to that described above. The process, however, requires a special enzyme called medium-chain acylthioesterase, which is present in the cytosol of the mammary gland. When this enzyme is present, chain-termination on fatty acid synthetase occurs when the growing chain is only 8 to 10 carbons long; in the absence of medium-chain acylthioesterase, fatty acid synthetase continues until the acyl chain is 16 carbons long (Dils, 1983). The cleavage and release of medium-chain fatty acids suggests that medium-chain acylthioesterase may have a more hydrophilic binding site than the long-chain acylthioesterase. In ruminants, medium-chain acylthioesterase activity results in the release of acyl-CoA esters (rather than free fatty acids as in nonruminants) that can be incorporated directly into triacylglycerols without prior activation (Grunnet and Knudsen, 1981; see section on triacylglycerol synthesis).

FORMATION OF UNSATURATED FATTY ACIDS

Most organisms are capable of desaturating palmitic and stearic acids to their corresponding monoenes, palmitoleic and oleic acids. The double bond is inserted in the 9,10 position relative to the carboxyl group; hence, the notation for palmitoleic acid is 16:1 (*n*-7) and 18:1 (*n*-9) for oleic acid. In plants and animals the introduction of the double bond is effected by a desaturase enzyme that requires molecular oxygen, NADH or NADPH, and an electron carrier, usually ferredoxin in plants and cytochrome b_5 in animals. A similar aerobic system also operates in some microorganisms (yeast), although an anaerobic pathway for formation of monoenoic fatty acids also has been elucidated.

Polyunsaturated fatty acids (i.e., containing two or more double bonds) also occur in all higher organisms (they are absent in bacteria); however, major differences exist between plants and animals in their ability to synthesize *de novo* various groups. Four families of polyunsaturated acids are recognized, based on the distance between the terminal methyl group and the nearest double bond.

Palmitoleic family	CH_3—$(CH_2)_5$—CH=CH—	*n*-7
Oleic family	CH_3—$(CH_2)_7$—CH=CH—	*n*-9
Linoleic family	CH_3—$(CH_2)_4$—CH=CH—	*n*-6
Linolenic family	CH_3—CH_2—CH=CH—	*n*-3

All polyunsaturated fatty acids are formed from these precursors by further elongation and/or desaturation mechanisms. Animals, with the possible exception of a terrestrial snail (Allen, 1976) and certain insects (Gilmour, 1965), cannot synthesize *de novo* linoleic acid (18:2) or linolenic acid (18:3) and, therefore, must obtain these precursors from their diet in order to manufacture polyunsaturated fatty acids in the *n*-3 and *n*-6 families. Both linoleic and linolenic acids are produced in abundance by plants. The various pathways used to synthesize polyunsaturated fatty acids in mammals are outlined in Figure 2.1.

SYNTHESIS OF TRIACYLGLYCEROLS

As indicated in Chapter 1, only small quantities of free fatty acids are found in the tissues and fluids of plants and animals. Instead the fatty acids and their CoA esters that are synthesized *de novo* or obtained from the diet rapidly react with glycerol-3-phosphate to form triacylglycerols, phosphoglycerides, or glycolipids. Glycerol-3-phosphate is derived primarily from dihydroxyacetone phosphate that is produced during glycolysis, or it may be formed from the direct phosphorylation of glycerol. In the formation of triacylglycerols, glycerol-3-phosphate first reacts with one molecule of fatty acyl-CoA to form lysophosphatidic acid, which in turn reacts with a second fatty acyl-CoA to form phosphatidic acid (see Chapter 1). Hydrolysis of the phosphatidic acid yields a diacylglycerol, which then reacts with a third

$$
\begin{array}{ccc}
CH_2OH & & CH_2OH \\
| & & | \\
C\!\!=\!\!O & + NADH + H^+ \longleftrightarrow HO\!-\!C\!-\!H & + NAD \\
| & & | \\
CH_2OPO_3H_2 & & CH_2OPO_3H_2 \\
\end{array}
$$

Dihydroxyacetone phosphate   Glycerol-3-phosphate

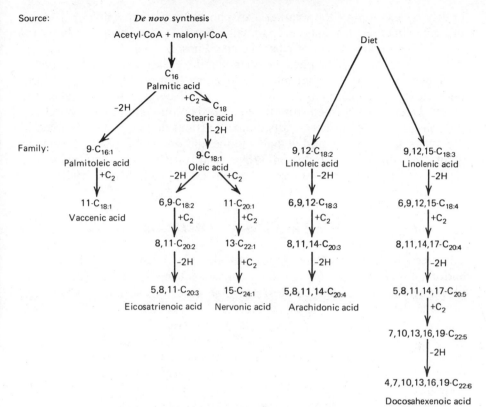

FIGURE 2.1. Pathways of polyunsaturated fatty acid synthesis in mammals. The large numbers indicate the position of the double bonds counting from the carboxyl carbon; the subscripts indicate the number of carbon atoms, with the number of double bonds to the right of the colon. Modified from E. L. Smith et. al., *Principles of Biochemistry: General Aspects,* 7th ed. Copyright 1983 by McGraw-Hill Book Co.

fatty acyl-CoA to form a triacylglycerol. Most cells are capable of triacylglycerol synthesis, but the primary sites are the liver and fat cells in animals and the oil seeds of plants. In higher animals, the intestinal mucosa also synthesizes triacylglycerols from monoacylglycerols formed following digestion of triacylglycerols in intestinal lumen. This synthetic pathway is unique in that it does not require phosphatidic acid as an intermediate.

SYNTHESIS OF PHOSPHOGLYCERIDES

Phosphatidic acid is also an important intermediate in the synthesis of phosphoglycerides. In addition, the synthetic pathways, which are basically the same in plants and higher animals, require the presence of cytidine nucleotides (e.g., cytidine triphosphate, CTP), which serve as carriers of either the polar head group or of phosphatidic acid. The process can be illustrated briefly by examining the synthesis of phosphatidylethanolamine, a major phosphoglyceride in plants and animals. The first step in its synthesis involves the phosphorylation of ethanolamine. Phosphoethanolamine then reacts with CTP to form CDP-phosphoethanolamine. The latter is an activated compound that transfers phosphoethanolamine to a diacylglycerol to form phosphatidylethanolamine. As in the case of triacylglycerol synthesis, the diacylglycerol was formed from the hydrolysis of the phosphate group from phosphatidic acid. Further details on the reactions and enzymes involved in the synthesis of phosphatidylethanolamine and other phosphoglycerides in plants and animals are found in White et al. (1973), Kates and Marshall (1975), Lehninger (1975), Gurr and James (1975), and Mudd (1980).

THE PROCESSING OF LIPIDS

The preceding sections have emphasized the *de novo* synthesis of saturated and unsaturated fatty acids and their subsequent incorporation into triacylglycerols and phosphoglycerides. In some cases, these end-products are stored at the site of synthesis, where they can provide a source of energy and carbon atoms for future metabolic needs (triacylglycerols) or become integral parts of cellular and subcellular structures (phosphoglycerides). In most instances, however, the synthesis site and the final site of storage or usuage are different, and it is necessary to transport these lipids to new localities within the organism. Furthermore, animals that depend heavily on exogenous lipid sources for essential precursors or have to supplement those lipids generated by *de novo* synthesis must first digest, absorb, and resynthesize these compounds before transporting them to various tissues in the body. Although the pathways by which lipids are synthesized by plants and animals are basically similar, there are significant differences in the means by which these lipids are transported. These differences will be examined in this section, beginning with the processes that precede lipid transport in animals. Further comparisons of sites and forms of lipid storage are made in the final section of this chapter.

ANIMALS

Digestion

Although the bulk of dietary lipid is triacylglycerol, smaller quantities of phosphoglycerides, sphingolipids, sterol esters, and fat-soluble vitamins also are consumed. The first step in their processing is the digestion of these large molecules into smaller and more polar molecules that can be absorbed from the digestive tract. Digestion requires the enzymatic action of lipases and carboxylic esterases, plus bile salts that promote emulsification and solubilization of the lipids. The secretion, function, and importance of bile salts are discussed in detail in Chapter 10.

The major enzyme involved in the digestion of triacylglycerol in vertebrates is pancreatic lipase, which converts triacylglycerol into 2-monoacylglycerol and two free fatty acids. A similar cleavage is accomplished by gastric and lingual lipases, but their significance is uncertain (Masoro, 1977). The pancreatic juice also contains other lipases that catalyze the hydrolysis of phosphoglycerides, cholesterol esters, and short-chain fatty acid esters. One such enzyme, phospholipase A, liberates lysophosphatidylcholine (Chapter 1) from phosphatidylcholine. Lysophosphatidylcholine is itself a good detergent and aids bile salts in the emulsification of the dietary lipids (White et al., 1973). Simple esterases appear to be widely distributed among the invertebrates as well, including representatives of phyla in which intracellular digestion predominates.

Absorption

Before free fatty acids, monoacylglycerols, and lysophosphoglycerides are absorbed into the mucosal cells of the small intestine, they first diffuse out of the emulsion particles and into micelles. Micelles are spherical bodies comprised of bile salts, phosphatidylcholine, and cholesterol arranged so that the hydrophilic heads are exposed to the aqueous medium, while the hydrophobic hydrocarbon chains project inward (Figure 2.2; see also Figure 10.6). It is now generally agreed that the digestion products first dissociate from the micelle before entering the cell. Inside the mucosal cell, the fatty acid, monoacylglycerols, and lysophosphoglycerides are transported to the endoplasmic reticulum, where they are resynthesized into triacylglycerols and phosphoglycerides. Small amounts of free fatty acids also combine with cholesterol, which also diffuses from the intestinal lumen into the mucosal cell, to form cholesterol esters.

Transport

Lipids absorbed from the digestive tract are transported to various organs in the body via the circulatory system. Unesterified fatty acids containing less than 10 carbon atoms enter the portal system, bind to plasma albumin, and are transported directly to the liver. Longer chain fatty acids tend to become incorporated into triacylglycerol molecules. The regenerated triacylglycerols and other lipids, however, are hydrophobic and cannot be effectively transported by the aqueous blood plasma. Their transport requires that they first become solubilized by combining with lipoproteins present in the plasma. Four major lipoprotein classes are recognized in humans, each discrete with respect to the type and amount of lipid present (Table 2.1). In addition, each lipoprotein class contains one or more apoproteins. The latter are the specific lipid binding proteins of the lipoproteins, and are

FIGURE 2.2.   Micelles in aqueous phase of lumen.

usually designated using as nomenclature A, B, C, and so forth (Simons and Gibson, 1980).

Chylomicrons are large particles formed in the intestinal mucosa. Their primary role is to transport dietary triacylglycerols to storage sites (adipose tissue) and other peripheral tissues. Chylomicrons enter the lymphatic capillaries and reach the blood via the thoracic duct. Once in the general

TABLE 2.1

Major classes of human plasma lipoproteins. Composition data are in weight percent

	Chylomicrons	Very low density lipoproteins (VLDL)	Low density lipoproteins (LDL)	High density lipoproteins (HDL)
Density (g ml^{-1})	0.94	0.94–1.006	1.006–1.063	1.063–1.21
Particle size (nm)	75–1000	30–50	0–20	7.5–10
Triacylglycerols	85–90	50–55	6–10	3–6
Cholesterol esters	3–4	14–16	35–45	12–18
Cholesterol	2–3	6–8	8–12	2–4
Phospholipids	6–8	16–20	20–25	25–30
Protein	1–2	8–10	18–22	47–52

SOURCE: Data from Lehninger (1975) and Simons and Gibson (1980).

circulation, the apoprotein C, derived from high density lipoproteins (HDL), is added to the chylomicron surface. Removal of triacylglycerol from the chylomicron involves the enzyme lipoprotein lipase, which is bound to the plasma membrane of endothelial cells lining the capillaries (Masoro, 1977). Lipoprotein lipase hydrolyzes the triacylglycerol into diacylglycerol and fatty acid, which are taken up by the extrahepatic tissue; the remnants of the chylomicron particles are removed by the liver (Allen, 1976). These processes are summarized in Figure 2.3.

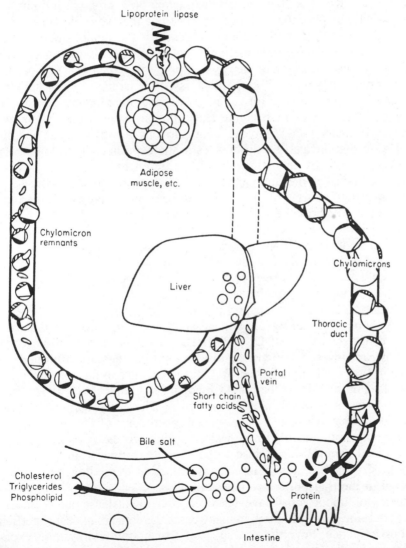

FIGURE 2.3. Schematic representation of chylomicron metabolism. Reprinted, by permission, from Levy et al. (1971).

The metabolic functions of very low density lipoproteins (VLDL) are similar to those of chylomicrons in that they both transport triacylglycerols. A major distinction here is that VLDL are produced primarily by the liver and transport triacylglycerols derived from endogenous (i.e., liver) rather than dietary sources. The intestine also synthesizes a triacylglycerol-containing VLDL during postabsorptive and fasting states (Simons and Gibson, 1980). Low density and high density lipoproteins (LDL and HDL), in contrast, are responsible for the transport of cholesterol to peripheral tissues for membrane synthesis. Cholesterol and cholesterol esters are most abundant in LDL, whereas HDL are approximately 50% protein, and phospholipids (especially phosphatidylcholine) constitute the major lipid class (Table 2.1). HDL are of particular interest because of a demonstrated inverse relationship between their circulating levels and cardiovascular disease risk.

The previous discussion of the physical properties, chemical composition, and metabolic function of lipoproteins is based largely on the human lipoprotein system. Except for a few avian and mammalian species, our knowledge of lipoprotein structure and function in other animals is much more poorly understood. Moreover, comparisons between human and other animals are complicated by differences in the experimental techniques and terminology used by investigators, plus differences in the age, sex, and nutritional status of the animals tested. These problems are addressed by Chapman (1980) in his extensive review of the comparative aspects of animal lipoproteins. Much of the information presented in the following synopsis is taken from this review.

Data on lipoproteins in invertebrates are limited to crustaceans and insects (Arthropoda). Relatively complex HDL with special lipid-binding proteins have been detected in the hemolymph of several species of crab and lobster. These hemolymph lipoproteins apparently transport lipid to sites of storage and utilization as well as transport pigments, lipid, and protein to maturing oocytes in females during vitellogenesis. Soluble lipoproteins that correspond to human LDL, HDL, and VHDL in terms of density have been isolated from insect hemolymph. Insects (and perhaps all arthropods) are unique in that lipids transported by LDL are mainly in the form of diacylglycerols (Figure 2.4) (Gilbert and Chino, 1974; Chino et al., 1981). Insect LDL also transport sterols and hydrocarbons. These two compounds are of particular interest because both are major constituents of the cuticular lipids and play an important role in the waterproofing process (Chapter 6). Thus, the chemical composition of insect LDL is quite distinct from human LDL. High density lipoproteins of insects also function in vitellogenesis and are involved in the transport of certain hormones.

Studies of lipoproteins have been reported for at least one member of each vertebrate class, although in many cases the data are at best superficial. Existing evidence, nevertheless, suggests that lipoprotein systems even in the most primitive vertebrates resemble those of humans in complexity and

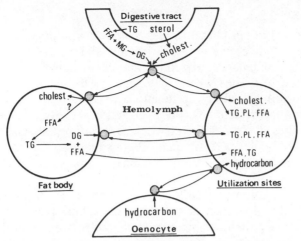

FIGURE 2.4. A scheme proposed for lipid transport in insects. Small circles, diacylglycerol-carrying lipoprotein (DGLP) molecule; TG, triacylglycerol; DG, diacylglycerol; MG, monoacylglycerol; FFA, free fatty acid; PL, phospholipid. Insect DGLP is formed primarily in the fat body and released into hemolymph. It loads various lipids, probably at the surface of different tissues, and unloads these lipids at the site of utilization. Reprinted, by permission, from Chino and Kitazawa (1981).

molecular arrangement (Chapman, 1980). In fish, the intraluminal hydrolysis of triacylglycerols is complete, resulting in the absorption of free fatty acids. Earlier studies indicated that these fatty acids were transported directly to the tissues via the portal system; however, Sire et al. (1981) demonstrated that the long-chain fatty acids are esterified into triacylglycerols by the intestinal cells and transferred as VLDL-like particles to the lymph. Sire and Vernier (1981) also have shown that, depending on the nature of the ingested fatty acids, trout are also capable of synthesizing chylomicrons of various sizes. Lipid transport in amphibians and reptiles has received little study. The amphibians examined thus far have low concentrations of all lipoprotein classes, whereas concentrations are much greater in reptiles, with LDL predominating. Birds, which in contrast have received considerable attention, are unique in that triacylglycerols are transported via the portal system as large VLDL rather than as chylomicrons, which enter the lymphatic system in mammals. Birds also increase the synthesis and circulating concentrations of VLDL in response to demands for egg formation. This is in marked contrast to the situation in arthropods, in which HDL or VHDL with relatively low lipid content are able to provide sufficient lipid for vitellogenesis (Chapman, 1980). Lipid transport processes in nonhuman mammals basically mirror those of humans despite major differences in diets and digestive physiology of the various groups; however, lipoprotein profiles and the chemical composition of specific lipoprotein classes are highly variable, even between members of a given taxon.

Storage

Lipids transported to peripheral tissues may be incorporated into structural elements (e.g., cuticle, membranes) or immediately oxidized to provide energy for cell maintenance. If food is abundant, however, excess dietary lipids are stored in tissues where they serve as a fuel reserve and, at the same time, perform a variety of functions of adaptive significance to the animal. Among the vertebrates, the liver and especially adipose tissue are the principal sites of lipid storage, whereas in arthropods and many other invertebrates, fat bodies perform an analogous function. The dominant lipid class in these tissues, with some exceptions, is triacylglycerol, which typically accounts for 98 to 99% of the lipid present.

Adipose tissue, commonly called fat, is a specialized form of loose connective tissue. It is a whitish or yellowish tissue that is widely distributed throughout the body, being especially abundant between muscles, beneath skin (subcutaneous), on the surface of the heart and kidneys, behind the eyeballs, in the abdominal mesenteries, and around certain joints (Hole, 1981). The cells within the adipose tissue that function as lipid reservoirs are called adipocytes. The adipocytes accumulate lipid (primarily triacylglycerol), which coalesces into a single droplet, forcing the cytoplasm and nucleus to the periphery of the cell (Figure 2.5a). The fat droplet, cytoplasm, and nucleus are surrounded by a plasma membrane that separates the adipocyte from the extracellular fluid. Fat globules usually range between 10 and 120 μm in diameter, but may reach 200 μm in the adipose tissue of obese animals (Gurr, 1980).

Another type of adipose tissue, brown fat, also occurs in mammals, including humans, and is especially prominent in hibernators and young mammals. The largest deposit of brown fat is located in the interscapular region. Structurally, brown fat is quite different from white adipose tissue. The cells of brown fat have a central nucleus and an abundance of cytoplasm that contains many fat droplets approximately 1 to 3 μm in diameter (Figure 2.5b) (Masoro, 1968). The tissue is highly vascularized and the cells contain numerous mitochondria. Unlike white adipose tissue, in which the reserve fat is mobilized via the blood and oxidized in cells of other tissues, the smaller fat globules in brown fat provide fuel for mitochondria in the same cell for heat production for hibernating and neonatal animals (Gurr, 1980). The degradation and mobilization of stored fat and other processes related to its utilization as an energy source and as a producer of heat for both types of adipose tissue are discussed further in Chapters 6 and 7.

PLANTS

The digestion, absorption, transport, and storage of dietary lipid are, of course, processes unique to animals. Each of these processes, nevertheless,

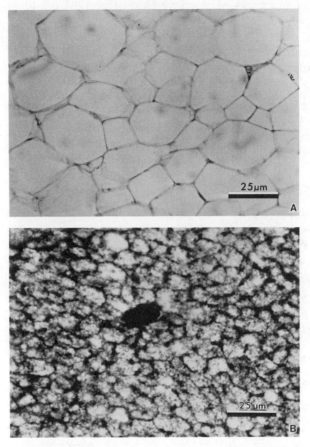

FIGURE 2.5. (A) Photomicrograph of white adipose tissue from an adult rabbit. The lipid has been extracted, leaving only a thin rim of cytoplasm for each cell. (B) Photomicrograph of brown adipose tissue from four-day old rabbit. These polygonal cells contain more cytoplasm than the white adipose tissue and have many small lipid droplets. The dark circle is a blood vessel cut in cross-section.

has its counterpart at some stage in the *de novo* production and metabolism of lipids in plants. For example, photosynthetic sugars not required for the immediate energy or structural demands of cells within the leaves can be transported to all parts of the plant. Sugars are taken up from the vascular tissue (phloem) and converted into the precursors necessary for the synthesis of palmitic acid and ultimately triacylglycerols. Note that plants, in contrast to animals, do not transport actual lipid molecules, but instead synthesize these from transported nonlipoidal precursors. Seeds, which are the principal storage site of triacylglycerols in certain plants, also contain lipases that hydrolyze these into glycerol and fatty acids. Oxidation of the latter produces acetyl-CoA, which must be converted into sugars so that the carbon can be transported to growing tissues of the seedling, where it provides structural compounds and energy. The conversion of acetyl-CoA to sugars involves a complex sequence of enzymatic reactions that constitute the glyoxylate cycle. This pathway is discussed in detail in Chapter 6. The remainder of this chapter is devoted to the accumulation of lipids during seed development and the nature of storage sites within the seed.

Seed Morphology and Lipid Content

Each seed contains an embryo, which grows into a mature plant after germination, endosperm, which provides nourishment for the growing seedling, and a protective seed coat (testa) (Figure 2.6). The embryo contains the cotyledons, or seed leaves. In monocots (e.g., corn), the single seed leaf is small and occupies a small part of the seed, whereas in dicots (e.g., peanuts, sunflower) the two seed leaves occupy the greatest portion of the seed (Bidwell, 1979). The endosperm, depending on the species, may accumulate primarily carbohydrate (starch) or lipid during seed development. This tissue may remain as a main tissue in the embryo until germination (monocots) or contribute its contents to the embryo during development and be highly reduced in the mature seed (most dicots). In the latter case, the cotyledons develop into photosynthesizing leaves following seedling growth. Thus, the triacylglycerols of oil-rich seeds (e.g., peanuts, sunflower, coconut) may be stored in endosperm, cotyledons, or both, depending on the species (Gurr, 1980). There also appears to be a relationship between the primary storage site within the seed and the fatty acid composition of the triacylglycerol. Seeds that have cotyledons as major storage organs usually contain triacylglycerols that are rich in linolenic acid (Appelqvist, 1975).

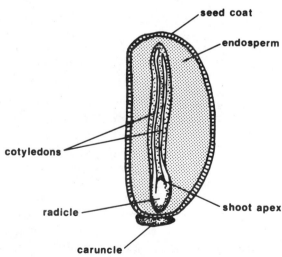

FIGURE 2.6. Drawing of section through castor bean, *Ricinus communis,* seed showing edge view of embryo embedded in endosperm.

The maturing seed is surrounded by the fruit, which may be "dry" or "fleshy" (Figure 2.7). In olive, oil palm, and avocado, the fleshy fruit also contains substantial amounts of triacylglyccrol (Hitchcock and Nichols, 1971). The triacylglycerol and other foodstuffs contained in the fruit are unavailable to the enclosed seed; however, they function biologically in wild plants by promoting seed dispersal as a result of their consumption by animals.

Lipid Accumulation in Developing Seeds

Seed development begins with the fertilization of the ovule in the flower. The maturation of the seed and fruit usually requires several weeks to months after flowering, and in oil-rich seeds, is associated with an increasing lipid content. The rate at which lipid accumulates, however, is seldom constant. In *Crambe abyssinica* (Cruciferae), whose pattern of lipid accumulation is typical of most oil-rich species, polar lipids, such as phosphoglycerides and galactosylglycerides, predominate during the initial stage of development (Gurr et al., 1972). The principal fatty acid at this time is linolenic acid (18:3). Triacylglycerols, which are present in only small amounts during this initial stage, accumulate rapidly during the second phase (8 to 30 days after flowering in *C. abyssinica*), and are accompanied by an increase in erucic acid (22:1). Linolenic acid, in contrast, decreases markedly when expressed as a percentage of total fatty acids present, but actually increases slightly when expressed on a per seed basis (Gurr et al., 1972). During the final phase, which ends with seed maturation, lipid accumulation ceases. Developmental patterns and associated changes in lipid composition for other oil-rich species are described by Appelqvist (1975) and the references cited therein.

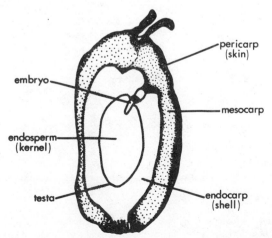

FIGURE 2.7. Section through mature oil palm fruit showing enclosed seed. Redrawn from photograph in Gurr (1980).

Oil Structure Organelles

A major portion of the lipids in mature oil-rich seeds are localized in specialized spherical structures within the cytoplasm. Although referred to here simply as lipid bodies (Figure 2.8), they have also been called spherosomes, oleosomes, oil bodies, and reserve oil droplets (Appelqvist, 1975; Gurr, 1980). Oil bodies are rich in triacylglycerols, with smaller amounts of phospholipid and protein also present. Their numbers and diameter vary according to the plant species and developmental stage of seed. It is generally agreed that oil bodies originate as small vesicles that become detached from endoplasmic reticulum, but the exact mechanism by which this takes place is not totally clear. The nature of the membrane surrounding each oil body has also been a subject of controversy. The failure of oil bodies to coalesce into one large lipid droplet indicates the presence of a limiting membrane. When viewed with an electron microscope, however, the thickness of this membrane is only one-half that of a typical unit membrane (see Chapter 3).

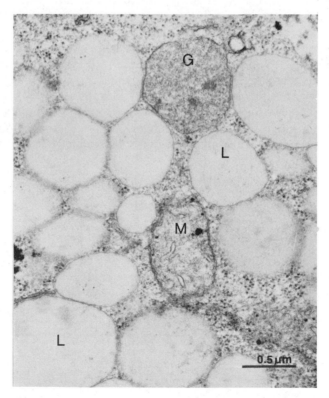

FIGURE 2.8. Electron micrograph of a section through a cotyledon of a germinated (24 hr) cotton seed. Numerous lipid bodies (L) occur in the cytoplasm around a glyoxysome (G) and mitochondrion (M). Note the single membrane around the lipid bodies and glyoxysome as opposed to two membranes bounding the mitochondrion. Courtesy of Richard N. Trelease; micrograph by Diane C. Doman.

Evidence suggests that the oil body membranes are indeed ''half membranes'' whose polar, hydrophobic surfaces face the triacylglycerols contained on the inside (Salisbury and Ross, 1978). This hypothesis supports the idea that triacylglycerols accumulate between the two halves of an endoplasmic reticulum membrane, forcing the halves apart, and thus, producing a mature oil body that ultimately detaches to become an independent particle.

There are both similarities and differences between the lipid storage organelles of plants and animals. As described earlier, the adipocytes of animals are large diameter cells composed almost entirely of a single oil droplet; each cell is enclosed by a typical unit membrane. Seed oil bodies, in contrast, are much smaller in diameter, and although they pack closely together, never coalesce. Moreover, the membrane that surrounds the oil bodies, when visible, does not appear to have the classic tripartite structure. As Gurr (1980) has noted, there is a closer parallel between the cotyledon oil bodies and the cells of brown adipose tissue, which are packed with discrete smaller diameter (1 to 3 μm) oil droplets. Further research may result in findings that reflect a more basic structural theme than is currently recognized.

REFERENCES

Allen, W. V. (1976). Biochemical aspects of lipid storage and utilization in animals. *Amer. Zool.* **16**:631–647.

Appelqvist, L.-Å. (1975). Biochemical and structural aspects of storage and membrane lipids in developing oil seeds. In: *Recent Advances in the Chemistry and Biochemistry of Plant Lipids* (Galliard, T. and E. I. Mercer, Eds.), pp. 247–286. Academic Press, New York.

Bidwell, R. G. S. (1979). *Plant Physiology,* 2nd ed. MacMillan, New York.

Chapman, M. J. (1980). Animal lipoproteins: Chemistry, structure, and comparative aspects. *J. Lipid Res.* **21**:789–853.

Chino, H. and K. Kitazawa (1981). Diacylglycerol-carrying lipoprotein of hemolymph of the locust and some insects. *J. Lipid Res.* **22**:1042–1052.

Chino, H., J. Katase, R. G. H. Downer, and K. Takahasi (1981). Diacylglycerol-carrying lipoprotein of hemolymph of the American cockroach: Purification, characterization, and function. *J. Lipid Res.* **22**:7–15.

Davies, D. T., C. Holt, and W. W. Christie (1983). The composition of milk. In: *Biochemistry of Lactation* (Mepham, T. B., Ed.), pp. 71–117. Elsevier, Amsterdam.

Dils, R. R. (1983). Milk fat synthesis. In: *Biochemistry of Lactation* (Mepham, T. B., Ed.), pp. 141–157. Elsevier, Amsterdam.

Gilbert, L. I. and H. Chino (1974). Transport of lipid in insects. *J. Lipid Res.* **15**:439–456.

Gilmour, D. (1965). *The Metabolism of Insects.* W. H. Freeman, San Francisco.

Green, D. E. (1960). The synthesis of fat. *Sci. Am.* **202**:46–51.

Grunnet, I. and J. Knudsen (1981). Direct transfer of fatty acids synthesised 'de novo' from fatty acid synthetase into triacylglycerols without activation. *Biochem. Biophys. Res. Commun.* **100**:629–636.

Gurr, M. I. (1980). The biosynthesis of triacylglycerols. In: *The Biochemistry of Plants*, Vol. 4 (Stumpf, P. K., Ed.), pp. 205–248. Academic Press, New York.

Gurr, M. I. and A. T. James (1975). *Lipid Biochemistry: An Introduction*, 2nd ed. Chapman and Hall, London.

Gurr, M. I., J. Blades, and R. S. Appleby (1972). Studies on seed-oil triglycerides. The composition of *Crambe abyssinica* triglycerides during seed maturation. *Eur. J. Biochem.* **29**:362–368.

Hitchcock, G. and B. W. Nichols (1971). *Plant Lipid Biochemistry*. Academic, Press, New York.

Hole, J. W., Jr. (1981). *Human Anatomy and Physiology*, 2nd ed. Wm. C. Brown, Dubuque, Iowa.

Kates, M. and M. O. Marshall (1975). Biosynthesis of phosphoglycerides in plants. In: *Recent Advances in the Chemistry and Biochemistry of Plant Lipids* (Galliard, T. and E. I. Mercer, Eds.), pp. 115–159. Academic Press, New York.

Lehninger, A. L. (1975). *Biochemistry*. Worth, New York.

Levy, R. I., D. W. Bilheimer, and S. Eisenberg (1971). The structure and metabolism of chylomicrons and very low density lipoproteins (VLDL). In: *Structure and Metabolism of Chylomicrons and VLDL* (Smellie, R. M. S., Ed.), pp. 3–17. Biochem. Soc. Symp., Vol. 33.

Masoro, E. J. (1968). *Physiological Chemistry of Lipids in Mammals*. W. B. Saunders, Philadelphia.

Masoro, E. J. (1977). Lipids and lipid metabolism. *Ann. Rev. Physiol.* **39**:301–321.

Mudd, J. B. (1980). Phospholipid biosynthesis. In: *The Biochemistry of Plants,* Vol. 4 (Stumpf, P. K., Ed.), pp. 249–282. Academic Press, New York.

Nelson, D. R. and R. W. Rinne (1975). Citrate cleavage enzyme from developing soybean cotyledons. Incorporation of citrate carbon into fatty acids. *Plant Physiol.* **55**:69–72.

Nelson, D. R. and R. W. Rinne (1977a). Citrate metabolism in soybean cotyledons. *Plant Cell Physiol.* **18**:405–412.

Nelson, D. R. and R. W. Rinne (1977b). The role of citrate in lipid synthesis in developing soybean cotyledons. *Plant Cell Physiol.* **18**:1021–1027.

Salisbury, F. B. and C. W. Ross (1978). *Plant Physiology,* 2nd ed. Wadsworth, Belmont, California.

Simons, L. A. and J. C. Gibson (1980). *Lipids: A Clinician's Guide*. University Park Press, Baltimore.

Sire, M.-F. and J.-M. Vernier (1981). Etude ultrastructurale de la synthese de chylomicrons au cours de l'absorption intestinale des lipides chez la Truite. Influence de la nature des acides gras ingeres. *Biol. Cell* **40**:47–62.

Sire, M.-F., C. Lutton, and J.-M. Vernier (1981). New views on intestinal absorption of lipids in teleostean fishes: An ultrastructral and biochemical study in the rainbow trout. *J. Lipid Res.* **22**:81–94.

Smith, E. L., R. L. Hill, I. R. Lehman, R. J. Lefkowitz, P. Handler, and A. White (1983). *Principles of Biochemistry: General Aspects*, 7th ed. McGraw-Hill, New York.

Stumpf, P. K. (1980). Biosynthesis of saturated and unsaturated fatty acids. In: *The Biochemistry of Plants,* Vol. 4 (Stumpf, P. K., Ed.), pp. 177–204. Academic Press, New York.

White, A., P. Handler, and E. L. Smith (1973). *Principles of Biochemistry*, 5th ed. McGraw-Hill, New York.

3

BIOMEMBRANES

DISTRIBUTION AND FUNCTIONS

Membranes are ubiquitous structures in all living organisms. The surface of every cell (plant, animal, bacterium) is covered by a plasma membrane that separates it from its environment and enables it to selectively control the entry and exit of substances (Figure 3.1). Plants and bacteria, in addition, have a cell wall that surrounds the plasma membrane. In higher plants, the cell wall consists largely of cellulose chains interwoven to form complex structures. Bacterial cell walls consist of layers that differ chemically from one another and from species to species. The major chemical constituent is usually a mycopeptide, but certain bacterial strains (gram-negative) also have a protective lipid layer as part of the cell wall. Although animal cells generally lack such a cell wall, they often possess extraneous coats (e.g., collagen) that function similarly to cell walls of plants and bacteria. In addition to the plasma membrane, virtually all subcellular organelles are made of or surrounded by membranes or pieces of membrane. Examples here include the nucleus, mitochondria, endoplasmic reticula, lysosomes, Golgi complexes, chloroplasts, and vacuoles (Figure 3.1). The membranes surrounding or composing the cell organelles separate the organelles from the surrounding cytoplasm and provide barriers to the movement of molecules and ions between the various compartments within the cell.

Membranes perform many other important functions in addition to compartmentation and controlling the extracellular and intracellular exchange of materials. They provide a structural framework for the attachment of enzymes, insuring that only certain types of chemical reactions occur in association with particular cell organelles, and that the enzymes are oriented or aligned in a manner that provides for maximum effective interaction (e.g., enzymes associated with oxidative phosphorylation in mitochondria).

45

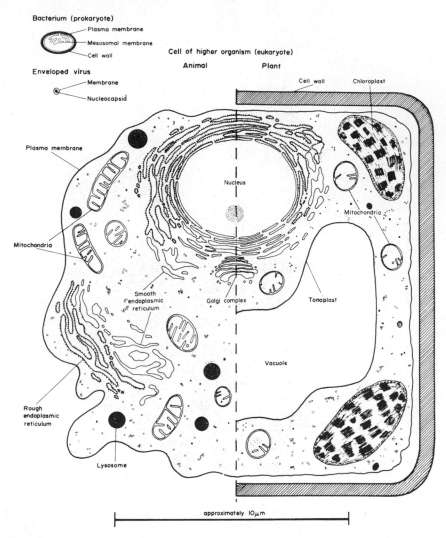

Bacterium (prokaryote)

Plasma membrane

Mesosomal membrane

Cell wall

Enveloped virus

Membrane

Nucleocapsid

Cell of higher organism (eukaryote)

Animal   Plant

Cell wall   Chloroplast

Plasma membrane

Nucleus

Mitochondria

Mitochondria

Smooth endoplasmic reticulum

Golgi complex   Tonoplast

Rough endoplasmic reticulum

Vacuole

Lysosome

approximately 10μm

FIGURE 3.1. Diagrammatic representation of eukaryote (plant and animal) and prokaryote (bacterium) cell. Reprinted, by permission, from Finean et al. (1978).

The outer surface of plasma membranes also contains specific binding sites that function as receptors for hormones and other chemical messengers. Thus, the plasma membrane controls not only materials but information received and transmitted across the cell boundary. Membranes also provide a matrix for reactions involved in the transduction of energy. Chloroplast membranes convert light energy from the sun into chemical energy, whereas mitochondria subsequently convert the chemical energy into ATP and other high energy compounds. Finally, processes such as motility, irritability, transmission of impulses, recognition and adhesion, all of which are characteristic of living systems, are intimately associated with membrane function.

MEMBRANE COMPOSITION

Biomembranes are composed primarily of lipids and proteins. Other constituents include carbohydrates, metal ions, and water. This rather general statement concerning chemical composition is necessitated by the fact that membrane constituents, particularly the types of proteins and lipids and their ratios, vary according to the type of organism, the type of membrane, and the type of tissue in which the membrane occurs (Table 3.1). Further,

TABLE 3.1
Composition of membranes

Membrane	mg lipid/mg protein	Major lipids[a]	Phospholipid/cholesterol ratio (mole/mole)
Plasma	0.5–1.0	Ch, PC, PS, PE, Sph	0.4–1.0
Myelin	3.5–4.0	PC, PE, Sph, Ch	0.7–1.2
Golgi	1.2	PC, PE, Sph, Ch	0.45–0.5
Rough endoplasmic reticulum	0.2–0.5	PC, PE, PI	0.06–0.1
Smooth endoplasmic reticulum	1.2	PC, PE, PI	0.1–0.2
Lysosomal	0.3	PC, PE, Sph, Ch	0.5
Outer mitochondrial	0.4	PC, PE, PI	0.1–0.2
Inner mitochondrial	0.3	PC, PE, DPG	0.06
Nuclear	0.2–0.6	PC, PE, PI	0.1
Chloroplast	0.6	GalL, PG, Chl, PC	
Gram + bacteria	0.3–0.5	PE, DPG	
Mycoplasma	0.3	Ch, TG, PG	

SOURCE: Reprinted, by permission, from Robinson (1975).

[a]PL :Phosphatidyl choline
 PS :Phosphatidyl serine
 PE :Phosphatidyl ethanolamine
 PI :Phosphatidyl inositol
Sph :Sphingomyelin
 Ch :Cholesterol
DPG:Diphosphatidyl glycerol
GalL:Galactolipid
 PG :Phosphatidyl glycerol
Chl :Chlorophyll
 TG :Triglyceride

membrane composition may be altered by age, diet, or environmental factors, often in a manner that appears to be adaptive. Although the emphasis here is on lipids, proteins and carbohydrates (where they occur) are also discussed, as these components are likely to be important in determining how membrane lipids perform their roles.

Lipids

Chemical analysis and permeability characteristics indicate that biomembranes are rich in lipids. As shown in Table 3.1, the amount (25 to 70% by weight) and composition of membrane lipids is quite variable. Phosphoglycerides are usually the dominant lipid class, especially in animal cell membranes, with sphingomyelins and cholesterol next in abundance. The phosphoglycerides are represented primarily by phosphatidylcholine (PC), and phosphatidylethanolamine (PE); smaller quantities of phosphatidylserine (PS), phosphatidylinositol (PI), and lysophosphatidylcholine (LPG) are also present in most membranes (Figure 3.2). Sphingomyelins and cholesterol are found largely in the plasma, Golgi, and erythrocyte membranes. In the latter, cholesterol content is approximately equimolar to the phospholipids (phosphoglycerides + sphingomyelin). The lipid composition of plant cell membranes is basically similar to those described for animal cells, except for the presence of significant quantities of sugar-containing glycosylglycerols in chloroplast membranes. A similar occurrence of glycosylglycerols in blue-green algae has been used to support the argument that chloroplasts

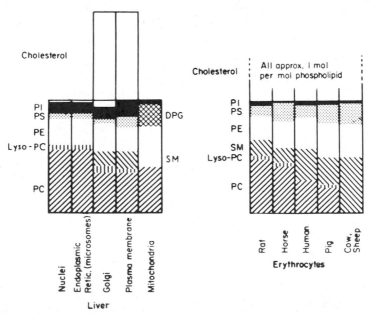

FIGURE 3.2. Lipid composition of liver membranes and erythrocytes (SM = sphingomyelin, DPG = diphosphatidylglycerol). Reprinted, by permission, from Finean et al. (1978).

and blue-green algae share a common ancestry (Finean et al., 1978). The lipid composition of bacterial membranes varies with the strain. Phosphatidylglycerol, which is a nitrogen-free derivative of glycerol phosphate, is the dominant phosphoglyceride in gram-positive bacteria, but in gram-negative bacteria PE predominates.

The fatty acids (acyl chains) of membrane phospholipids are also variable. The fatty acids are typically even-numbered chains ranging from 14 to 24 carbon atoms in length. Both saturated and unsaturated (one to six double bonds) are present, with the unsaturated moiety usually occupying position two on the glycerol molecule. Membrane phosphoglycerides contain both saturated, short-chain (C_{16} and C_{18}) fatty acids and longer chain (C_{20} to C_{24}) polyunsaturated fatty acids. In contrast, sphingomyelins tend to have more long-chain (C_{20} to C_{24}) saturated acyl residues. The fatty acid constituents of phospholipid molecules also vary, depending upon the type of membrane. For example, the fatty acids in the plasma membrane of liver cells have longer chains and are more saturated than comparable molecules found in the remainder of the cell (Nystrom, 1973).

Despite considerable variation in composition, membrane lipids all share one important feature—they are amphipathic, that is, one end of the molecule is water-insoluble or hydrophobic (the fatty acyl chains) and the other end is water-soluble or hydrophilic (the polar head group). The polarity of the head group is due to the presence of electrical charges associated with the phosphate moiety and nitrogenous base (if present) and/or a high number of hydroxyl groups. The amphipathic nature of membrane lipids can be clearly illustrated by reexamining the structure of the most abundant membrane phosphoglyceride, phosphatidylcholine (PC):

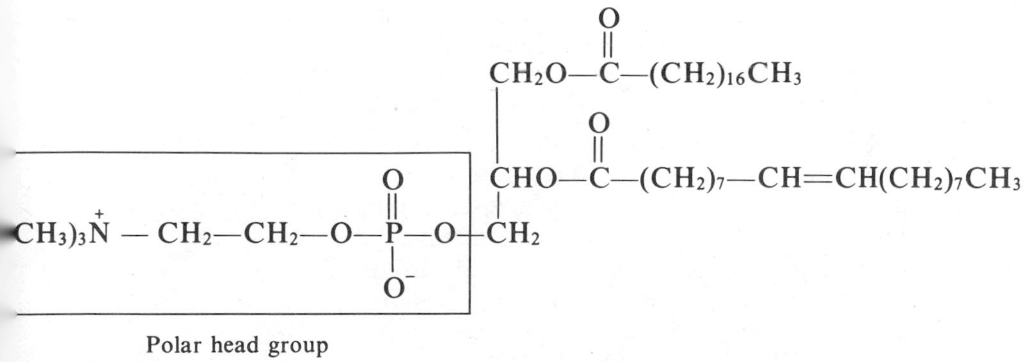

$$CH_2O-\overset{\overset{\displaystyle O}{\|}}{C}-(CH_2)_{16}CH_3$$

$$CHO-\overset{\overset{\displaystyle O}{\|}}{C}-(CH_2)_7-CH{=}CH(CH_2)_7CH_3$$

$$(CH_3)_3\overset{+}{N}-CH_2-CH_2-O-\overset{\overset{\displaystyle O}{\|}}{\underset{\underset{\displaystyle O^-}{|}}{P}}-O-CH_2$$

Polar head group

Sphingomyelin is physically similar to PC and other phosphoglycerides; however, cholesterol is a sterol with much of its surface hydrophobic. Nevertheless, the hydroxyl group on carbon three of ring A makes the molecule polar at one end, thus giving it an amphipathic character (Beck, 1980). As I discuss shortly, many of the special properties and functions of biological membranes, including lipid-protein interactions, are due to the amphipathic nature of the constituent lipids.

Proteins

Two categories of membrane proteins are generally recognized: extrinsic (or peripheral) and intrinsic (or integral). This classification is based largely on the ease by which the protein can be removed from the membrane. Extrinsic proteins are water-soluble and can be solubilized by simple aqueous salt solutions. The ease by which they can be removed suggests that extrinsic proteins are weakly bound to the surface of the membrane. Intrinsic proteins, in contrast, are extracted only by strong detergents or solvents. They cannot be removed without disrupting the lipid components of the membrane. Moreover, lipids often remain associated with the extracted protein. Intrinsic proteins are believed to be closely associated with the interior or the nonpolar portion of the membrane. The relatively high percentage of hydrophobic amino acid residues found in many of the intrinsic proteins studied is consonant with such an association (Harrison and Lunt, 1975). Intrinsic proteins also have regions that contain mainly ionized and polar amino acids. Thus, like membrane lipids, intrinsic proteins are also amphipathic. The location, interaction, and function of extrinsic and intrinsic proteins are discussed further when models of membrane structure are examined.

Carbohydrates

Although lipids and proteins are the major constituents of biomembranes, the plasma membrane and perhaps some of the intracellular membranes also contain some carbohydrates. The uncertainty about their occurrence in some intracellular membranes (e.g., Golgi complex, endoplasmic reticulum) stems from the possible contamination of membrane extracts with carbohydrates employed in the synthesis and storage of protein molecules. Some of the carbohydrates are free polysaccharides loosely attached to the membrane surface, but the majority are covalently bound to membrane lipids (glycolipids) and especially to proteins (glycoproteins). The carbohydrate moieties of glycolipids and glycoproteins are typically oligosaccharide chains containing from 2 to 15 sugar residues (Roseman, 1975). Some of the simple sugars in these chains include galactose, mannose, fucose, glucose, N-acetylglucosamine, and N-acetylgalactosamine. All of the carbohydrates are asymmetrically distributed on the external surface of the membrane.

STRUCTURAL ORGANIZATION OF BIOMEMBRANES

The presence of lipids and proteins as the principal components of biomembranes is firmly established; however, their arrangement is not and continues to stimulate much research and speculation. Our knowledge of the structural organization of biomembranes has benefited greatly from the development of techniques such as high resolution electron microscopy, x-ray crystallography, freeze-fracture and freeze-etching, electron spin resonance, and nuclear magnetic resonance. From investigations using these techniques, a model has evolved that is consistent with observed functional properties of biomembranes and that takes into account two important attributes stressed in the discussion of chemical constituents: the hydrophobic-hydrophilic nature of the lipid molecules, which permits their arrangement in a bilayer, and the presence of protein molecules with amphipathic properties, which enables them to interact with the lipid components. The following section briefly outlines the major steps and developments leading up to the currently accepted membrane model, and also provides background information for understanding the physical properties of biomembranes. Comprehensive historical accounts of the development of membrane models can be found in Danielli (1975), Harrison and Lunt (1975), Robinson (1975), and Karp (1979).

Early Studies

The first indication of the lipid nature of biomembranes, specifically the plasma membrane, was based on studies of diffusion rates of different substances in and out of cells such as root hairs. Experimental results showed that materials that were soluble in lipids generally entered the cell more rapidly than nonlipid-soluble substances. Artificial systems also played an important role in these early investigations. Chemists and physicists showed that when amphipathic lipids (e.g., phospholipids) were placed in an aqueous solution, they arranged themselves so that their hydrophilic "heads" remained in contact with the water, while the hydrophobic "tails" oriented themselves upwards toward the air (Chapter 1). Commonly the phospholipids would arrange themselves in a double or bilayer with the polar heads at the two surfaces and the nonpolar tails facing one another. Evidence for the existence of a lipid bilayer in an actual plasma membrane was suggested by Gorter and Grendel in 1925, who spread lipids extracted from an erythrocyte membrane on a water surface and found that the area occupied was approximately twice the total surface area of the intact erythrocyte. Although this finding was fortuitous in that two errors in their techniques and calculations canceled one another (Harrison and Lunt, 1975), subsequent studies indeed did show that sufficient lipid was present to cover the erythrocyte surface completely as a bilayer if the bilayer was highly expanded. Other studies being conducted at this time, however, demonstrated that lipid solubility was not the sole factor controlling permeability, but that molecular size and electrical charge were also important. Obviously some other membrane component(s) must be present in addition to amphipathic lipids.

Davson-Danielli-Robertson Model

The first model proposed to explain on a structural basis the properties of the plasma membrane was provided by Davson and Danielli in 1935 without the aid of direct observation. Their original model featured a centrally located lipid bilayer sandwiched between two layers of adsorbed globular protein. Revised versions of this model in later years included the "unrolling" of the globular protein onto the surface and the presence of protein-lined pores, which provided a route through which water, ions, and polar solutes could pass. This model dominated discussion of membrane model systems for the next 30 years and served as the basis for the concept of membrane structure. With the development of the electron microscope in the 1950s, investigators were able to resolve the plasma membrane for the first time and to compare their observations with the arrangement postulated by Davson and Danielli. The appearance of a trilaminar structure (two electron dense bands surrounding an electron lucent layer, Figure 3.3) was consistent with the protein-lipid-protein sandwich proposed by Davson and Danielli. The universal nature of this arrangement in a variety of membranes examined led Robertson to propose the concept of the "unit membrane." This model extended the one proposed by Davson and Danielli by stating that the outside and inside of the membrane consisted of a monolayer of protein fully spread on the surface, and that differences existed in the protein present on the two surfaces. A diagrammatic representation of the Davson-Danielli-Robertson model is shown in Figure 3.4.

Fluid Mosaic Model

The electron microscopic data offered in support of the Davson-Danielli-Robertson model applied more specifically to the lipid bilayer than to any particular arrangement of protein. Subsequent studies of molecular transport across membranes plus optical measurements suggested that the protein probably did not uniformly coat the external surface of the lipid bilayer.

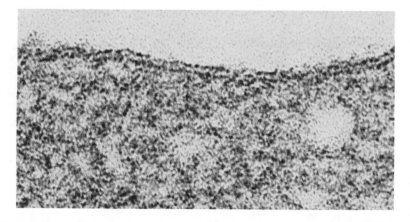

FIGURE 3.3.   Electron micrograph of plasma membrane surrounding a neutrophil. ×560,000. Courtesy of D. Chandler.

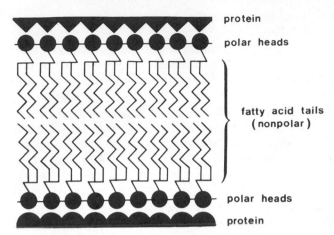

FIGURE 3.4.   Schematic diagram of the Davson-Danielli-Robertson model.

These findings led to the formulation of the "fluid mosaic model" proposed by Singer and Nicolson (1972). In their model, the lipid bilayer is retained, but the proteins (intrinsic) are now pictured as discontinuous globular particles that may penetrate the interior of the membrane or span the entire structure (Figure 3.5). It is believed that the polar groups of lipid and protein are in direct contact with the aqueous surroundings, while the nonpolar portions of both molecules are associated with the central core of the membrane. The model does not portray the loosely bound extrinsic proteins that

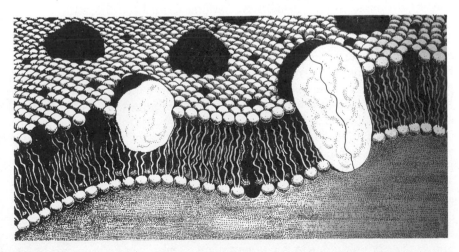

FIGURE 3.5.   Current model of plasma membrane structure. The hydrophilic ends of the proteins protrude from the membrane, while their hydrophobic portions are embedded in the bilayer of phospholipids (clear) and cholesterol (black). Reprinted, by permission, from Singer (1975).

are probably associated with the polar head groups of the lipid bilayer (Robinson, 1975).

The fluid mosaic model is the most commonly accepted model and serves as the basis for the discussion of the physical properties of membranes, as well as for alterations in the structure and composition of membrane lipids, alterations that are believed to have functional significance. There is strong experimental evidence for the existence of globular protein embedded in a lipid bilayer, especially from data derived from freeze-fracture electron microscopic examinations of various membranes. Nevertheless, acceptance of this model does not preclude the existence of other arrangements, permanent or temporal, of the major membrane constituents. In organelles such as mitochondria and chloroplasts, the lipid may be organized into globular microspheres (spherical micelles) rather than in a bilayer (Karp, 1979). Also, lateral movements of proteins within the membrane have been demonstrated, which suggests the possibility that proteins can exhibit some form of short-term order for specific functional purposes. The idea of an universal membrane structure may be unrealistic in view of the great number of lipid and protein constituents present in membranes and the enormous possibilities for molecular association between lipid/lipid, lipid/protein, and protein/protein.

ARTIFICIAL MEMBRANES

Much of our knowledge concerning the general properties of biomembranes has been derived from experiments using artificial membrane systems that simulate the structural organization of plasma and other cellular membranes. Two of the most important of these systems have been the artificial lipid bilayer and the liposome. The artificial bilayer is typically formed by dipping a brush into a dilute solution of lipid dissolved in a hydrocarbon solvent and applying it across a hole that is surrounded on both sides by water. The bubble that forms initially soon thins to form a single lipid bilayer having approximately the same thickness and many of the same properties as those described for the lipid bilayer in biological membranes. The lipid bilayer, also known as the black lipid membrane because of the color generated by the film due to light interference, can remain stable for days if provided the appropriate environment (Bangham, 1975). Liposomes consist of a single lipid membrane surrounding an aqueous compartment. Liposomes are often prepared from phospholipids extracted from cells, especially the plasma membrane of erythrocyte "ghosts." When sufficient water is added to the dried lipid extract, the phospholipids reorganize into a series of relatively large droplets several microns across and consisting of numerous concentric bilayers surrounding aqueous compartments. These can be reduced, by exposure to ultrasonic radiation, to very small vesicles (200 to 500 Å diameter) with only one or two layers of membranes (Fettiplace and Haydon, 1980).

PHYSICAL PROPERTIES OF MEMBRANES

Early membrane models pictured the membrane as consisting of a discrete package of static molecules arranged in a highly organized and uniform manner. Such an arrangement, however, was not consonant with observed properties and physiological functions of either biological or artificial membranes. Through experimentation it became apparent that membranes are, in contrast, labile structures, somewhat disordered and undergoing continuous changes in the living state (Nystrom, 1973). The popularity and acceptance of the fluid mosaic model is based largely on the fact that it attempts to account for the dynamic nature of biomembranes. Nevertheless, it is necessary to go beyond the pictorial arrangement of membrane components as envisaged in the fluid mosaic model to more fully understand the mechanical properties exhibited by biomembranes and to account for their diverse functions, which range from specialized transport to immunological interactions. Of particular significance here, especially as they pertain to the lipid moiety, are membrane fluidity, the asymmetry of membrane components, and the movement of components within the membrane.

Membrane Fluidity

Lipids in biological or artificial membranes can exist in either a highly ordered crystalline gel state or a less ordered liquid crystalline state of varying viscosity. For example, an artificial bilayer composed of PC with two acyl chains of myristic acid (14:0) connected to the glycerol moiety is in the gelated state at temperatures below 25°C. As the temperature is raised above 25°C, the bilayer lipids are converted from a crystalline gel to a much more fluid liquid crystal (Karp, 1979). In the latter state, the acyl chains melt and acquire considerably more mobility, with the flexing and twisting greatest at the methyl end of the chain. Although both the glycerol and the polar head group retain a fairly regular arrangement, there is evidence of increased motion of the polar group and some reorganization of water around the polar group during the phase change (Chapman, 1975a,b). The temperature at which the change from gel to liquid crystal occurs is referred to as the transition temperature. It can be measured using several physicochemical techniques including calorimetry, which detects the heat absorbed as bonds that hold the molecules in a fairly rigid structure are broken during the phase change (Figure 3.6).

Experiments on artificial systems and natural membranes have shown that transition temperature is dependent on the length of the fatty acid chains, the degree of saturation of these chains, the nature of the polar head group, the hydration state of the membrane, the presence or proportion of cholesterol, and the purity (homogeneity) of the lipid constituents (Chapman and Wallach, 1968; Chapman, 1975a). Perhaps the most important of these factors is the degree of unsaturation of the acyl chains. Highly unsaturated

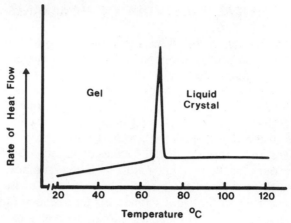

FIGURE 3.6. Diagrammatic representation of calorimetric scan of pure phospholipid preparation. The sharp peak represents heat flow that marks the transition from gel to liquid crystal.

lipids have a low transition temperature, whereas saturated lipids have a high transition temperature. This relationship is consistent with the melting point/saturation characteristics of fatty acids described in Chapter 1. The physical basis for this relationship can be further explained by packing differences between saturated and unsaturated acyl chains. Because saturated fatty acids are essentially straight in shape, they can pack together more tightly and tend to remain fixed or "frozen" at low temperatures. Unsaturated fatty acids, however, have crooks in the chain where the double bond(s) are located, crooks that permit them to remain more fluid as temperatures are decreased. In some organisms the acyl chains contain cyclopropyl groups or branched chains that produce similar kinks and help maintain membrane fluidity (Eisenberg and McLaughlin, 1976). The length of the acyl chains of the phospholipid molecules also influences the transition temperature, with membranes that contain a high proportion of long-chain fatty acids exhibiting higher transition temperatures. This relationship is also consistent with the melting properties of short- versus long-chain fatty acids discussed in Chapter 1.

The heterogeneity of lipid packing in membranes also influences the transition temperature. The "classical" phase transition (Figure 3.6), which occurs sharply at a single temperature, is valid only for homogeneous phospholipid preparations (e.g., PC). The lipid composition of biomembranes is complex, consisting of phospholipid molecules with fatty acyl chains of various lengths and degrees of saturation. Thus, the change from a gel to a fluid state occurs over an appreciable range of temperatures and produces a much more rounded peak on a calorimetric tracing. Because of the mixed composition and location of membrane lipids, it is possible to have one portion of a membrane in a solid state, while an adjacent portion is in a fluid state (Lee, 1977; Karp, 1979).

Although discussion thus far has centered on the hydrocarbon tails of phospholipid molecules, there is strong evidence that the polar head group also exerts an effect on transition temperature. Biophysical studies have shown that the temperature of transition from a gel to liquid crystalline state of a membrane is approximately 25°C higher in artificial membranes of phosphatidylethanolamine (PE) than phosphatidylcholine (PC) when fatty acyl chains are identical (Chapman and Wallach, 1968). Similarly, a decrease in fluidity of fibroblast membranes was observed when ethanolamine was substituted for choline, with no change in fatty acid composition (Esko et al., 1977). Since there is a tendency for unsaturated fatty acids to associate with PE, the increased stability of the head group may compensate for the increased fluidity characteristics of the constituent acyl chains.

Another variable that can affect the transition temperature of a membrane is the presence of cholesterol. Cholesterol is thought to intercalate in the lipid bilayer with the β-hydroxyl group near the aqueous interface and the fused rings and hydrocarbon tail oriented toward the inner half of the bilayer (Figure 3.5) (Huang, 1977). Cholesterol increases the fluidity of membranes that are below their transition temperature by interfering with van der Waals interaction between the apposed fatty acyl chains. In other words, cholesterol causes a membrane to remain in a liquid crystal state even when it is cooled well below its normal transition temperature. Moreover, high concentrations of cholesterol may abolish the transition temperature entirely (Figure 3.7). Conversely, cholesterol stabilizes membranes that are above their transition temperature by enhancing van der Waals interactions (Oldfield and Chapman, 1972; Singer, 1974). The effect of cholesterol is also dependent on the phospholipid composition of the membrane. For example, experimental studies have shown that the interaction between sphingomyelin and cholesterol is much stronger than it is between PC and cholesterol (Barenholz and Thompson, 1980).

Lipid Movements Within the Membrane

The fluidity of a membrane is directly related to the mobility of its phospholipid components. As a transition temperature is approached, there is an increase in methylene group mobility along the acyl chains toward the center of the bilayer. The increased mobility allows the adjacent chains to move across one another within the plane of the membrane. Lipid movements, however, can involve an entire phospholipid molecule. Most membrane lipids, for example, are capable of rapid lateral motion, changing places with neighbors about 10^7 times per second. This rate converts into a lateral movement of approximately 1 μm per second (Eisenberg and McLaughlin, 1976). It is also possible for a phospholipid to move transversely across the membrane. This process, known as flip-flop, is much slower, being measured in hours or days. The slow rate reflects in part the difficulty in moving the polar head group through the hydrocarbon core of the bilayer.

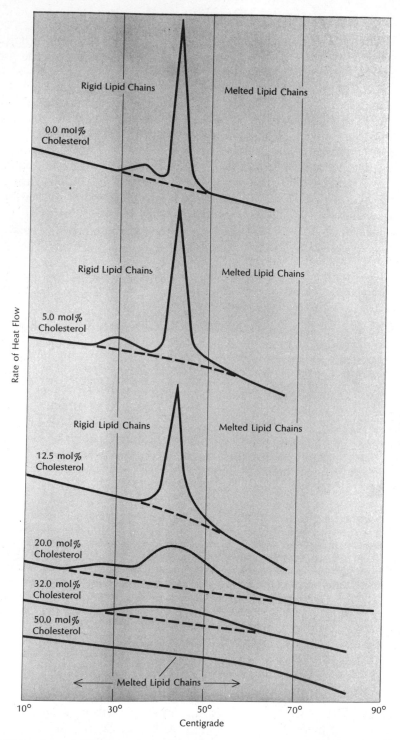

Rigid Lipid Chains   Melted Lipid Chains

0.0 mol%
Cholesterol

Rigid Lipid Chains   Melted Lipid Chains

5.0 mol%
Cholesterol

Rigid Lipid Chains   Melted Lipid Chains

12.5 mol%
Cholesterol

20.0 mol%
Cholesterol

32.0 mol%
Cholesterol

50.0 mol%
Cholesterol

Rate of Heat Flow

←——— Melted Lipid Chains ———→

10°   30°   50°   70°   90°

Centigrade

FIGURE 3.7.   Increasing the mol percent of cholesterol in a phospholipid preparation blurs the transition of the membrane from a gel (left side of peak) to the liquid crystal (right side of peak) state. Reprinted, by permission, from Chapman (1975a).

Asymmetry of Membrane Lipids

Most biomembranes not only contain more than one kind of lipid, but the lipids are typically asymmetrically distributed between the two surfaces. For example, PS and PE are more abundant on the inner monolayer of an erythrocyte membrane, whereas the outer monolayer contains most of the PC and sphingomyelin (Rothman and Lenard, 1977). This arrangement is consistent with the tendency for positively charged phospholipids to occur on the outside and negatively charged phospholipids to occur on the inside of the lipid bilayer. Another difference between the two membrane surfaces is the degree of unsaturation of the phospholipid molecules, with the inner monolayer containing a higher percentage of unsaturated acyl chains (Karp, 1979). The molecular mechanisms by which this asymmetry is achieved are unknown, but may involve the extremely low frequency of transmembrane movement of phospholipids described earlier.

MEMBRANE DYNAMICS: BIOLOGICAL CONSEQUENCES

Physical changes in the structural organization of membranes are well documented, but the functional significance of these changes is less clear. Although it is generally agreed that the physical state of the lipid bilayer is an important factor in determining the physiological performance of a membrane, how and to what degree this physical state alters a specific membrane function is still quite controversial. The strongest evidence, based primarily from direct measurements on artificial bilayers, relates to membrane fluidity and alterations that occur in conjunction with the transition temperature. A membrane containing a high percentage of unsaturated fatty acyl chains or one undergoing a transition from a gel to liquid crystal state might produce a number of functional changes. There could be a change in affinity between the membrane polar surface and various ligands (e.g., ions, proteins) that might result from the lateral expansion of the membrane and from the "loosening" of the polar head group arrangement (Träuble and Haynes, 1971). Träuble (1972) showed that the number of binding sites for certain dyes and ions on the membrane increased approximately threefold above the transition point. A second and very important change relates to increased membrane permeability to water, ions, and certain larger molecules. Experiments using artificial bilayers have demonstrated that water permeability increased when the degree of unsaturation of the acyl chains was increased and that water permeability decreased when the lipid constituents were hydrogenated or when analogues with saturated chains were substituted (see review by Fettiplace and Haydon, 1980). Increasing the length of the acyl chains (less fluid) also decreased permeability. When cholesterol was added to membranes formed from PC, thus reducing the fluidity, a three- to fivefold reduction in membrane permeability was observed (Fettiplace and Haydon,

1980). The change in permeability with increased membrane fluidity may reflect the decrease in membrane thickness, changes in the organization of the hydrocarbon chains (i.e., increased number of kinks), and changes in the arrangement of the polar head groups. Increased membrane fluidity may also enhance the role played by proteins in the transport of substances into and out of a cell by enabling them to move more freely within the plane of the membrane.

The functional significance of lipid asymmetry, in comparison, is much more obscure and speculative. One possible consequence of an asymmetric lipid composition is a membrane with differential fluidities in each half of the bilayer. A more rigid outer monolayer might be conducive to its function in terms of increasing the tolerance of a membrane to increased lateral compression. A higher proportion of negatively charged phospholipids on the outer monolayer may also facilitate interactions with extrinsic proteins and be a necessary condition for the activity of membrane-associated enzymes (Bergelson and Barsukov, 1977). Alternatively, lipid asymmetry may serve no biological purpose, but instead represent an incidental consequence of the topological constraints imposed on the mechanisms by which external lipids are synthesized (Rothman and Lenard, 1977). Answers to these and other possibilities must await further experimental analysis.

COMPENSATORY ADJUSTMENTS IN MEMBRANE LIPIDS AND MEMBRANE FUNCTION

In the previous section some evidence was presented indicating that certain enzyme or physiological functions are likely directly correlated with the molecular ordering or fluidity of membrane lipids. In view of this relationship, it is not surprising to find that organisms are able to alter their membrane composition so that membrane fluidity is maintained at a level that preserves normal membrane function. The key environmental variable in this process is temperature. Not only is fluidity determined by temperature, but changes in temperature can cause many organisms to restructure their biological membranes so that lipids with appropriate physical properties are matched to the prevailing temperature conditions. The restructuring includes changes in the degree of saturation of the fatty acid chains of the phospholipid molecules and changes in the relative proportions of specific phospholipids. In this final section, examples of functional adjustments in membrane lipids and fluidity are presented for microorganisms, plants, and animals. Although the emphasis is on temperature-induced alterations, compositional changes resulting from diet and other parameters are included where appropriate.

Microorganisms

Much of the strongest evidence demonstrating the importance of membrane fluidity in controlling membrane function has come from studies of microorganisms. The eubacterium *Escherichia coli* exhibits many of the advantages of using microbial systems for such investigations. In *E. coli* phospholipids account for over 90% of the lipids present, and these are located exclusively in the cytoplasmic and outer membranes of the cell envelope (Silbert, 1975). This organism grows readily on simple defined media under aerobic or anaerobic conditions. Moreover, its lipid composition can be systematically adjusted by altering the growth medium. For example, there are mutants of *E. coli* (K_{1060}) that cannot synthesize unsaturated fatty acids but can incorporate into their membrane phospholipids fatty acids with various chain structures if supplied to the growth medium. Although many measurements made in these studies are largely indirectly related to *in vivo* membrane properties (i.e., observing membrane function in liposomes comprised of phospholipids extracted from bacteria), recent studies have shown altered membrane function in bacterial cells measured directly.

Many workers have reported that as the temperature of growth is decreased, *E. coli* increases the ratio of unsaturated to saturated fatty acids in its membrane phospholipids and vice versa (Marr and Ingraham, 1962; Haest et al., 1969; Sinensky, 1971). An even more defined variation in fatty acid pattern is observed in the *E. coli* K_{1060} mutants when they are grown at different temperatures (Haest et al., 1972). Similar results have been obtained in studies of *Tetrahymena pyriformis*, a facultative protozoan parasite, and *Acholeplasma laidlawii* B cells, a free-living bacterium that lacks a cell wall (Wunderlich et al., 1973; Nozawa et al., 1974; McElhaney, 1974). Since lowering the temperature of the environment of a membrane renders a membrane of given fatty acid composition less fluid, the increased unsaturation may be viewed as a means to ensure that the proper degree of membrane fluidity is maintained. This ability to synthesize phospholipids so that membranes have identical viscosities at a given growth temperature has been termed homeoviscous adaptation (Sinesky, 1974). There is accumulating evidence that both transport processes and biochemical reactions in microorganisms are affected by membrane lipid viscosity. Haest et al. (1969) showed that the permeability of liposomes decreased markedly when the latter were manufactured from bacteria grown at high temperatures. Utilizing *E. coli* cells, Davis and Silbert (1974) found that permeability remained low until the monoenoic fatty acid content of membrane phospholipids exceeded 82%, but thereafter increased sharply as the level rose from 82 to 92%. Conversely, passive permeability eventually decreased when the saturated fatty acid content of cells was increased by growing them in a stearic acid–rich medium. Altered fatty acid composition of membranes may also affect enzymatic activities. Poon et al. (1981) found that the activity of two enzymes, adenylate cyclase and Na^+/K^+-ATPase, in the murine T lymphocyte tumor EL4 was inhibited by saturated fatty acids, while unsaturated fatty acids had a moderate enhancing effect.

In comparison to rather extensive studies of temperature-induced alterations in the fatty acyl composition of phospholipids and membrane function in microorganisms, there have been only a few investigations of changes in phospholipid polar head groups or the interactions of other membrane lipids with phospholipids. Van Deenen and co-workers were able to incorporate sterols into the membranes of *A. laidlawii*, an organism that does not require sterols for growth. Certain sterols decreased the permeability of *A. laidlawii* cells to nonelectrolytes; however, sterol incorporation was neither influenced by nor changed the fatty acid composition of the membrane phospholipids (De Kruyff et al., 1973; McElhaney et al., 1973). Recently changes in fluidity and composition of total lipids during temperature acclimation from three membrane fractions—mitochondria, pellicles, microsomes—of *Tetrahymena* were examined by Yamauchi et al. (1981). The usual increase in the ratio of unsaturated to saturated fatty acids occurred in all three fractions following a temperature shift from 39° to 15°C, but membrane fluidity increased only in the microsomes and pellicles; the fluidity of mitochondrial lipids was constant up to 10 hours after the temperature shift. The authors proposed that cardiolipin, which is localized almost exclusively in the inner mitochondrial membranes, might act as a membrane stabilizer to keep fluidity constant during temperature acclimation. Removal of cardiolipin caused no change in the fluidity of mitochondrial lipids from 39°C-grown cells, whereas it reduced fluidity of mitochondrial lipids from 15°C-acclimated cells (Table 3.2). When cardiolipin was readded to the mitochondrial lipids depleted of cardiolipin, fluidity was restored to the initial level, thereby confirming the rigidifying effect of cardiolipin in cold-acclimated cells. The physiological significance of constant fluidity in mitochondrial membranes at the lower temperature is uncertain, nor has it been determined if cardiolipin exhibits a similar stabilizing effect in the mitochondrial membranes of other cells.

Recent studies of microsomal phospholipids isolated from *Tetrahymena* suggest that the precise location of fatty acids in specific phospholipid molecules may be even more important than the degree of fatty acid unsaturation in enabling the organism to achieve rapid homeoviscous adaptation in response to low temperature stress. Dickens and Thompson (1982) found only a comparatively small increase in fatty acid unsaturation one

TABLE 3.2

Effects of cardiolipin on fluidity of mitochondrial lipids from *Tetrahymena*. A lowered order parameter (S), which is calculated from electron spin resonance spectra, indicates decreased membrane fluidity

	Mitochondrial lipids from:		Cardiolipin-free mitochondrial lipids from:	
	39°C-grown cells	15°C-acclimated cells	39°C-grown cells	15°C-acclimated cells
S	0.651	0.651	0.651	0.637

SOURCE: Reprinted, by permission, from Yamauchi et al. (1981).

hour after *Tetrahymena* were transferred from 39 to 15°C, yet the fluidity of the lipids, as measured with fluorescence depolarization, had increased almost to the level found in cells fully acclimated to 15°C. Using innovative GC-MS techniques, the authors were able to show that the increased fluidity resulted from an extensive redistribution of both preexisting and newly modified fatty acids into different positions on the phospholipid molecules. Previous investigations had shown that merely reversing fatty acids between the *sn*-1 and the *sn*-2 positions of a mixed-acid phospholipid molecule markedly affects the physical properties of the lipid in bilayers (Roberts et al., 1978; De Bony and Dennis, 1981). At present the mechanisms by which this reorganization takes place in *Tetrahymena* are unknown, nor is there experimental evidence that similar response occurs in other organisms. Nonetheless, it is an exciting find in that it supports the concept that different mechanisms contribute to the homeoviscous response at different times following a change in temperature.

Plants

It is clear from the previous discussion that molecular ordering of membrane lipids can be altered by variations in fatty acyl composition in microbial systems. Whether similar changes in fatty acyl composition have comparable influence on molecular ordering of lipids in the more complex membranes of higher plants and, if so, what effects the changes in membrane fluidity have on the function of these membranes has been a subject of much investigation and controversy. Most research to date has focused on the responses of plants to low temperature and what constitutes low temperature resistance. Many plants of tropical and subtropical origin, including economically important crop species such as corn, tomato, cotton, and soybean, are sensitive to low temperatures in the range of 20°C down to 0°C (Lyons et al., 1979). At some critical temperature within this range, the plant suffers physiological damage that may be reflected by restricted germination, growth, reproduction and post-harvest longevity. The physiological damage is referred to as "chilling injury" as there is no actual freezing of plant tissues. Plant species of temperate origin, in contrast, do not exhibit this sensitivity to low temperature, and may even have temperature optima in this range (Lyons et al., 1979).

Plants that are chilling sensitive can be conditioned or hardened by exposure to temperatures just slightly above the chilling range for some period of time before they experience the injurious temperature. Presumably this pre-exposure period allows the plant to alter the fluidity of its membranes so that enzymatic and other activities associated with membranes remain functional. As illustrated in Figure 3.8, the temperature change must be gradual to ensure that the plant cell has ample time to initiate changes in lipid composition that prevent membrane fluidity from decreasing to a suboptimal level. Sudden chilling can decrease membrane fluidity so rapidly

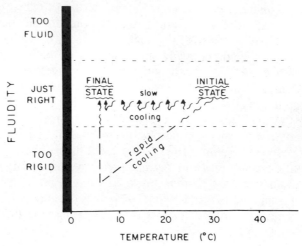

FIGURE 3.8. Representation of an organism's ability to maintain its membrane fluidity within an optimum range during a slow temperature reduction but not a fast one. Reprinted, by permission, from Thompson (1979).

that irreversible damage occurs before membrane fluidity can be restored to optimal conditions.

An increase in unsaturated fatty acids is most often suggested as being primarily responsible for the maintenance of proper membrane function at low and even freezing temperatures. The evidence for this, which is largely indirect, is based on a rather consistent correlation observed between the degree of unsaturation of the lipids of plants and the environmental temperature at which they are grown (Hilditch and Williams, 1964) and on numerous studies of principally herbaceous plants that demonstrate a distinct increase in the unsaturation of phospholipid fatty acyl residues, especially linolenic acid (18:3), in species resistant to chilling (see Lyons, 1973, for review). A general increase in phospholipids, changes in the relative proportions of specific phospholipids (i.e., PC, PE), and changes in sterol concentration and composition may also accompany changes in unsaturation, but their contribution to the cold hardening process has received far less study (Willemot, 1979).

It has been generally assumed that increases in lipid unsaturation not only increase membrane fluidity, but also lower the temperature of the order-disorder transition (i.e., change in liquid crystalline to gel phase), a phenomenon that can be determined using electron spin resonance (esr) spectroscopic techniques. The lower temperature limit of the order-disorder transition is referred to as T_s and the upper limit as T_f. Below T_s the lipids are relatively rigid, while above T_f the lipids are relatively fluid. Because plants exposed to temperatures below T_s develop symptoms of injury, T_s is considered the critical temperature for plant growth (Raison and Chapman, 1976; Chapman et al., 1979).

A similar relationship between the temperature limits of the order-disorder transition and the Arrhenius activation energy (E_a) of enzyme activity has been cited as further evidence that compositional changes in membrane lipids (i.e., increased unsaturation) are integral to the cold hardening process in plants. The E_a represents the energy the enzyme-substrate complex must gain for the reaction to proceed. The change in rate of succinate-oxidase activity of mitochondria from maize root tissue as a function of temperature is shown in Figure 3.9. There is an abrupt vertical displacement in rates at about 27°C (T_f) and 12°C (T_s). The E_a above T_f is 1 kcal per mole, above T_s 10 kcal per mole, and below T_s 25 kcal per mole (Raison et al., 1979). Moreover, the authors report that the temperatures obtained for T_f and T_s using the plots of enzyme activity and fitting straight lines coincide with those obtained from spin-label (esr) motion. In contrast, the E_a for this reaction for membranes of chilling-tolerant plants, which do not exhibit changes in molecular ordering of lipids, is constant between 0 and 30°C. These results suggest that the temperature-induced change in temperature-sensitive membranes is an intrinsic property of the membrane lipid and that a direct correlation exists between the physical state of the membrane component and enzyme activity.

Despite these impressive correlations, experimental data also exist that are contrary to the general hypothesis that an increase in unsaturation of membrane lipids is essential for cold-hardening. Woody plants such as black locust and poplar trees, which are tolerant of extreme freezing, are no more unsaturated in winter than they are in summer (Siminovitch et al., 1975). Strains of wheat differing widely in freezing resistance show the same increase in their fatty acid unsaturation upon hardening (de la Roche et al.,

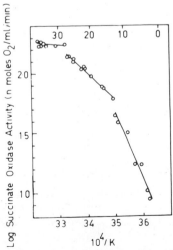

FIGURE 3.9. Changes in the succinate-oxidase activity of maize root mitochondria as a function of temperature. Reprinted, by permission, from Raison et al. (1979).

1975). Singh et al. (1977) found no significant lowering of the phase transition in rye phospholipids even though the linolenic acid content of the membranes increased during hardening. The validity of order-disorder transitions based on data generated from Arrhenius plots and esr probes has also been questioned. Bishop et al. (1979) concluded that, because of the highly unsaturated nature of chloroplast membranes of algae and higher plants, no transition would be expected to occur above 0°C in the bulk lipids of either chilling-sensitive or chilling-resistant species and that measurements of such transitions obtained from Arrhenius plots of spin label motion must be viewed with caution. These findings do not necessarily mean that lipids are not involved in frost hardening, but do indicate the need for further investigation and reexamination. It may be that certain plants have alternate means of maintaining a functional level of membrane fluidity in those cases where increased fatty acid unsaturation is not indicated. In such cases, conformational changes in membrane proteins that initiate changes in the molecular ordering of discrete domains of membrane lipids may be a key factor.

The effects of high temperature and other environmental factors on plant membrane lipids and membrane function have received far less study. Pearcy (1978) determined the fatty acid composition of leaf lipids originating predominantly in the chloroplasts of *Atriplex lentiformis*, a shrub that grows in both coastal and desert habitats. He found a decrease in linolenic acid (18:3) and increases in the more saturated fatty acids at high growth temperatures (43°C day/30°C night). Although increased lipid saturation is correlated with a greater thermostability of the photosynthetic apparatus at high growth temperatures, any cause and effect relationship is uncertain. There is no evidence, for example, that the increased saturation would extend phase transitions in chloroplasts to between 35 and 45°C where thermal inactivation of photosynthesis occurs. Raison (1980) also states that there is no apparent correlation between T_f and the upper temperature limit or optimum temperature for plant growth. Because *Atriplex* grows well on soils with high salt content, it would appear to be an ideal species on which to examine the relationship between salinity and membrane lipids. Kuiper (1968) found that both PE and PC are inversely related to chloride accumulation in grapes and that long-chain saturated fatty acids in phospholipids may contribute to membrane stability and a consequent low permeability for chloride.

ANIMALS

The restructuring of biological membranes in response to changes in environmental temperature has been extensively investigated in animals, especially in poikilotherms, which often must survive large fluctuations in body temperature. These studies have centered on the ability of animals to make compensatory adjustments in their membrane lipid composition so that membrane fluidity is maintained in a more or less constant condition, a process referred to earlier as homeoviscous adaptation (Sinensky, 1974). Because the integrity and functional properties of cellular membranes are vital to numerous cellular processes, it is presumed that the strategy of maintaining membrane function independent of temperature changes by modulating the effective viscosity of the hydrophobic compartment of membranes would have survival value during crisis periods and would also be of considerable importance for the fitness of species at normothermic temperatures (Cossins, 1981).

The most thoroughly documented and widespread response of animals to control membrane fluidity at low environmental temperatures is to increase the level of unsaturation of membrane fatty acids. Because unsaturated fatty acids possess lower melting points (Chapter 1), and thus are more fluid at reduced temperatures than their saturated homologues, the incorporation of unsaturated fatty acids into membrane lipids counteracts the consequences of reduced kinetic energy at low temperature. The increase in unsaturation is most commonly achieved by increasing the proportion of long-chain polyunsaturated fatty acids, accompanied by either a reduction or no change in the content of monoenes (Hazel and Sellner, 1980). This pattern is illustrated by data in Table 3.3 depicting the fatty acid composition of hepatocytes prepared from thermally acclimated trout (Hazel and Sellner, 1980). These data indicate that during the process of cold acclimation, fatty acids belonging to the oleic (n-9) acid family are replaced by the more highly unsaturated fatty acids of the linolenic (n-3) acid family. Increased amounts of polyunsaturated fatty acids in the phospholipids of cold-exposed poikilotherms have also been detected in the muscle, gill, and hepatopancreas tissue of aquatic crustaceans (Chapelle, 1978; Farkas and Nevenzel, 1981), gill and muscle tissue of the goldfish (Caldwell and Vernberg, 1970; Cossins et al., 1978), liver, adipose tissue, and eggs of frogs (Baranska and Wlodawer, 1969; Marinetti et al., 1981), and the intestinal membranes of turtles (Chapelle and Gilles-Baillien, 1981). A second pattern of adaptation, which is less common, involves an elevation in monoene content with little or no change in the content of polyunsaturated fatty acids (Hazel and Sellner, 1980). Brain synaptosomes of goldfish (Cossins, 1977) and carp muscle mitochondria (Wodtke, 1978) exhibit this pattern of change in their phospholipid fatty acids following acclimation to cold.

The metabolic adjustments responsible for the increased unsaturation of membrane lipids with cold adaptation are complex and vary among different animal groups. Several possible mechanisms for increasing the degree of unsaturation are discussed in Hazel et al. (1983) and Hazel (1984). These include: (1) altering the fatty acyl-CoA pool by selectively increasing unsaturated fatty acid biosynthesis (involves increased activity of desaturase enzymes); (2) selectively removing only unsaturated fatty acids from the acyl-CoA pool during the *de novo* synthesis of membrane phospholipids; (3) substituting unsaturated acyl chains for saturated chains on an otherwise

TABLE 3.3

Effects of acclimation temperature upon the fatty acid composition of hepatocytes from warm (20°C)- and cold (5°C)-acclimated rainbow trout

Fatty acid class	Cold-acclimated	Warm-acclimated
Saturated fatty acids	25.8 ± 1.16[a]	28.7 ± 2.81
Total unsaturated fatty acids	75.2 ± 1.19	71.1 ± 3.12
Monoenes	23.9 ± 1.99	36.2 ± 2.34[b]
Dienes	8.5 ± 1.06	7.8 ± 0.42
Trienes	2.6 ± 0.16	2.1 ± 0.43
Tetra- and pentaenes	11.6 ± 1.33	6.7 ± 1.30[b]
22:6n3	24.3 ± 0.75	20.8 ± 1.79
Polyunsaturates (:4–6)	35.9 ± 1.55	27.5 ± 1.65[b]
n-3 Family	34.5 ± 1.56	27.0 ± 1.59[b]
n-6 Family	15.4 ± 1.08	14.6 ± 0.79
n-9 Family	21.9 ± 1.88	24.3 ± 1.76
$\dfrac{\text{Total unsaturates}}{\text{Total saturates}}$ (U/S)	3.13 ± 0.15	2.44 ± 0.35
Unsaturation index[c]	251.5 ± 7.88	201.3 ± 9.43[b]

Source: Reprinted, with permission, from J. R. Hazel and P. A. Sellner, The regulation of membrane lipid composition in thermally-acclimated poikilotherms. In: *Animals and Environmental Fitness* (Gilles, R., Ed.). Copyright 1980, Pergamon Press, Ltd.

[a]Data presented as weight percentage ± standard error for six experiments.
[b]Significant difference ($p < 0.05$) between acclimation groups.
[c]The summed product of weight percentage and the number of double bonds.

stable glycerol skeleton (involves increased activity of deacylation-reacylation cycle enzymes); and (4) direct desaturation of intact phospholipid moieties. Although there is evidence that each of these mechanisms for restructuring membrane lipid composition is operative in poikilotherms, the relative importance of each cannot be stated with certainty. It is likely that certain of the mechanisms listed above contribute to the homeoviscous adaptation at different times during the acclimation process. For example, the rearrangement of fatty acid pairs on the phospholipid molecule resulting from increased activity of phospholipase and acyltransferase enzymes appears to be an initial and fairly rapid response for maintaining optimal membrane fluidity in an altered thermal environment, whereas the selective metabolism and incorporation of monoene and polyunsaturated fatty acids into the biomembrane occurs at later stages in the acclimation process.

Changes in relative phosphatide composition in response to altered temperature can also result in the restructuring of biological membranes. Membranes from cold-acclimated animals usually contain more PE than membranes from warm-acclimated counterparts [e.g., carp (Wodtke, 1978), trout (Hazel, 1979), locust (Kallapur et al., 1982)]. Moreover, in crayfish (Cossins, 1976) and trout (Hazel, 1979), the increase in PE is accompanied by a decrease in sphingomyelin. Such changes in relative phosphatide composition may constitute an additional means of regulating membrane unsaturation, for PE is generally the most highly unsaturated of animal phosphatides, whereas sphingomyelin contains predominantly saturated fatty acids (Schmidt et al., 1977). In addition, ethanolamine phosphoglycerides, because of their small headgroup size and high proportion of unsaturated fatty acids, adopt a wedge-shaped conformation that is not easily packed into lamellae (Hui et al., 1981).

Most studies of temperature effects on poikilotherms have focused on the end results of acclimation rather than on the kinetics of the process; hence, there are few data concerning the rates at which higher animals accomplish the homeoviscous restructuring of membranes (White and Somero, 1982). Cossins et al. (1977), however, have provided a detailed investigation of the period of time necessary for homeoviscous adjustments in the synaptosomal membranes of cold- versus warm-acclimated goldfish. Changes in the fluidity of these membranes were determined using fluorescence polarization, a technique in which a fluorescent molecule is intercalated into the hydrophobic region of the membrane and the rotation that occurs during its fluorescence lifetime is monitored by measuring the polarization of fluorescent light. A high value of polarization indicates a low membrane fluidity. When goldfish that had been pre-acclimated to 5°C were transferred to 25°C (Figure 3.10a), membrane fluidity decreased rapidly, until after approximately 14 days it was identical to the value for fish previously acclimated to 25°C. In contrast, transfer from 25 to 5°C (Figure 3.10b) resulted in no significant change in fluidity for 20 days, after which there was a rapid change to the value characteristic of 5°C-acclimated fish. The lag in lipid restructuring in response to cold acclimation likely reflects the dependence of lipid biosynthesis on temperature (Chapter 2). A good correlation was found between the adjustment of fluidity and the ratio of saturated to unsaturated fatty acids in the acyl group composition of synaptosomal membranes during the cold transfer experiment (Cossins et al., 1977).

The goldfish and many other poikilothermic animals experience dramatic seasonal changes in environmental temperature. Data presented for the goldfish show that they can modify their synaptosomal membrane fluidity to partially compensate for the effects of a changed environmental temperature. Other fish species, such as those inhabiting polar seas or hot springs, live in relatively extreme but thermally stable environments. It has been suggested that selection pressure for cellular adaptation to an extreme but stable thermal environment would be strong, and that the potential for seasonal flexibility need not be maintained. Cossins and Prosser (1978) examined this hypothesis by comparing the fluidity of the synaptosomal membranes of thermally acclimated goldfish with two relatively stenothermal fish species, the arctic sculpin, *Myoxocephalus verrucosus*, and the desert pupfish, *Cyprinodon nevadensis*. Data obtained from the synaptosomal membranes

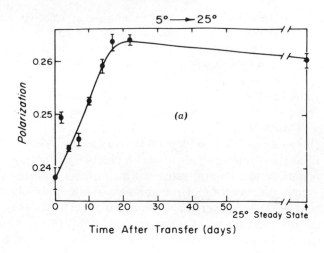

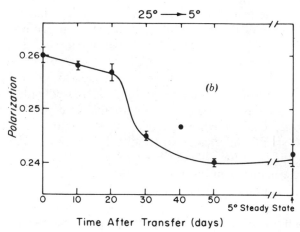

FIGURE 3.10. The time-course of changes in the membrane fluidity of brain synaptosomes of goldfish during reacclimation. (*a*) From 5 to 25°C. (*b*) From 25 to 5°C. Reprinted, by permission, from Cossins et al. (1977).

of the rat are included for comparison. Arrhenius plots of polarization for the various preparations are presented in Figure 3.11. The curves clearly show that membrane fluidity decreased as the measurement temperature was lowered, and that the curves shifted upwards and to the left with higher acclimation, habitat, or body temperature. Thus, over the entire temperature range, synaptosomal membranes of the arctic sculpin were more fluid than those of 5°C-acclimated goldfish, and the synaptosomal membranes of the desert pupfish had a fluidity between those of 25°C-acclimated goldfish and the rat. Furthermore, the curve for the arctic sculpin was shifted approximately 5°C with respect to the 5°C-treated goldfish, which corresponds to the difference in acclimation temperatures, indicating complete fluidity com-

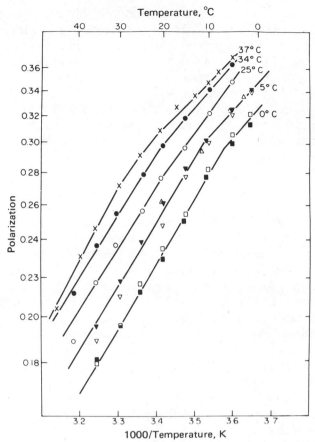

FIGURE 3.11. A comparison of membrane fluidity between 3° and 42°C of brain synaptosomes isolated from the arctic sculpin (*Myoxocephalus* sp., 0°C, □ ■), 5°C-acclimated goldfish (▽ ▼), 25°C-acclimated goldfish (○), the desert pupfish (34°C, ●) and the rat (37°C, ×). Each symbol represents a separate preparation. Reprinted, by permission, from Cossins and Prosser (1978).

pensation. Similarly, the desert pupfish and rat curves were shifted 9 and 12°C with respect to the 25°C-treated goldfish, again corresponding to their respective differences in acclimation or body temperature. However, the fact that polarization values obtained for the arctic sculpin and 5°C-treated goldfish at their respective cellular temperatures were different from those obtained from 25°C-treated goldfish, desert pupfish, and rat indicates that all species did not adapt to similar levels of fluidity. It appears that species inhabiting relatively unchanging thermal environments permit a more complete specialization of their membranes to their respective cellular temperatures than is possible in eurythermal species, which must tolerate and adapt to large seasonal fluctuations in environmental temperature (Cossins and Prosser, 1978).

Whereas poikilothermic animals can experience substantial changes in body temperature, most birds and mammals are obligate homeotherms with a euthermic range of 38 ± 4°C (Hazel and Prosser, 1974). It is well known that the tissues of homeothermic mammals often fail if the temperature is reduced 10 to 15°C. An exception among mammals are the hibernators, which during this period of torpor and reduced metabolism permit their body temperature to drop to between 1 and 5°C. Although the activity of organs such as the heart and lung is considerably reduced during hibernation, they still remain functional. Consequently, the membranes of these organs must be adapted to function at low body temperatures (Aloia, 1978). This fact has prompted several recent investigations of the lipid composition/fluidity characteristics of membranes from hibernating species to determine if a homeoviscous response occurs.

The data obtained thus far indicate that many hibernating mammals alter the lipid composition of their membranes, but the patterns of change are not as sharply defined or consistent as those observed for poikilotherms acclimated to cold temperatures. For example, the fatty acids of phospholipids of brain tissues exhibited increased levels of unsaturation in hibernating hamsters, *Mesocricetus auratus* (Goldman, 1975), and hibernating ground squirrels, *Citellus lateralis* (Aloia, 1979); however, membrane lipids from whole heart tissue of *C. lateralis* (Aloia and Pengelley, 1979) and mitochondria of heart and liver from *C. tridecemlineatus* (Platner et al., 1976) either exhibited no change in saturation or became more saturated during the hibernating state. In the case of phospholipid composition, both Aloia (1979) and Robert et al. (1982) found a decrease in the ratio of ethanolamine phosphoglycerides to choline phosphoglycerides in hibernating squirrels and hamsters, respectively. In cold-acclimated poikilotherms, PE levels typically increase while PC decreases.

At present it is difficult to correlate the complex and sometimes conflicting data on membrane fatty acid and phospholipid alterations that occur during hibernation with continued membrane function. The lack of a consistent trend toward increased unsaturation of fatty acids in hibernating mammals suggests that this is not a primary factor in maintaining membrane fluidity, at least on a bulk or general scale. In fact, there is recent evidence indicating that membranes of hibernating mammals in their active (summer) physiological condition may be sufficiently fluid to sustain a liquid-crystalline state at reduced temperatures without adjustments being necessary. Even some nonhibernators apparently possess cell membranes that remain in a fluid state down to approximately 0°C (Aloia, 1980). In view of these findings, Aloia (1980) has proposed that any homeoviscous adjustments resulting from alteration of fatty acid levels may be restricted to specific regions or compartments of membranes and that the nature of the alteration would be dictated by the viscosity or internal milieu necessary for that specific region of the membrane to continue functioning under the new physiological state. This might explain why different organs and individual phospholipid classes experience different alterations in fatty acid composition during hibernation.

CONCLUDING REMARKS

The studies reported here confirm that microorganisms, plants, and animals can compensate for the effect of temperature upon the fluidity of their constituent membranes by modifying the chemical composition of the membrane lipids. Although increasing the unsaturation of fatty acyl chains of phospholipids is the most commonly observed response to lowered temperature, changes in the degree of branching and length of acyl chains, retailoring of phospholipid molecular species, substitution of different polar headgroups, and insertion of spacer molecules such as cholesterol can also render a membrane more disordered or fluid. It is presumed that the observed changes in membrane fluidity permit continued membrane function (e.g., carbohydrate transport, mitochondrial oxidation and ATP production, catalytic activity of membrane-bound enzymes, nonelectrolyte permeability) at the new temperature. Except for Hazel (1979), however, few studies have demonstrated that the thermally induced alteration in membrane composition actually altered the functional properties of the membrane in a compensatory manner. Some membranes exhibit little or no measurable fluidity adjustments during thermal acclimation even though the acyl groups of the phospholipids become more unsaturated (Cossins et al., 1978). In these cases it is possible that changes in the fatty acid composition of phospholipids intimately associated with membrane-bound proteins may affect their functional properties without having any noticeable effect upon the bulk membrane fluidity. It is also very likely that the homeoviscous response is only one of several adaptive responses that must occur if membranes are to retain an optimal function under a new thermal regime. For example, Sidell (1977) found that fish increased the concentration of electron transport enzymes in mitochondrial membranes by up to 50% in response to cold acclimation. Obviously much additional study is needed to establish with certainty the functional significance of membrane lipids, especially as it pertains to the concept of homeoviscous adaptation. Nonetheless, tremendous advances have been made in understanding the complexity of organization of cellular membranes and the dynamic and unique role of lipids in membrane function. Continued research in this area will undoubtably open up new and exciting horizons.

REFERENCES

Aloia, R. C. (1978). Phospholipid composition of hibernating ground squirrel (*Citellus lateralis*) kidney and low-temperature membrane function. *Comp. Biochem. Physiol.* **60B**:19–26.

Aloia, R. C. (1979). Brain lipid composition of the hibernating and active ground squirrel, *Citellus lateralis. J. Thermal Biol.* **4**:223–231.

Aloia, R. C. (1980). The role of membrane fatty acids in mammalian hibernation. *Federation Proc.* **39**:2974–2979.

Aloia, R. C. and E. T. Pengelley (1979). Lipid composition of cellular membranes of hibernating mammals. In *Chemical Zoology* (Florkin, M. and B. T., Scheer, Eds.), Vol. 11, pp. 1–48. Academic Press, New York.

Bangham, A. D. (1975). Models of cell membranes. In: *Cell Membranes: Biochemistry, Cell Biology and Pathology* (Weissmann, G. and R. Claiborne, Eds.), pp. 24–34. HP Publishing, New York.

Baranska, J. and P. Wlodawer (1969). Influence of temperature on the composition of fatty acids and on lipogenesis in frog tissue. *Comp. Biochem. Physiol.* **28**:553–570.

Barenholz, Y. and T. E. Thompson (1980). Sphingomyelins in bilayers and biological membranes. *Biochim. Biophys. Acta* **604**:129–158.

Beck, J. S. (1980). *Biomembranes: Fundamentals in Relation to Human Biology.* McGraw-Hill, New York.

Bergelson, L. D. and L. I. Barsukov (1977). Topological asymmetry of phospholipids in membranes. *Science* **197**:224–230.

Bishop, D. G., J. R. Kenrick, J. Bayston, A. S. Macpherson, S. R. Johns, and R. I. Willing (1979). The influence of fatty acid unsaturation on fluidity and molecular packing of chloroplast membrane lipids. In: *Low Temperature Stress in Crop Plants* (Lyons, J. M., D. Graham, and J. K. Raison, Eds.), pp. 375–389. Academic Press, New York.

Caldwell, R. S. and J. F. Vernberg (1970). The influence of acclimation temperature on the lipid composition of fish gill mitochondria. *Comp. Biochem. Physiol.* **34**:179–191.

Chapelle, S. (1978). The influence of acclimation temperature on the fatty acid composition of an aquatic crustacean (*Carcinus meanas*). *J. Exp. Zool.* **204**:337–346.

Chapelle, S. and M. Gilles-Baillien (1981). Variation in the lipids in the intestinal membranes of active and hibernating tortoises. *Biochem. System. Ecol.* **9**:233–240.

Chapman, D. (1975a). Lipid dynamics in cell membranes. In: *Cell Membranes: Biochemistry, Cell Biology and Pathology* (Weissmann, G. and R. Claiborne, Eds.), pp. 13–22. HP Publishing, New York.

Chapman, D. (1975b). Fluidity and phase transitions of cell membranes. *Biomembranes* **7**:1–9.

Chapman, D. and D. F. H. Wallach (1968). Recent physical studies of phospholipids and natural membranes. In: *Biological Membranes* (Chapman, D., Ed.), pp. 125–202. Academic Press, New York.

Chapman, E., L. C. Wright, and J. K. Raison (1979). Seasonal changes in the structure and function of mitochondrial membranes of artichoke tubers. *Plant Physiol.* **63**:363–366.

Cossins, A. R. (1976). Changes in muscle lipid composition and resistance adaptation to temperature in the freshwater crayfish, *Austropotamobius pallipes*. *Lipids* **11**:307–316.

Cossins, A. R. (1977). Adaptation of biological membranes to temperature—The effect of temperature acclimation of goldfish upon the viscosity of synaptosomal membranes. *Biochim. Biophys. Acta* **470**:395–411.

Cossins, A. R. (1981). The adaptation of membrane dynamic structure to temperature. In: *Effects of Cold on Biological Membranes* (Morris, G. and A. Clarke, Eds.), pp. 83–106. Academic Press, New York.

Cossins, A. R. and C. L. Prosser (1978). Evolutionary adaptation of membranes to temperature. *Proc. Nat. Acad. Sci. USA.* **75**:2040–2043.

Cossins, A. R., M. J. Friedlander, and C. L. Prosser (1977). Correlations between behavioral temperature adaptations of goldfish and the viscosity and fatty acid composition of their synaptic membranes. *J. Comp. Physiol.* **120**:109–121.

Cossins, A. R., J. Christiansen, and C. L. Prosser (1978). Adaptation of biological membranes to temperature—The lack of homeoviscous adaptation in the sarcoplasmic reticulum. *Biochim. Biophys. Acta* **511**:442–454.

Danielli, J. F. (1975). The bilayer hypothesis of membrane structure. In: *Cell Membranes:*

Biochemistry, Cell Biology and Pathology (Weissmann, G. and R. Claiborne, Eds.), pp. 3–11. HP Publishing, New York.

Davis, M. B. and D. F. Silbert (1974). Changes in cell permeability following a marked reduction of saturated fatty acid content of *Escherichia coli* K-12. *Biochim. Biophys. Acta* **373**:224–241.

De Bony, J. and E. A. Dennis. 1981. Magnetic nonequivalence of the two fatty acid chains in phospholipids of small unilamellar vesicles and mixed micelles. *Biochemistry* **20**:5256–5260.

De Kruyff, B., W. J. De Greef, R. V. W. Van Eyk, R. A. Demel, and L. L. M. Van Deenen (1973). The effect of different fatty acid and sterol composition on the erythritol flux through the cell membrane of *Acholeplasma laidlawii*. *Biochim. Biophys. Acta* **298**: 479–499.

de la Roche, I. A., M. K. Pomeroy, and C. J. Andrews (1975). Changes in fatty acid composition in wheat cultivars of contrasting hardiness. *Cryobiology* **12**:506–512.

Dickens, B. F. and G. A. Thompson, Jr. (1982). Phospholipid molecular species alterations in microsomal membranes as an initial key step during cellular acclimation to low temperature. *Biochemistry* **21**:3604–3611.

Eisenberg, M. and S. McLaughlin (1976). Lipid bilayers as models of biological membranes. *BioScience* **26**:436–443.

Esko, J. D., J. R. Gilmore, and M. Glaser (1977). Use of a fluorescent probe to determine the viscosity of LM cell membranes with altered phospholipid compositions. *Biochemistry* **16**:1881–1890.

Farkas, T. and J. C. Nevenzel (1981). Temperature acclimation in the crayfish: Effects on phospholipid fatty acids. *Lipids* **16**:341–346.

Fettiplace, R. and D. A. Haydon (1980). Water permeability of lipid membranes. *Physiol. Rev.* **60**:510–550.

Finean, J. B., R. Coleman, and R. H. Michell (1978). *Membranes and Their Cellular Functions*, 2nd ed. Halsted Press, New York.

Goldman, S. S. (1975). Cold resistance of the brain during hibernation. III. Evidence of a lipid adaptation. *Am. J. Physiol.* **228**:834–838.

Haest, C. W. M., J. De Gier, and L. L. M. Van Deenen (1969). Changes in the chemical and the barrier properties of the membrane lipids of *E. coli* by variation of the temperature of growth. *Chem. Phys. Lipids* **3**:413–417.

Haest, C. W. M., J. De Gier, G. A. Van Es, A. J. Verkheij, and L. L. M. Van Deenen (1972). Fragility of the permeability barrier of *Escherichia coli*. *Biochim. Biophys. Acta* **288**:43–53.

Harrison, R. and G. G. Lunt (1975). *Biological Membranes: Their Structure and Function*. Wiley, New York.

Hazel, J. R. (1979). Influence of thermal acclimation on membrane lipid composition of rainbow trout liver. *Am. J. Physiol.* **236**:R91–R101.

Hazel, J. R. (1984). Effects of temperature upon the structure and metabolism of cell membranes in fish. *Am. J. Physiol.* **346**:R460–R470.

Hazel, J. R. and C. L. Prosser (1974). Molecular mechanisms of temperature compensation in poikilotherms. *Physiol. Rev.* **54**:620–677.

Hazel, J. R. and C. L. Prosser (1978). Incorporation of 1-^{14}C-acetate into fatty acids and sterols by isolated hepatocytes of thermally acclimated rainbow trout (*Salmo gairdneri*). *J. Comp. Physiol.* **134**:321–329.

Hazel, J. R. and P. A. Sellner (1980). The regulation of membrane lipid composition in thermally-acclimated poikilotherms. In: *Animals and Environmental Fitness* (Gilles, R., Ed.), pp. 541–560. Pergamon Press, Oxford.

Hazel, J. R., P. A. Sellner, A. F. Hagar, and N. P. Neas (1983). Thermal adaptation in biological membranes. The acylation of *sn*-glycerol-3-phosphate by liver microsomes of thermally acclimated trout (*Salmo gairdneri*). *Molec. Physiol.* **4**:125–140.

Hilditch, T. P. and P. N. Williams (1964). *The Chemical Composition of Natural Fats*, 4th ed. Wiley, New York.

Huang, C. (1977). Configurations of fatty acyl chains in egg phosphatidylcholine-cholesterol mixed bilayers. *Chem. Phys. Lipids* **19**:150–158.

Hui, S. W., T. P. Stewart, P. L. Yeagle, and A. D. Albert (1981). Bilayer to non-bilayer transition in mixtures of phosphatidylethanolamine and phosphatidylcholine: Implications for membrane properties. *Arch. Biochem. Biophys.* **207**:227–240.

Kallapur, V. L., R. G. H. Downer, J. C. George, and J. E. Thompson (1982). Effect of environmental temperature on the phase properties and lipid composition of flight muscle mitochondria of *Schistocerca gregaria*. *Insect Biochem.* **12**:115–121.

Karp, G. (1979). *Cell Biology*. McGraw-Hill, New York.

Kuiper, P. J. C. (1968). Lipids in grape roots in relation to chloride transport. *Plant Physiol.* **43**:1367–1371.

Lee, A. G. (1977). Lipid phase transitions and phase diagrams. II. Mixtures involving lipids. *Biochim. Biophys. Acta* **474**:285–344.

Lyons, J. M. (1973). Chilling injury in plants. *Ann. Rev. Plant Physiol.* **24**:445–466.

Lyons, J. M., J. K. Raison, and P. L. Steponkus (1979). The plant membrane in response to low temperature: An overview. In: *Low Temperature Stress in Crop Plants* (Lyons, J. M., D. Graham, and J. K. Raison, Eds.), pp. 1–24. Academic Press, New York.

Marinetti, G. V., J. T. Bagnara, and K. Cattieu (1981). Changes in the phospholipid composition and in fatty acids of phosphatidylcholine and phosphatidylethanolamine of various tissues of the developing tadpole of *Agalychnis dacnicolor*. *Comp. Biochem. Physiol.* **70B**:783–785.

Marr, A. G. and J. L. Ingraham (1962). Effect of temperature on the composition of fatty acids in *Escherichia coli*. *J. Bacteriol.* **84**:1260–1267.

McElhaney, R. N. (1974). The effect of alterations in the physical state of the membrane lipids on the ability of *Acholeplasma laidlawii* B to grow at various temperatures. *J. Mol. Biol.* **84**:145–157.

McElhaney, R. N., J. De Gier, and E. C. M. Van Der Neut-kok (1973). The effect of alterations in fatty acid composition and cholesterol content on the nonelectrolyte permeability of *Acholeplasma laidlawii* B cells and derived liposomes. *Biochim. Biophys. Acta* **298**:500–512.

Nozawa, Y., H. Iida, H. Fukushima, K. Ohki, and S. Ohnishi (1974). Studies on *Tetrahymena* membranes: Temperature-induced alterations in fatty acid composition of various membrane fractions in *Tetrahymena pyriformis* and its effect on membrane fluidity as inferred by spin-label study. *Biochim. Biophys. Acta* **367**:134–147.

Nystrom, R. A. (1973). *Membrane Physiology*. Prentice-Hall, New Jersey.

Oldfield, E. and D. Chapman (1972). Dynamics of lipids in membranes: Heterogeneity and the role of cholesterol. *FEBS Lett.* **23**:285–297.

Pearcy, R. W. (1978). Effect of growth temperature on the fatty acid composition of the leaf lipids in *Atriplex lentiformis* (Torr.) Wats. *Plant Physiol.* **61**:484–486.

Platner, W. S., D. G. Steffen, G. Tempel, and X. J. Musacchia (1976). Mitochondrial membrane fatty acids of liver and heart of the euthermic and hibernating ground squirrel (*Citellus tridecemlineatus*). *Comp. Biochem. Physiol.* **53A**:279–283.

Poon, R., J. M. Richards, and W. R. Clark (1981). The relationship between plasma membrane lipid composition and physical-chemical properties. II. Effect of phospholipid fatty acid modulation on plasma membrane physical properties and enzymatic activities. *Biochim. Biophys. Acta* **649**:58–66.

Raison, J. K. (1980). Membrane lipids: Structure and function. In: *The Biochemistry of Plants*, Vol. 4 (Stumpf, P. K., Ed.), pp. 57–83. Academic Press, New York.

Raison, J. K. and E. A. Chapman (1976). Membrane phase changes in chilling-sensitive *Vigna radiata* and their significance to growth. *Aust. J. Plant Physiol.* **3**:291–299.

Raison, J. K., E. A. Chapman, L. C. Wright, and S. W. L. Jacobs (1979). Membrane lipid transitions: Their correlation with the climatic distribution of plants. In: *Low Temperature Stress in Crop Plants* (Lyons, J. M., D. Graham, and J. K. Raison, Eds.), pp. 177–186. Academic Press, New York.

Robert, J., D. Montaudon, L. Dubourg, G. Rebel, J.-L. Miro, and B. Canguilhem (1982). Changes in lipid composition of the brain cellular membranes of an hibernating mammal during its circannual rhythm. *Comp. Biochem. Physiol.* **71B**:409–416.

Roberts, M. F., A. A. Bothner-By, and E. A. Dennis (1978). Magnetic nonequivalence within the fatty acyl chains of phospholipids in membrane models: [1]H nuclear magnetic resonance studies of the α-methylene groups. *Biochemistry* **17**:935–942.

Robinson, G. B. (1975). The isolation and composition of membranes. In: *Biological Membranes* (Parsons, D. S., Ed.), pp. 8–32. Clarendon press, Oxford.

Roseman, S. (1975). Sugars of the cell membrane. In: *Cell Membranes: Biochemistry, Cell Biology and Pathology* (Weissmann, G. and R. Claiborne, Eds.), pp. 55–64. HP Publishing, New York.

Rothman, J. E. and J. Lenard (1977). Membrane asymmetry. *Science* **195**:743–753.

Schmidt, C. F., Y. Barenholz, and T. E. Thompson (1977). A nuclear magnetic resonance study of sphingomyelin in bilayer systems. *Biochemistry* **16**:2649–2656.

Sidell, B. D. (1977). Turnover of cytochrome C in skeletal muscle of green sunfish (*Lepomis cyanellus*, R.) during thermal acclimation. *J. Exp. Zool.* **199**:233–250.

Silbert, D. F. (1975). Genetic modification of membrane lipid. *Ann. Rev. Biochem.* **44**:315–339.

Siminovitch, D., J. Singh, and I. A. de la Roche (1975). Studies on membranes in plant cells resistant to extreme freezing. I. Augmentation of phospholipids and membrane substance without changes in unsaturation of fatty acids during hardening of black locust bark. *Cryobiology* **12**:144–153.

Sinensky, M. (1971). Temperature control of phospholipid biosynthesis in *Escherichia coli*. *J. Bacteriol.* **106**:449–455.

Sinensky, M. (1974). Homeoviscous adaptation—A homeostatic process that regulates the viscosity of membrane lipids in *Escherichia coli*. *Proc. Nat. Acad. Sci. USA* **71**:522–525.

Singer, S. J. (1974). The molecular organization of membranes. *Ann. Rev. Biochem.* **43**:805–834.

Singer, S. J. (1975). Architecture and topography of biologic membranes. In: *Cell Membranes: Biochemistry, Cell Biology and Pathology* (Weissmann, G. and R. Claiborne, Eds.), pp. 35–44. HP Publishing, New York.

Singer, S. J. and G. L. Nicolson (1972). The fluid mosaic model of the structure of cell membranes. *Science* **175**:720–731.

Singh, J., A. de la Roche, and D. Siminovitch (1977). Differential scanning calorimeter analyses of membrane lipids isolated from hardened and unhardened black locust bark and from winter rye seedlings. *Cryobiology* **14**:620–624.

Thompson, G. A., Jr. (1979). Molecular control of membrane fluidity. In: *Low Temperature Stress in Crop Plants* (Lyons, J. M., D. Graham, and J. K. Raison, Eds.), pp. 347–363. Academic Press, New York.

Träuble, H. (1972). Phase transitions in lipids. In: *Biomembranes* (Kreuzer, F. and Slegers, J. F. G., Eds.), pp. 197–227. Plenum Press, New York.

Träuble, H. and D. H. Haynes (1971). The volume change in lipid bilayer lamellae at the crystalline-liquid crystalline phase transition. *Chem. Phys. Lipids* **7**:324–335.

White, F. N. and G. Somero (1982). Acid-base regulation and phospholipid adaptations to temperature: Time courses and physiological significance of modifying the milieu for protein function. *Physiol. Rev.* **62**:40–90.

Willemot, C. (1979). Chemical modification of lipids during frost hardening of herbaceous species. In: *Low Temperature Stress in Crop Plants* (Lyons, J. M., D. Graham, and J. K. Raison, Eds.), pp. 411–430. Academic Press, New York.

Wodtke, E. (1978). Lipid adaptation in liver mitochondrial membranes of carp acclimated to different environmental temperatures—Phospholipid composition, fatty acid pattern, and cholesterol content. *Biochim. Biophys. Acta* **529**:280–291.

Wunderlich, F., V. Speth, W. Batz, and H. Kleinig. 1973. Membranes of *Tetrahymena*. III. The effect of temperature on membrane core structures and fatty acid composition of *Tetrahymena* cells. *Biochim. Biophys. Acta* **298**:39–49.

Yamauchi, T., K. Ohki, H. Maruyama, and Y. Nozawa (1981). Thermal adaptation of *Tetrahymena* membranes with special reference to mitochondria—role of cardiolipin in fluidity of mitochondrial membranes. *Biochim. Biophys. Acta* **649**:385–392.

4

BIOELECTRICAL PHENOMENA

Electricity underlies almost every facet of biological existence. Every cell possesses an electrical charge across its membrane, and hence may be considered as an electrical unit. Moreover, in higher animals the processing and transmission of information, as well as the contraction of muscle, require either electrical stimuli or stimuli that elicit events in which there are electrical phases. As indicated in the beginning of Chapter 3, irritability and the transmission of impulses are functions intimately associated with cell membranes. In this chapter we consider the structure, chemical composition, and function of membranes of nerve cells, emphasizing particularly the role of constituent lipids in generating and controlling electrical activities associated with neural function. Coverage includes a comparison of lipid distribution within the nervous system of invertebrates and vertebrates, alterations in composition that occur with development and aging, and the functional significance of changes that occur in neural lipids that may help an animal better adapt to a changing environment. Before addressing these topics, however, it is useful to review some basic concepts in electricity that have application in biological systems.

ELECTRICITY AND BIOLOGY—SOME BASIC CONCEPTS

Current can be defined as the movement of an electric charge. In metallic conduction (e.g., copper wire), current involves the movement of free electrons. These electrons are negatively charged and can be transferred from one point to another if energy is provided, the measure of which is the voltage. In biological systems, current flow involves the movement of ions rather than electrons. Because ions have a much greater mass than electrons, under the same voltage conditions ions will move more slowly. Furthermore, membranes act as barriers to the free diffusion of all ions. The rate at which ions traverse a membrane is as much as 10^8 times lower than the average rate at which they diffuse across an equivalent distance (50 to 100 Å) of cytoplasm or extracellular fluid (Eckert and Randall, 1983). The tendency to prevent the movement of ions across a membrane is referred to as resistance—the more permeable a membrane is to an ion the less will be its resistance. The relationship between current (I), voltage (E), and resistance (R) is given by Ohm's law:

$$I(\text{current}) = \frac{E(\text{voltage})}{R(\text{resistance})}$$

Materials like glass and rubber are known as good insulators (high electrical resistance) because only a small amount of current flows through them even when high voltages are applied. In the membranes of living cells, the lipid bilayer provides a similar region of high electrical resistance because the constituent molecules are nonionizing and under most conditions bear only weak electrostatic charges.

   Another important physical concept in electricity that has application in biological systems is capacitance, or the ability to store a charge. A capacitor consists basically of three layers—two good conducting layers that sandwich a poor conductor (Figure 4.1). The poor conductor is often referred to as the dielectric material. The capacity of such a structure depends upon the thickness, surface area, and dielectric constant of the nonconducting material. As a general rule, the thicker the dielectric material, the poorer the capacitance or layer ability to store energy.

   In living systems, the lipid bilayer of the cell membrane has the property of capacitance. The bilayer is relatively impermeable to ions, but is thin (ca. 50 Å), and thus, can separate and store the charges resulting from the electrostatic interaction of positive ions on the outside of the cell membrane with negative ions on the inside. The dielectric constant of membrane lipids is only approximately 2 to 3 times that of a vacuum, as compared with 80 times for water (Needham, 1965). Thus, lipid acts as an excellent dielectric

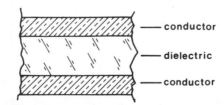

   conductor
   dielectric
   conductor

FIGURE 4.1.   Basic structural arrangement of a capacitor.

medium for membranes such as those found in nerve cells (neurons) where sudden large changes in electrical potential accompany the propagation of a nerve impulse.

The functional relationship between resistance and capacitance in a living cell can be shown by considering an electrical circuit connected across a plasma membrane (Figure 4.2). In the circuit shown, a capacitor is wired in parallel with a resistor. The lipid bilayer, which is relatively impermeable to ions, is essentially a nonconductor, and hence provides high resistance to current flow across the membrane at potential differences that normally exist. The lipid bilayer does serve as the membrane capacitor. Current applied to the membrane passes into the bilayer, where the charges accumulate, causing the potential across the capacitor to increase until a steady-state level is reached. The time required for the charge to reach 63% of this steady-state value is referred to as the time constant and is proportional to both the resistance and capacitance of the membrane. Actual current flow (i.e., movement of ions) across the membrane occurs through pores or channels in the lipid bilayer. The rate of passage of various ions through these openings in the bilayer is inversely dependent on the resistance of the membrane pores. The importance of membrane resistance and capacitance in determining the way in which membrane potentials and currents spread with distance along a nerve axon is discussed in the next section.

Because the hydrocarbon interior of a biomembrane has a low dielectric constant, considerable energy is required to transport a small ion such as potassium across the membrane. Such transport requires special mechanisms by which the activation energy of the ion transport is drastically reduced. One solution to this problem is the employment of mobile carrier molecules that bind the ion (K^+) at one membrane-solution interface and then migrate through the membrane to the opposite interface where the ion is released (Läuger, 1972). Several ion carrier molecules have been identified—valinomycin, enniatin, monactin. These are all macrocyclic molecules that

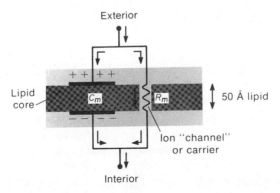

FIGURE 4.2. Equivalent circuit for a cell membrane showing membrane capacitance (C_m) and membrane resistance (R_m). Arrows indicate direction of current. After *Animal Physiology*, 2nd ed., by R. Eckert and D. Randall. W. H. Freeman and Company. Copyright © 1983.

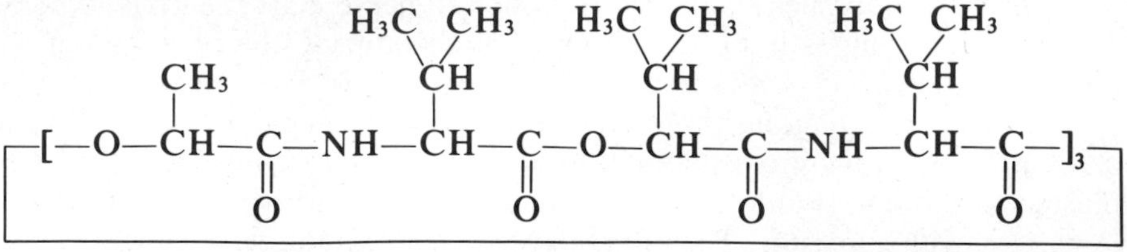

$$H_3C \quad CH_3 \qquad H_3C \quad CH_3 \qquad H_3C \quad CH_3$$

Valinomycin

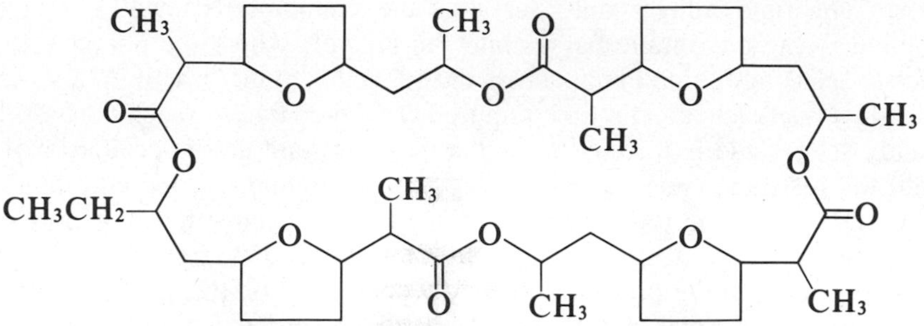

Monactin

feature a strongly hydrophobic exterior and a hydrophilic interior that offers to the cation an environment that is similar to the hydration shell of the ion in aqueous solution. The conformation of these carrier molecules may be a key factor in overcoming the high electrostatic energy barrier in the center of the membrane. Läuger (1972) proposed that these macrocyclic compounds probably are adsorbed at the membrane-solution interface, with the carbonyl groups of the molecule pointing toward the aqueous phase and the apolar side chains oriented toward the hydrocarbon interior of the membrane. When the carrier molecule binds with the cation, the carbonyl groups are believed to be turned toward the interior of the complex, and the surface activity of the molecule is lost.

STRUCTURAL BASIS FOR NERVOUS SYSTEM FUNCTION

All organisms must be capable of responding quickly and appropriately to changes in their environment. In single cell organisms, most internal structures are close enough to one another and to the surface so that communication depends solely on simple biophysical and biochemical mechanisms. No specialized communication network is required. In large multicellular organisms, however, the surface area to volume ratio is much less favorable. Although these organisms can compensate to some extent by modifying their form as they grow (e.g., branching systems of plant roots; intestinal villi and respiratory alveoli of mammals) (Beck, 1971), the more evolutionarily advanced animals still require some means of regulating and co-ordinating the activities of their cells. It is the nervous system, in association with the endocrine system, that provides these functions.

Neurons

The basic structural and functional component of the nervous system is the nerve cell, or neuron (Figure 4.3*a*). The neuron consists of the cell body, which contains a single nucleus and various cytoplasmic organelles, and two types of processes (dendrites and axons) for transmitting impulses to and from the cell body. Dendrites are highly branched extensions of the cell body that conduct an impulse toward the cell body, whereas the axon is a single, elongated extension of the cell body that conducts an impulse away from the cell body. The axon usually arises from a thickened area on the cell body called the axon hillock. Each neuron has only one axon, but side branches called collaterals may arise along the course of an axon. The axon and its collaterals terminate by branching into many fine filaments called telodendria or synaptic knobs. These knobs contain granules or vesicles in which the synaptic transmitter secreted by the nerve is stored.

Many axons in vertebrate nervous systems are surrounded by a multilay-ered, lipid-rich sheath called myelin (Figure 4.3*b*). In the peripheral nervous system, the myelin sheath is produced by flattened Schwann cells, which are located along the axon. During the myelination process, each Schwann cell wraps itself around the axon several times, pushing the cytoplasm and nucleus of the Schwann cell to the periphery, where they eventually form the neurilemma (Figure 4.4). The inner portion, consisting of several layers of Schwann cell membrane, is the myelin sheath. When viewed in cross

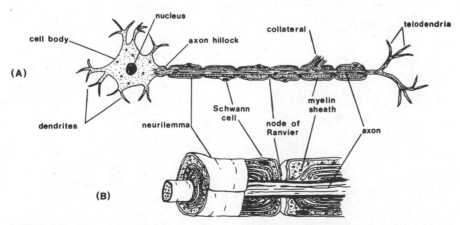

FIGURE 4.3. The neuron. (*A*) Structure of an entire multipolar neuron. (*B*) Cross section and longi-tudinal section of myelinated nerve fiber. Redrawn, by permission, from Tortora and Anagnostakos (1984).

BIOELECTRICAL PHENOMENA

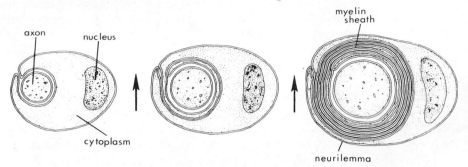

FIGURE 4.4. Stages in the formation of the myelin sheath.

section with an electron microscope, a pattern of alternating dark and light lines forming concentric rings is seen (Figure 4.5). The dark lines result from the apposition of the inner faces of the Schwann cell membrane when the membrane compacts on itself, while the lighter lines result from the coming together of the outer membrane of the Schwann cell during successive wraps. At regular intervals along the length of a myelinated axon there are constrictions, called the nodes of Ranvier, where myelin is absent.

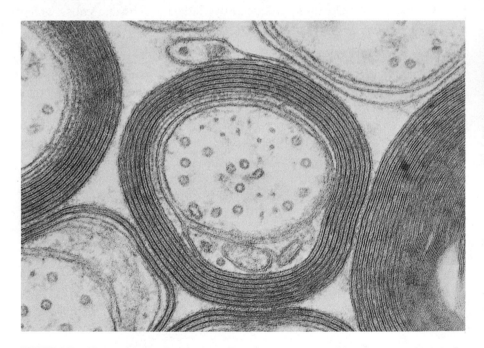

FIGURE 4.5. Electron micrograph of single myelinated axon from the central nervous system. × 150,000. From P. Morell and W. T. Norton, Myelin. Copyright © 1980 by Scientific American, Inc. All rights reserved.

Myelinated fibers are also present in the central nervous system (CNS), being responsible for the white matter of the brain and spinal cord. In the CNS, the myelin sheath is formed from cells called oligodendrocytes rather than Schwann cells, which are absent. Each oligodendrocyte sends out processes (ca. 40) that spiral around an adjacent axonal segment. Nodes of Ranvier are present along the myelinated fibers of the CNS, but they are not as numerous, and there is no neurilemma. A detailed account of the lipid composition of myelin and its relationship to the insulative and protective functions of myelin during nerve impulse conduction will be presented following a brief review of the physiological basis for the transmission of action potentials along the axon.

Resting and Action Potentials

The outside of most, if not all, cells is usually electrically charged or polarized with respect to the inside. The charge is the result of an unequal distribution of ions across the cell membrane. In neurons, the concentration of Na^+ is approximately 14 times greater on the outside than on the inside of the membrane, while K^+ is approximately 30 times more abundant inside the membrane. In addition, the inside of a nerve cell also contains large numbers of nondiffusable, negatively charged ions (anions), most of which are protein. The nerve membrane is only slightly permeable to Na^+, but very permeable to K^+; thus, K^+ tends to diffuse outward quite readily while Na^+ diffuses inward more slowly. To prevent these ions from reaching equilibrium by diffusion, the cell actively pumps Na^+ outward and K^+ inward. However, because K^+ tends to diffuse outward again at a higher rate than the inward diffusion of Na^+ and because the membrane is impermeable to most anions inside the cell, the outside of the membrane becomes positively charged with respect to the inside. The difference in charge on either side of the membrane when the cell is at rest (not conducting impulses) is called the "resting" or "membrane" potential. This potential varies from one cell to another but generally is of the order of 70 to 100 mV (millivolts) (Eyzaguirre and Fidone, 1975).

As long as the neuron is undisturbed it remains in this polarized state. However, when a stimulus of sufficient strength (threshold) is applied to the neuron, the membrane undergoes a sudden change in permeability. Channels or pores in the membrane that are highly selective for Na^+ now open, allowing Na^+ to rapidly diffuse inward. This depolarizes the membrane and eventually reverses the membrane potential so that the inside of the cell is briefly positive. A return to the normal resting potential (repolarization) is brought about by a reduced drive for sodium entry as the inside of the membrane becomes more positive and a simultaneous increase in potassium permeability, which causes K^+ to rapidly move out of the cell. This sequence of events constitutes an "action potential" and is the basis for the transmission of nerve impulses along an axon. Because so few Na^+

and K$^+$ move in or out of the cell during a single action potential, there is virtually no change in their concentration gradients. Yet continued flux of these ions without restoration would eventually eliminate the concentration gradient and render the membrane inexcitable. To prevent this imbalance, the distribution of Na$^+$ and K$^+$ is restored rapidly by the membrane active transport system. This restoration occurs mainly after the action potential is completed, and thus plays no direct role in the production or recovery of an action potential.

Action Potential Propagation—Myelinated Versus Unmyelinated Fibers

In order for the electrochemical change produced by the stimulus to serve as a communication signal, it must travel from one part of the neuron to another. The action potential itself does not travel along the membrane; instead, each action potential triggers a new one at an adjacent area of the membrane, producing a wave of depolarization, or a nerve impulse. In a nonmyelinated fiber (Figure 4.6), the propagation of action potentials is achieved by the electrotonic spread of local circuit current from the region of sodium influx (i.e., point of stimulation) to nearby regions of inactive membrane. The current spreads longitudinally along the inside of the axon and out of the membrane, depolarizing the unexcited portions of the membrane ahead of the region of inward Na$^+$ flux. The reduced potential across the membrane once again increases Na$^+$ conductance and initiates the regenerative sequence that produces a new action potential. These processes repeat themselves until the end of the axon is reached. A current will also flow into regions behind the advancing action potential to complete the circuit, but no restimulation of the axon is produced because the membrane here is in the refractory state.

The process by which action potentials are conducted in myelinated fibers is somewhat different (Figure 4.6). The axonal membrane of myelinated fibers is exposed to the extracellular fluid only at the nodes of Ranvier. The rest of the fiber is covered by the myelin sheath, which has a much higher

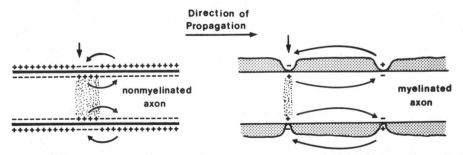

FIGURE 4.6. Current flow in a nonmyelinated and myelinated axon. Diameter of myelinated axon greatly enlarged for clarity.

electrical resistance and a much lower capacitance than the axonal membrane itself. Because of this structural arrangement and the electrical properties of myelin, resting and action potentials are generated only at the nodes. When a stimulus depolarizes the membrane at a node, the current generated has to travel along the inside of the fiber until it reaches the next node. Here the current leaves the fiber and returns along the outside of the myelin sheath to the original depolarized region. The current leaving the second node causes its depolarization and the development of another action potential. These events repeat themselves at each subsequent node along the axon. Thus, instead of the action potential being propagated with continuity along the length of the axon as it occurs in nonmyelinated fibers, a series of discontinuous action potentials are produced, one at each node. Propagation by this means is called saltatory conduction (from the Latin *saltare*, to leap or dance).

Saltatory conduction along a myelinated nerve fiber has several advantages over the continuous conduction exhibited by nonmyelinated fibers. Most important is the fact that saltatory conduction is much more rapid. As discussed above, the capacity or the amount of charge stored across the myelin sheath is considerably smaller than the charge at a node. Since the length of time taken for an active node to depolarize an adjacent inactive node to threshold is determined by the amount of charge that must be removed, the conduction speed of a myelinated nerve fiber is many times greater than that of a nonmyelinated fiber having the same diameter and membrane properties. Increased velocity of impulse propagation can be achieved by increasing the diameter of the axon, which also results in reduced internal longitudinal resistance. This is the evolutionary course adopted by invertebrates, and is exemplified by the giant axons of the squid and crayfish. Still, a $10\text{-}\mu$ myelinated fiber of a frog has the same conduction velocity as a $500\text{-}\mu$ unmyelinated squid axon. Moreover, nearly 2500 $10\text{-}\mu$ fibers can be packed into the same volume occupied by the giant squid axon (Ruch et al., 1965). Thus, myelination permits the thousands of nerve fibers that make up a nerve trunk to take up comparatively little space, yet conduct impulses at velocities reaching 120 m per sec. The latter is obviously a significant contributory factor to the evolutionary success of higher vertebrates, particularly in the development of the complex nervous system and its control of muscle activity. A final consideration is that myelination conserves energy for the neuron. The Na^+ and K^+ that depolarize the membrane must be pumped back to their respective sides only at the nodes rather than along the entire axonal membrane, as is the case in a nonmyelinated fiber. Thus, only a modest amount of energy is required, allowing the nerve to conduct action potentials continuously with very little energy consumption. If there is a disadvantage to myelination it would be that the loss or destruction of myelin through such diseases as multiple sclerosis has very deleterious effects on the functioning of the nervous system (Morell and Norton, 1980).

LIPIDS OF THE NERVOUS SYSTEM

Studies of the quantitative distribution of lipids in the nervous systems of invertebrates and vertebrates continues to be an active area of research. New methods for lipid isolation combined with advances in analytical instrumentation have resulted in the discovery of several different kinds of lipids, some of which are unique to neural tissue. By using innovative microsurgical techniques, investigators have been able to link many of these lipids with specific nervous structures or materials such as the myelin sheath or synaptic membranes. The composition and concentration of nervous system lipids at the gross, cellular, and subcellular levels have been reviewed in some detail by McMurray et al. (1964), Eichberg et al. (1969), Rouser and Yamamoto (1969), and Bowen et al. (1974). This synopsis draws upon information contained in these papers, but emphasizes those regions or structures of the nervous system where it has been shown that modification of constituent lipids may influence function in a manner that is adaptive.

Myelin

Myelin has been chosen as a starting point because its functional significance as an electrical insulator has been described in the previous section and because its presence or absence largely dictates the nature of the lipid composition of a specific tissue or region of the nervous system. Although myelin consists of successive wrappings of plasma membrane from Schwann cells, its molecular composition is quite different from that of other plasma membranes (Figure 4.7). The most conspicuous feature is its high proportion of lipid to protein (see also Table 3.1). Lipids constitute approximately 70%

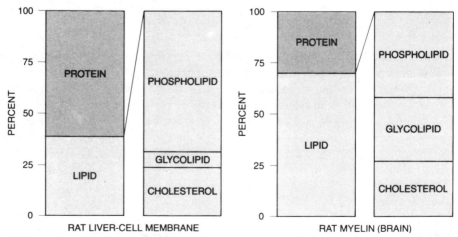

FIGURE 4.7. Comparison (by weight) of the molecular content of myelin with that of a liver cell plasma membrane. From P. Morell and W. T. Norton, Myelin. Copyright © 1980 by Scientific American, Inc. All rights reserved.

of the dry weight of myelin from the CNS and nearly 80% of myelin from the peripheral nervous system, compared with 40 to 50% of the plasma membranes of other animal cells. The high lipid content of myelin effectively prevents the passage of water-soluble ions such as sodium and potassium, thus contributing to its effectiveness as an electrical insulator for neurons. The protein composition of myelin also reflects its rather stable and metabolically inert nature. Not only does myelin have much less protein than other cell surface membranes, but there are fewer types of protein (Morell and Norton, 1980). One predominant protein is soluble in organic solvents and is believed to be located within the lipid bilayer of myelin, where it possibly complements the role played by lipids. The ionic impermeability of myelin is in part due to the absence of proteins (ion channels) present in other cell membranes that facilitate the passage of ions through the lipid bilayer (Morell and Norton, 1980).

The principal lipid classes present in myelin are phospholipids, glycolipids, and cholesterol (Figure 4.7). Major groups of phospholipids include phosphatidylethanolamine (PE), phosphatidylcholine (PC), phosphatidylserine (PS), phosphatidylinositol (PI), sphingomyelin, and ethanolamine plasmalogens (see Chapter 1). The principal glycolipids (-glycosphingolipids) are cerebrosides, which contain the sugar galactose. The presence of cerebrosides in nervous tissue is essentially restricted to the myelin sheath. The sterol content of mature myelin is almost entirely cholesterol, although other sterols and derivatives do exist (Mokrasch, 1969).

A comparison of the lipid content of brain myelin from selected vertebrate species is shown in Table 4.1. The quantitative lipid patterns, with some exceptions, are quite similar to each other. Cholesterol accounts for about 35 to 50% of the myelin lipid in the species listed, while total phospholipids

TABLE 4.1
Lipid content of brain myelin from various vertebrate species[a]

Lipid	Goldfish	Frog	Pigeon	Rabbit	Rat	Human
Total phospholipids	57.9	49.1	31.6	40.3	33.0	36.2
Phosphatidylethanolamine	18.5	19.6	14.2	17.6	16.6	14.4
Phosphatidylcholine	24.0	19.2	8.1	9.5	10.5	11.5
Phosphatidylserine	1.9	3.7	2.9	5.4	2.9	5.4
Sphingomyelin + phosphatidylinositol	2.2	5.7	3.3	7.3	2.7	4.9
Phosphatidic acid	1.7	0.8	2.8		0.3	
Cholesterol	35.9	42.3	49.3	39.5	40.3	42.9
Cerebrosides	5.2	8.6	19.1	20.2	23.2	20.7

SOURCE: Modified from D. P. Selivonchick and B. I. Roots, Variation in myelin lipid composition induced by change in environmental temperature of goldfish (*Carassius auratus* L.). *J. Therm. Biol.* **1**. Copyright 1976, Pergamon Press, Ltd.

[a]All lipid data are expressed as mole %.

are appreciably higher in the lower vertebrates and cerebrosides much more abundant in birds and mammals. The low cerebroside content of fish and amphibian myelin appears to be compensated for by an increased amount of PC. Although not shown in Table 4.1, ethanolamine plasmalogen is also uniformly present in all vertebrate species (Eichberg et al., 1969), with concentrations for the goldfish, rat, and guinea pig approaching mole % values for their respective PE (Selivonchick and Roots, 1976). Among mammalian species there is a tendency for myelin from the peripheral nerves to have a higher proportion of sphingomyelin and a lower proportion of cerebrosides than myelin from the brain or spinal cord, but thus far no physiological significance has been ascribed to these differences.

Gray Versus White Matter

The brain and spinal cord can be divided into regions of white and gray matter. White matter is composed primarily of myelinated processes of neurons, whereas the gray area is composed of nerve cell bodies, unmyelinated internuncial neurons, and supporting nerve cells called neuroglia. In the brain, the gray matter covers the surface of the cerebrum with the white matter lying beneath; in the spinal cord, white matter surrounds the butterfly-shaped gray matter. Chemical analyses of white matter indicate that it contains more lipid per gram tissue than does gray matter, with cholesterol, cerebrosides, sphingomyelin, and ethanolamines typically abundant. These findings undoubtedly reflect the presence of myelin in the white matter. Gray matter, in contrast, exhibits a high proportion of gangliosides, a complex group of glycosphingolipids, which are discussed in greater detail in conjunction with synaptic membranes. There are also differences in the fatty acid moieties of the constituent phospholipids and glycolipids of these regions of the nervous system. Gray matter contains more polyunsaturated fatty acids than white matter, whereas the latter, especially the myelin fraction, contains a high percentage of long-chain fatty acids and hydroxy derivatives of fatty acids (Clausen, 1969). A higher percentage of saturated fatty acids in white matter may also contribute to the high stability of myelin.

Invertebrates

Information on the lipid composition of invertebrate nervous tissue is relatively meager in comparison with vertebrates. In general, invertebrate nervous systems do not contain as much lipid per unit weight, nor is the variety of lipid classes present. Again, myelin is a determinant factor in this pattern. Invertebrate neurons lack a true myelin sheath, although the giant squid axon and other invertebrate nerve fibers do possess an axonal membrane and/or covering that is rich in lipids. This covering differs from vertebrate myelin in having a high content of PC but a relatively low level of PE (McMurray et al., 1964). That cerebrosides, including gangliosides, are either absent or only present in low concentrations in invertebrate nervous systems is borne out by Lee and Gonsoulin's (1979) study of the lipids from nerve tissue of the horseshoe crab, *Limulus polyphemus*. Ethanolamine plasmalogens, which contain an ether rather than a fatty acid in the one position of glycerol, are a significant component of the phospholipid fraction of the horseshoe crab nerve tissue lipids and appear to be widely distributed in the nervous systems of invertebrates (Eichberg et al., 1969).

Synaptic Membranes

For an impulse to be transmitted from one neuron to another, it must cross over a minute gap (200 Å) between the two neurons. This anatomical area, which serves as the functional junction between the two neurons, is called the synapse, and the neuron located before the synapse is the presynaptic neuron. By using appropriate fractionation and homogenization techniques, investigators are able to remove the endings of the presynaptic neurons (synaptic terminals) and recover these as discrete particles composed of the presynaptic membrane and enclosed vesicles and mitochondria. These structures, known as synaptosomes, have been the subject of much recent electrophysiological and biochemical investigation. Lipid analyses of the synaptic membrane have revealed a high accumulation of gangliosides, a complex group of glycosphingolipids containing different numbers of sialic acid (-N-acetylneuraminic acid, abbreviated NeuAc or NANA) linked to one or more sugar residues (see Chapter 1 for structure). A model depicting their believed location on the outer surface of a synaptic terminal is shown in Figure 4.8. The close association of gangliosides with the presynaptic membrane and knowledge of their ability to form complexes with Ca^{2+}, which

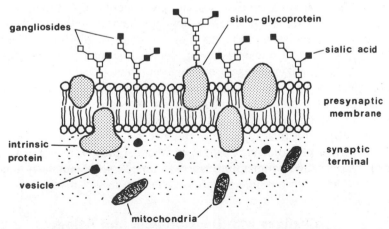

FIGURE 4.8. Model of the localization of gangliosides and sialoglycoproteins in portion of presynaptic membrane. Modified, by permission, from Rahmann (1978).

is essential for the release of the neurotransmitter, strongly suggests their involvement in interneuronal communication (Rahmann, 1980).

The concentration and composition of brain gangliosides have been examined for more than 60 vertebrate species. The results of these studies have revealed a definite correlation between the quantity and types of gangliosides present and the level of nervous organization. In birds and mammals, the ganglioside concentration ranges from 500 to 1000 μg ganglioside bound NANA per gram fresh weight, while in reptiles, amphibians, and fish, the range is about 110 to 500 μg (Rahmann and Hilbig, 1981a). Furthermore, the ganglioside pattern of homeotherms consists mainly of mono- and disialogangliosides with the nonpolar fraction prevailing, whereas in ectothermic vertebrates multisialogangliosides predominate, and the overall pattern is much more polar. In addition to gangliosides, synaptic membranes contain lipid molecules also present in myelin and other membranes. The sphingomyelins of synaptic membranes, however, differ from those of myelin in having predominantly shorter chain (C_{16}, C_{18}) fatty acids rather than C_{24} acids (Breckenridge et al., 1972). Synaptic membranes are also reported to have a much higher content of polyunsaturated acid in their PE and PS moieties, a feature that may be important in the excitability of the synaptic membranes (Nielson et al., 1970).

MODIFICATION OF LIPID COMPOSITION AND NERVOUS FUNCTION

Because initial studies of lipids associated with the nervous system indicated that these chemical compounds were relatively stable, it was thought that lipids played a rather static role in nervous system function. However, as more physiological and biochemical research was conducted on the nervous system, it became apparent that this view was incorrect and that the role of lipids required reevaluation. There is now good evidence that lipids perform a dynamic and diversified function. For example, in addition to their roles in the generation and transmission of electrical signals, they actively participate in many enzyme reactions, in membrane translocation processes, and in intermediary metabolism (Bowen et al., 1974). Recent studies have also demonstrated that several physical and biological factors can produce changes in the concentration and composition of nervous system lipids, often in a direction that appears to enhance function. In this final section on bioelectrical phenomena, changes in neural lipids in response to development, aging, and temperature are examined, with emphasis again on alterations that have possible adaptive value for an organism.

Changes with Development and Aging

Marked changes occur in the concentration and composition of lipids during growth and development of the nervous system. Most information on the nature of these changes has been obtained from studies of mammals, especially rats and mice, and more recently humans. Changes in total lipid composition and total phospholipids with age in human whole brain are summarized in Figure 4.9. The curves indicate that lipid and phospholipid composition change throughout life, with the greatest increase in amounts of both occurring early in life, followed by a decrease associated with aging (Rouser and Yamamoto, 1969). The overriding event responsible for the accretion of lipids is the elaboration of myelin. Analyses of isolated myelin during different stages of an animal's development show changes in lipid composition that are reflected in extracts from whole brain tissue. Rat brain myelin exhibits a dramatic decrease in the molar ratio of total phospholipid to cholesterol and a marked increase in cerebroside content when animals at 10 days of age are compared with adults (Cuzner and Davison, 1968). Thus, "early" myelin is quite different from adult myelin, with the former being more similar to the oligodendroglial (Schwann cell) plasma membrane. The functional significance of these changes has not been explored experimentally; however, the insertion of cerebrosides into the myelin lamellae later in development is believed to give rise to a more stable structure (Davison et al., 1966).

Paralleling the changes in concentration and distribution of lipid classes during development are alterations in fatty acid patterns within lipid classes. The general pattern is a shift during development from medium-chain length fatty acids toward longer chains and greater unsaturation (Eichberg et al., 1969). This trend is clearly evident in sphingolipids of white matter where saturated fatty acids up to C_{22} (especially 18:0) decrease, while longer chain acids (up to C_{26}) increase, particularly 24:1.

There have been fairly extensive studies of changes in lipid composition of the human brain before and after birth (Yusuf and Dickerson, 1977). The latter investigation is especially informative in that they examined the effect of growth and development on the phospholipids of different regions (forebrain, cerebellum, brain stem) of the brain. The concentration of total lipid-P was highest in the brain stem and lowest in the cerebellum at any age. The forebrain exhibited a sharp increase in total lipid-P between 13 and 20 weeks of gestation, a period during which there is rapid neuronal multiplication. A second rapid increase in total lipid-P occurred between the 35th week of gestation and the 10th month after birth. The authors attributed this second spurt to the growth of extensive neuronal interconnections and the formation of myelin, which in humans does not begin until the 30th week of gestation. The cerebellum, in contrast, had a sharp phospholipid growth spurt between three months before and six months after birth. The phospholipid pattern was similar in all three parts of the fetal brain, with choline phosphoglycerides the predominant class. After birth, ethanolamine phosphoglycerides became the dominant phospholipid in the brain stem, while little change in pattern was noted for the other two parts of the brain (Yusuf and Dickerson, 1977). It is tempting to correlate changes in lipid composition during development of the various parts of the brain with the functions of

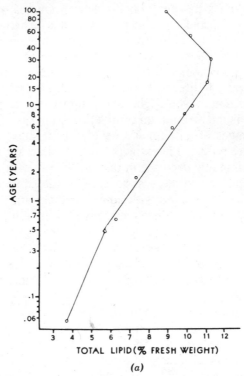

(a)

FIGURE 4.9. (a) Total lipid composition changes with age for human whole brain. (b) Total phospholipid changes with age in human whole brain (next page). Reprinted, with permission, from Rouser and Yamamoto (1969).

these regions and their importance at a specific developmental stage for the fetus and the newborn infant; however, additional research is needed to establish with certainty the validity of these correlations.

Temperature-Induced Changes

It is well established that changes in environmental temperature can lead to changes in the phospholipid composition of the nervous system of both poikilotherms and homeotherms. In Chapter 3, alterations in the phospholipids of synaptosomal membranes of cold-acclimated goldfish were cited as one example of the homeoviscous restructuring that takes place to ensure normal membrane function at low temperatures and the time required for these compositional changes (Cossins, 1977; Cossins et al., 1977). Earlier studies of the nervous system of the goldfish showed a trend toward increased unsaturation of fatty acids and fatty aldehydes of brain phospholipids (Johnston and Roots, 1964; Roots, 1968) and a decrease in brain plasmalogen content (Roots and Johnston, 1968; Driedzic et al., 1976) in cold-acclimated individuals. These changes are also consistent with expected alterations in

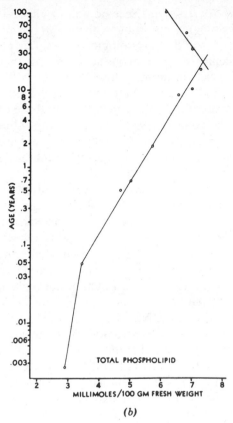

AGE (YEARS)

TOTAL PHOSPHOLIPID

MILLIMOLES/100 GM FRESH WEIGHT

(b)

FIGURE 4.9. (continued)

membrane fluidity and permeability. Myelin-specific lipids of goldfish showed little change in response to thermal acclimation except for an increase in plasmalogen content at the higher temperatures (Selivonchick and Roots, 1976). Higher plasmalogen levels might compensate for the increased thermal motion at higher temperatures, and thus help maintain stability and the normally low permeability of myelin. Consistently higher lipid weights were found in the optic tectum and optic nerve in goldfish acclimated to 5°C compared with 25°C-fish (Matheson et al., 1980). This finding is particularly interesting as it has also been shown that the mean myelinated axon diameter of the optic nerve at 5°C is considerably larger than at 25°C (1.99 μm vs. 1.48 μm; Matheson et al., 1978). Not only is the degree of myelination supported by the lipid weight changes, but the resulting increase in axon diameter is commensurate with the observed increase in conduction velocity of action potentials in some cold-acclimated poikilotherms (Lagerspetz, 1974).

The neuronal basis for thermal acclimation has been examined in detail in the land snail *Helix aspersa* (Zecevic and Levitan, 1980). Using intracellular electrophysiological techniques, these authors monitored the effects of temperature change on the physiological properties of an identified neuron. Both the frequency of spontaneous action potentials and the excitability of the neuron decreased in animals transferred from 20 to 5°C; however, after a two-week period of acclimation to 5°C, spontaneous impulse activity and excitability increased to levels comparable to those of warm-acclimated neurons at 20°C. The compensatory changes with acclimation are believed to be correlated with compensatory changes in passive membrane permeability, the latter in turn reflecting changes in membrane fluidity. The membrane fluidity of a neuron cooled to 5°C decreases as a continuous function of temperature, causing a change in disposition of membrane proteins relative to lipids. With acclimation to 5°C, the lipid composition of the membrane changes (i.e., increased unsaturation of phospholipid fatty acids), so that it becomes more fluid. The increased fluidity allows the disposition of membrane proteins and corresponding anionic field strength to return to warm-acclimated levels. Compensatory changes in neuronal properties enable poikilotherms such as snails to function normally in a wide range of environmental temperatures.

Unlike poikilotherms, the CNS of most homeothermic animals is relatively immune to changes in environmental temperature. In hibernating mammals, however, nerve conduction and brain activity must continue, despite body temperatures being reduced to 1 to 2°C above ambient. The effects of hibernation on the composition of membrane-bound lipids of brain tissue have been studied in the Syrian hamster, *Mesocricetus auratus* (Goldman, 1975; Blaker and Moscatelli, 1978), and the ground squirrel, *Citellus lateralis* (Aloia, 1979). The inconsistent and often conflicting patterns reported for membranes of hibernating mammals in general (Chapter 3) are also evident when compositional changes in lipids are examined with respect solely to nervous tissue. Still, a significant increase in the unsaturation of fatty acids in phosphoglycerides was noted in all three of the above studies. The increased unsaturation is thought to be essential for continued function of Na^+-K^+-ATPase, the enzyme primarily responsible for maintaining the proper ion gradient for nervous function. Na^+-K^+-ATPase requires phospholipids for optimal activity; furthermore, the fatty acyl chains of the phospholipids must be in a fluid state. The increase in unsaturation found in the above studies would tend to maintain lipid fluidity at the low temperatures encountered during hibernation. Goldman's (1975) finding that the phospholipids believed to be essential for ATPase activity—PS, PI, PE—had the highest levels of unsaturation further supports the thesis that continued nervous function during hibernation is dependent upon changes in membrane-bound lipids.

Electrophysiological data suggest that synapses within the nervous system are a primary site of processes associated with thermal adaptation (Lagerspetz, 1974). Although the molecular basis for such adaptation is still poorly known, a series of investigations by Rahmann and his colleagues have provided evidence that gangliosides, which are highly localized in synaptic membranes (Figure 4.8), play an important role in these adaptive processes. As discussed earlier, gangliosides are glycosphingolipids that contain different numbers of sialic or neuraminic acid (NANA) residues; the more sialyl groups present, the more polar the ganglioside. It has also been shown that the stability of the Ca^{2+}-ganglioside complex increases with increasing number of sialyl groups in the ganglioside (Behr and Lehn, 1973), and that lipid-Ca^{2+} complexes are less stable at lower temperatures (Love, 1970). This led Breer and Rahmann (1976) to propose that animals, in adapting to cold, form more polar gangliosides (i.e., polysialylated membranes) to ensure that synaptic transmission rates equal those achieved at the higher temperature with lower sialylated membranes.

Increased polysialylation of neuronal membranes in response to lowered environmental temperatures has been demonstrated for both poikilotherms and homeotherms under natural and experimental laboratory conditions. The results of some of these studies are summarized in Figure 4.10. The antarctic icefish, *Trematomus hansoni,* which survives subzero body temperatures, has the most polar ganglioside fraction known for any vertebrate; 45% of its gangliosides contain more than four NANA residues (Rahmann and Hilbig, 1980). In contrast, the warm-adapted zebrafish, *Pseudotrophaeus auratus*, exhibits more of a balance between mono- and disialogangliosides (nonpolar) and tetra- and pentasialogangliosides (polar). Seasonal adaptation to lower environmental temperatures in winter by the goldfish, *Carassius auratus*, is characterized by the formation of polar polysialoganglioside fractions. A similar pattern of change in the brain gangliosides is found in summer- and winter-collected carp and rainbow trout, and also following their thermal acclimation in the laboratory (Hilbig et al., 1979). Strict homeotherms such as the adult rat, whose body temperature varies little from the 37°C mean, have a brain ganglioside composition that is decidedly nonpolar. During neonatal development of rats, however, there is a heterothermic phase during which the newborn animal is unable to fully thermoregulate. It has been shown that simultaneous to the gradual acquisition of thermoregulation, the composition of brain gangliosides exhibits an increase in disialogangliosides and a concomitant decrease in the higher sialylated gangliosides until the pattern of the homeothermic adult is reached. Similarly, in hibernating mammals the brain gangliosides tend to be much more polar in comparison with their normothermic counterparts. Polysialylation occurs during the first three weeks of hibernation, and there is no change in ganglioside pattern with length of torpor. Furthermore, the most significant polysialylation effects occur in those brain parts—pons, brain stem, medulla—that are believed to be the most thermosensitive regions of the CNS (Geiser et al., 1981).

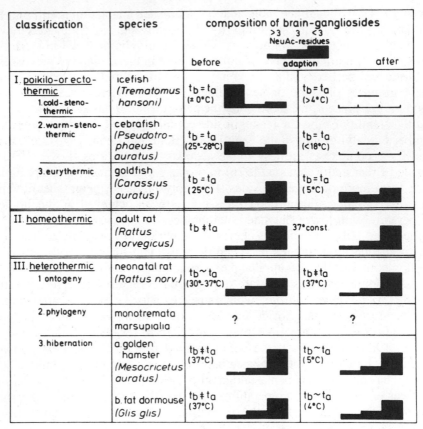

classification	species	composition of brain-gangliosides

FIGURE 4.10. Percent distribution of the three different polar groups of ganglioside fractions of several vertebrates according to their thermal classification, before and after adaptation to changes in the environmental temperature. Reprinted, with permission, from H. Rahmann and R. Hilbig, The possible functional role of neuronal gangliosides in temperature adaptation of vertebrates. *J. Therm. Biol.* **6.** Copyright 1981, Pergamon Press, Ltd.

CONCLUDING REMARKS

It is clear from the preceding discussion that lipids are integral to both the normal function of the nervous system and its ability to adapt to changing environmental conditions. The role of lipids in bioelectrical events centers around their involvement in neural membranes. The resistance and capacitance properties of the lipid bilayer are largely responsible for the generation and maintenance of resting potentials. The possibility that cell membranes with altered fatty acid composition might also exhibit altered electrical properties has received little research attention thus far, although Friedman (1977) reported that increased fatty acid unsaturation in the membrane of breadmold, *Neurospora crassa,* resulted in increased *in vivo* resistance.

Contributing to the dynamic role of the lipid bilayer is the lipid-rich myelin sheath, itself a type of biomembrane but with a unique structure and chemical composition. When present, myelin provides a mechanism for rapid transmission of nerve signals, while at the same time conserving space and energy for the animal. Although some compositional changes occur in the lipid components of myelin in response to developmental and temperature changes, these appear to be primarily associated with maintaining its stability and impermeability. Instead, changes in the rate of ionic conductance, and thus, propagation of action potentials in animals subjected to altered thermal environments, appear to be a function of the synapse, with modification of the lipid composition of synaptic membranes believed to be essential in the adaptive process. Changes in polar head groups, the saturation of fatty acyl moieties, and plasmalogen content have been cited as factors that might alter membrane fluidity and permeability, and modulate activity of membrane-bound enzymes (e.g., Na^+-K^+-ATPase) so that nervous function continues at an optimal level. Polysialylation of gangliosides associated with the synaptic membrane to ensure continued complexing of gangliosides with Ca^{2+} at low temperatures represents another important mechanism for modulating the thermosensitivity of the membrane-mediated process of transmission. Although the evidence that these changes in lipid concentration and composition are functionally significant is still largely indirect, the elucidation of the precise molecular events occurring in membrane during nerve conduction and synaptic transmission should establish a firm causal relationship.

REFERENCES

Aloia, R. C. (1979). Brain lipid composition of the hibernating and active ground squirrel, *Citellus lateralis. J. Therm. Biol.* **4**:223–231.

Beck, W. S. (1971). *Human Design.* Harcourt Brace Jovanovich, New York.

Behr, J. P. and J. M. Lehn (1973). The binding of divalent cations by purified gangliosides. *FEBS Lett.* **31**:297–300.

Blaker, W. D. and E. A. Moscatelli (1978). The effect of hibernation on the lipids of brain myelin and microsomes in the Syrian hamster. *J. Neurochem.* **31**:1513–1518.

Bowen, D. M., A. N. Davison, and R. B. Ramsey (1974). The dynamic role of lipids in the nervous system. In: *Biochemistry of Lipids: MTP International Review of Science,* Vol. 4 (Goodwin, T. W., Ed.), pp. 141–179. University Park Press, Baltimore.

Breckenridge, W. C., G. Gombos, and I. G. Morgan (1972). The lipid composition of adult rat brain synaptosomal plasma membranes. *Biochim. Biophys. Acta* **266**:695–707.

Breer, H. and H. Rahmann (1976). Involvement of brain gangliosides in temperature adaptation of fish. *J. Therm. Biol.* **1**:233–235.

Clausen, J. (1969). Gray-white matter differences. In: *Handbook of Neurochemistry* (Lajtha, A., Ed.), pp. 273–300. Plenum Press, New York.

Cossins, A. R. (1977). Adaptation of biological membranes to temperature—The effect of temperature acclimation of goldfish upon the viscosity of synaptosomal membranes. *Biochim. Biophys. Acta* **470**:395–411.

Cossins, A. R., M. J. Friedlander, and C. L. Prosser (1977). Correlations between behavioral temperature adaptations of goldfish and the viscosity and fatty acid composition of their synaptic membranes. *J. Comp. Physiol.* **120**:109–121.

Cuzner, M. L. and A. N. Davison (1968). The lipid composition of rat brain myelin and subcellular fractions during development. *Biochem. J.* **106**:29–34.

Davison, A. N., M. L. Cuzner, N. L. Banik, and J. Oxberry (1966). Myelinogenesis in the rat brain. *Nature* **212**:1373–1374.

Driedzic, W., D. P. Selivonchick, and B. I. Roots (1976). Alk-1-enyl ether-containing lipids of goldfish (*Carassius auratus* L.). Brain and temperature acclimation. *Comp. Biochem. Physiol.* **53B**:311–314.

Eckert, R. and D. Randall (1983). *Animal Physiology* 2nd ed., W. H. Freeman, San Francisco.

Eichberg, J., G. Hauser, and M. L. Karnovsky (1969). Lipids of nervous tissue. In: *The Structure and Function of Nervous Tissue* (Bourne, G. H., Ed.), pp. 185–287. Academic Press, New York.

Eyzaguirre, C. and S. J. Fidone (1975). *Physiology of the Nervous System,* 2nd ed. Year Book Medical Publishers, Chicago.

Friedman, K. J. (1977). Role of lipids in the *Neurospora crassa* membrane II. Membrane potential and resistance studies; The effect of altered fatty acid composition on the electrical properties of the cell membrane. *J. Membrane Biol.* **36**:175–190.

Geiser, F., R. Hilbig, and H. Rahmann (1981). Hibernation-induced changes in the ganglioside composition of dormice (*Glis glis*). *J. Therm. Biol.* **6**:145–151.

Goldman, S. S. (1975). Cold resistance of the brain during hibernation. III. Evidence of a lipid adaptation. *Am. J. Physiol.* **228**:834–838.

Hilbig, R., H. Rahmann, and H. Rösner (1979). Brain gangliosides and temperature adaptation in eury- and stenothermic teleost fish (carp and rainbow trout). *J. Therm. Biol.* **4**:29–34.

Johnston, P. V. and B. I. Roots (1964). Brain lipid fatty acids and temperature acclimation. *Comp. Biochem. Physiol.* **11**:303–309.

Lagerspetz, K. Y. H. (1974). Temperature acclimation and the nervous system. *Biol. Rev.* **49**:477–514.

Läuger, P. (1972). Carrier-mediated ion transport. *Science* **178**:24–30.

Lee, R. F. and F. Gonsoulin (1979). Lipids from nerve tissues of the horseshoe crab, *Limulus polyphemus*. *Comp. Biochem. Physiol.* **64B**:375–379.

Love, R. M. (1970). *Chemical Biology of Fishes*. Academic Press, New York.

Matheson, D. F., M. S. Diocee, S. T. Hussain, and B. I. Roots (1978). Microtubules in optic nerves of temperature acclimated goldfish. In: *Electron Microscopy* (Sturgess, J. M., Ed.), pp. 268–269. Microscopical Society of Canada, Toronto.

Matheson, D. F., R. Oei, and B. I. Roots (1980). Changes in the fatty acyl composition of phospholipids in the optic tectum and optic nerve of temperature-acclimated goldfish. *Physiol. Zool.* **53**:57–69.

McMurray, W. C., J. D. McColl, and R. J. Rossiter (1964). A comparative study of the lipids of the invertebrate and vertebrate nervous system. In: *Comparative Neurochemistry* (Richter, D., Ed.), pp. 101–107. MacMillen, New York.

Mokrasch, L. C. (1969). Myelin. In: *Handbook of Neurochemistry*, Vol. 1. (Lajtha, A., Ed.), pp. 171–193. Plenum Press, New York.

Morell, P. and W. T. Norton (1980) Myelin. *Sci. Am.* **242**:88–118.

Needham, A. E. (1965). *The Uniqueness of Biological Materials*. Pergamon Press, London.

Nielson, N. C., S. Fleischer, and D. J. McConnell (1970). Lipid composition of bovine retinal outer-segment fragments. *Biochim. Biophys. Acta* **211**:10–19.

Rahmann, H. (1978). Gangliosides and thermal adaptation in vertebrates. *Jap. J. Exp. Med.* **48**:85–96.

Rahmann, H. (1980). Gangliosides and thermal adaptation. In: *Structure and Function of Gangliosides* (Svennerholm, L., H. Dreyfush, and P.-F. Urban, Eds.), pp. 505–514. Plenum Press, New York.

Rahmann, H. and R. Hilbig (1980). Brain gangliosides are involved in the adaptation of antarctic fish to extreme low temperatures. *Naturwissenschaften* **67**:259.

Rahmann, H. and R. Hilbig (1981a). Involvement of neuronal gangliosides and thermal adaptation. In: *Survival in the Cold Hibernation and Other Adaptations* (Musaccia, X. Y. and J. Jansky, Eds.), pp. 177–189. Elsevier, Amsterdam.

Rahmann, H. and R. Hilbig (1981b). The possible functional role of neuronal gangliosides in temperature adaptation of vertebrates. *J. Therm. Biol.* **6**:315–319.

Roots, B. I. (1968). Phospholipids of goldfish (*Carassius auratus* L.) brain: The influence of environmental temperature. *Comp. Biochem. Physiol.* **25**:457–466.

Roots, B. I. and P. V. Johnston (1968). Plasmalogens of the nervous system and environmental temperature. *Comp. Biochem. Physiol.* **26**:553–560.

Rouser, G. and A. Yamamoto (1969). Lipids. In: *Handbook of Neurochemistry*, Vol. 1 (Lajtah, A., Ed.), pp. 121–169. Plenum Press, New York.

Ruch, T. C., H. D. Patten, J. W. Woodbury, and A. L. Towe (1965). *Neurophysiology*, 2nd ed. W. B. Saunders, Philadelphia.

Selivonchick, D. P. and B. I. Roots (1976). Variation in myelin lipid composition induced by change in environmental temperature of goldfish (*Carassius auratus* L.). *J. Therm. Biol.* **1**:131–135.

Tortora, G. J. and N. P. Anagnostakos (1984). *Principles of Anatomy and Physiology*, 4th ed. Harper and Row, New York.

Yusuf, H. K. M. and J. W. T. Dickerson (1977). The effect of growth and development on the phospholipids of the human brain. *J. Neurochem.* **28**:783–788.

Zecevic, D. and H. Levitan (1980). Temperature acclimation: Effects on membrane physiology of an identified snail neuron. *Am. J. Physiol.* **239**:C47–C57.

5

INTEGUMENTAL WATERPROOFING

Maintaining water balance is a critical problem for all organisms regardless of their habitat. The problem is particularly severe for terrestrial plants and animals, especially those species inhabiting desert regions where high temperatures combined with low humidity create a high desiccation potential, and access to free water is at best limited. Terrestrial plants and animals have evolved many complex and often elaborate mechanisms for conserving body water. Paramount among these is an integument that is relatively impervious to water. The nature of the integumental barrier varies, especially among the vertebrates (Hadley, 1980b); however, for many species lipids deposited on the surface and/or impregnated within the integumental layers are primarily responsible for the observed waterproofing. Although a lipid barrier to water flux is best known for plant and arthropod cuticle, where cuticular wax/permeability relationships provide an outstanding example of parallel evolution between two widely divergent groups (Hadley, 1981a), recent studies have also demonstrated a waterproofing function for lipids in the skin of some vertebrates, including humans.

This chapter examines the chemical composition, physical structure, and arrangement of the lipid deposits in the integument of plants and animals. Coverage includes discussion of the specific features of lipids that comprise the waterproofing barrier that are essential for surface function, how these lipids are synthesized and transported to the cuticle surface, and relationships between physicochemical properties of surface lipids and their effectiveness in restricting water loss. The chapter also includes examples of how lipids are modified by physical and biological factors to enhance barrier performance when environmental conditions require increased water conservation.

LIPID-WATER INTERACTIONS

The hydrophobic behavior of lipids essentially reflects differences between polar and nonpolar molecules. In the former, the shared electrons are attracted more strongly by one of the atoms forming the bond than they are by the other. Water is a highly polar molecule; the shared electrons of the oxygen-hydrogen bonds are held more tightly by the oxygen atom. As a result, the water molecule has a positive (hydrogen) and negative (oxygen) end, or in other words, behaves as a dipole. Because the majority of biological compounds (e.g., proteins, carbohydrates, nucleic acids, and salts) are either ionized or are polar materials, they can interact with the electrical field produced by the water molecules and enter into solution (Beament, 1976). Lipid molecules, in contrast, are typically nonpolar; the electrons of the carbon-hydrogen bonds that dominate these molecules are equally attracted by the atoms forming the bond so that no asymmetry or charge distribution occurs. As a result, lipids have little or no affinity for water, a feature implicit in their definition (Chapter 1). Still, the degree of nonpolarity (polarity) varies among the lipid classes. At one end of the spectrum are the *n*-alkanes and wax esters, both of which are very nonpolar, and thus ideal compounds for repelling water. At the other extreme are the phospholipids, the most polar of the major lipid classes. These have structural components that are ionizable or asymmetric (i.e., polar head groups), which allow phospholipids to at least electrostatically interact with water to form micelles or lipid bilayers. Despite their increased polarity, even phospholipids, when assembled, are not readily dispersed by water and, therefore, serve as durable barriers for separating the cell interior from the surrounding fluid (Chapter 3).

GENERAL DESIGN OF SURFACE LIPIDS (WAXES)

The lipids associated with the surface of plants and animals are collectively termed waxes. The word "wax" is derived from "weax" which describes honeycomb material, a substance with physical properties similar to those of surface lipids (Kolattukudy, 1976). Although waxes in a strict chemical sense refer only to long-chain alcohols esterified to long-chain fatty acids (i.e., wax esters), in a biological or functional sense waxes encompass a complex mixture of long-chain and cyclic compounds that belong to a variety of lipid classes. When referring to the integumental barrier to water flux, however, the terms "waxes" and "lipids" are often used interchangeably.

Surface lipids as a group exhibit a number of features that are designed to permit function under potentially adverse environmental conditions. The cuticular lipids of plants and arthropods, which have received the greatest study, usually have long carbon chains and are highly saturated compounds. As a result, these lipids are not easily lost through volatilization and are protected to some extent against degradation by atmospheric agents (e.g., oxygen) or microorganisms (Kolattukudy, 1976). Long-chain, saturated lipid compounds also provide thermal stability. For example, n-alkanes containing 23 to 35 carbon atoms have melting points that range from 47.6 to 75°C, respectively. These temperatures are well above those encountered by plants and animals under most natural conditions. Methyl branches, which are rare in internal lipids, are common in surface lipids. The functional significance of methyl branches is presently unclear; however, their presence may contribute to increased immunity against enzymatic breakdown by microbes, or they may be required to prevent premature release of long-chain molecules from the elongation system (Jackson and Blomquist, 1976). Because many of these same lipid features, which are essential for surface function, also influence the effectiveness of the waterproofing barrier, they are discussed further when lipid composition/permeability relationships are considered.

CHEMICAL COMPOSITION OF SURFACE WAXES

Although the chemistry of waxes has been studied for over 100 years, only in the past two decades have significant advances been made in our knowledge of the structure and composition of waxes associated with surfaces of plants and animals. Closely linked with these advances is the development of modern analytical instrumentation such as the gas chromatograph and mass spectrometer. With the aid of these research tools, investigators can, with reasonable ease and quickness, separate and identify the complex lipid mixtures that often characterize surface wax. The information contained in the virtual explosion of papers that have appeared on the subject fortunately has been synthesized in several comprehensive reviews (see below). The following survey of the major lipid components and their structural features in plant and animal surfaces is intended to reveal some general patterns and trends within and between various groups, with emphasis on those components believed to participate in the waterproofing process. More extensive coverage of chemical composition is found in the reviews of plant waxes by Kolattukudy and Walton (1972), Kolattukudy (1975), and Tulloch (1976) and insect waxes by Jackson and Blomquist (1976), Blomquist and Jackson (1979), and Blomquist and Dillwith (1983). A survey of the lipid classes found in the skin of mammals is provided by Nicolaides (1974) and Downing (1976).

The principal components comprising the surface waxes extracted from plants and animals are presented in Table 5.1. They can be divided into three categories: hydrocarbons, oxygenated compounds, and cyclic compounds. Hydrocarbons are one of the most common and abundant of the cuticular lipids of higher plants and arthropods (insects and arachnids). They are also the most intensively investigated surface lipid, a fact that reflects not only their abundance, but also the relative ease with which they can be

TABLE 5.1
Major components of surface waxes in plants and animals

I. HYDROCARBONS

(A) n alkanes $CH_3—(CH_2)_n—CH_3$

(B) n alkenes $CH_3—(CH_2)—CH=CH—(CH_2)_m—CH_3$

(C) Branched alkanes

    (1) 2 methyl alkanes $CH_3—\overset{\displaystyle CH_3}{\overset{|}{CH}}—(CH_2)_n—CH_3$

    (2) 3 methyl alkanes $CH_3—CH_2—\overset{\displaystyle CH_3}{\overset{|}{CH}}—(CH_2)_n—CH_3$

    (3) internally branched $CH_3—(CH_2)_n—\overset{\displaystyle CH_3}{\overset{|}{CH}}—(CH_2)_m—CH_3$
        alkanes

II. KETONES

    (A) Monoketones $CH_3—(CH_2)_n—\overset{\displaystyle O}{\overset{||}{C}}—(CH_2)_m—CH_3$

    (B) β-Hydroxyketones $CH_3—(CH_2)_n—\overset{\displaystyle O}{\overset{||}{C}}—CH_2—\overset{\displaystyle OH}{\overset{|}{CH}}—(CH_2)_m—CH_3$

    (C) β-Diketones $CH_3—(CH_2)_n—\overset{\displaystyle O}{\overset{||}{C}}—CH_3—\overset{\displaystyle O}{\overset{||}{C}}—(CH_2)_m—CH_3$

III. WAX ESTERS

$CH_3—(CH_2)—\overset{\displaystyle O}{\overset{||}{C}}—O—(CH_2)_n—CH_3$

IV. FREE FATTY ACIDS

$CH_3—(CH_2)_n—C{\overset{\displaystyle O}{\diagdown\!\!\!OH}}$

V. ALDEHYDES

$CH_3—(CH_2)_n—C{\overset{\displaystyle O}{\diagdown\!\!\!H}}$

VI. ALCOHOLS

    (A) Primary $CH_3—(CH_2)_n—CH—OH$

    (B) Secondary $CH_3—(CH_2)_m—\overset{\displaystyle OH}{\overset{|}{CH}}—(CH_2)_n—CH_3$

VII. CYCLIC COMPOUNDS

    (A) Sterols   (B) Triterpenes

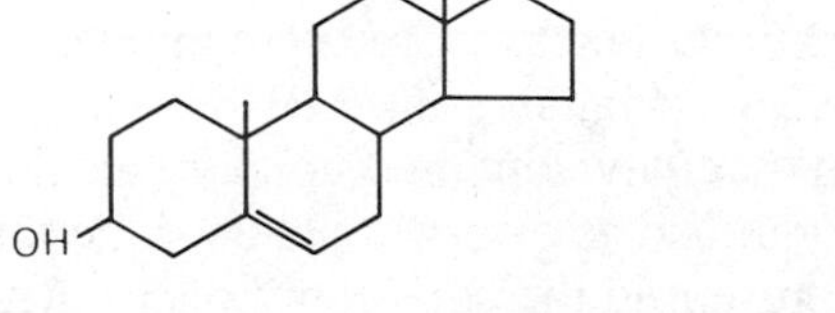

cholesterol

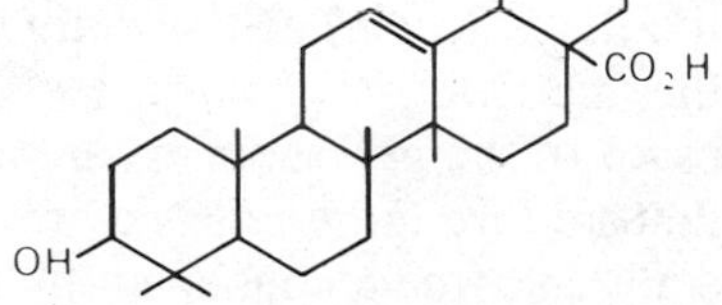

oleanolic acid

SOURCE: Reprinted from Hadley (1981a).

extracted, separated, and analyzed. In plants, *n*-alkane (straight-chain) hydrocarbons constitute the principal hydrocarbon fraction. These range in length from 21 to 37 carbon atoms, with odd-numbered chains predominating (especially C_{27}, C_{29}, C_{31}, and C_{33}). Branched alkanes, particularly 2-methyl and 3-methyl compounds, have also been reported. The percentage of branched alkanes in plants is usually small, although in tobacco and some grasses, they form up to 50% of the hydrocarbons (Tulloch, 1981). Dimethylalkanes, internally branched alkanes, and unsaturated (alkene) hydrocarbons are more rare in plants, but their apparent absence may be due in part to the more difficult procedures for separation and analysis (Tulloch, 1976). The cuticular alkane pattern of insects and arachnids is quite similar to that found in plants; however, branched alkanes are more common in arthropods and can account for over 90% of the total hydrocarbon fraction (Hadley, 1977). Branched hydrocarbons also constitute the hydrocarbons with the longest chain lengths, as neither insects nor arachnids can apparently synthesize, at least in significant quantities, *n*-alkanes containing more than 34 carbon atoms. Unsaturated hydrocarbons (primarily *n*-alkenes) are also found in the cuticular lipids of approximately 50% of the insects studied thus far (Lockey, 1980), but these are usually minor components and the position of the double bond has been determined for only a few species. Hydrocarbons are usually present in those vertebrates for which surface lipids have been analyzed; however, unlike plants and arthropods, they are seldom a principal component, and in mammalian skin they are thought to be of exogenous origin (Downing, 1976).

A second important group of cuticular lipids includes the oxygenated derivatives of alkanes—ketones, wax esters, free fatty acids, alcohols, and fatty aldehydes (Table 5.1). A variety of ketones with chain lengths similar to alkanes have been extracted from plant cuticle. These compounds do not appear to be characteristic components of arthropod cuticular lipids, but in the Florida indigo snake, methyl ketones are a principal integumentary lipid (Ahern and Downing, 1974). Wax esters are a common constituent in most plants. They consist of fatty acids with carbon chains of C_{20} to C_{24} esterified to alcohols having an even number of carbon atoms in the range C_{12} to C_{32} (Kolattukudy, 1975). In insects and arachnids, wax esters are usually reported as minor or trace components, but I suspect they often have been overlooked because analyses have concentrated solely on the hydrocarbon fraction. In the black widow spider, wax esters are five times more abundant than hydrocarbons (Hadley, 1978a), and they also are significant components in the surface lipids of some grasshopper species (Soliday et al., 1974), the desert cockroach, *Arenivaga investigata* (Jackson, pers. comm.), and the tree frog, *Phyllomedusa sauvagei* (McClanahan et al., 1978). The wax esters of the grasshoppers are unique in that they contain secondary alcohols that are rare in insects and are never combined as esters in plants. Free fatty acids and free alcohols are invariably present in the surface waxes of plants and animals, although amounts may be small. The alcohols are usually straight-chain, even-numbered compounds containing 22 to 34 carbon atoms. Free fatty acids may be saturated, unsaturated, and even branched, with long-chain molecules (28:0 and 30:0) found in some scorpions (Hadley and Jackson, 1977). Triacylglycerols are not included in this section because their presence is usually attributed to internal contamination. Nevertheless, trace amounts are often detected in the cast skins of insects and arachnids, indicating that in some species triacylglycerols represent a true cuticular component.

Cyclic compounds represent a third category of cuticular lipids. In plants these include the pentacyclic triterpenes (β-amyrins, taraxerone), tetracyclic phytosterols (sitosterol, campesterol), flavones, and aromatic hydrocarbons. Older literature reported that triterpenes were abundant in the cuticular wax of fruits and berries (Martin and Juniper, 1970), but that none of these cyclic compounds were major constituents in the cuticular waxes of most plant species. Recent investigations, however, have shown that these cyclic compounds occur more frequently than had been previously supposed (Manheim and Mulroy, 1978; Barthlott and Wollenweber, 1981), and that terpenoids and flavonoids can even be identified by their micromorphological structure when viewed with a scanning electron microscope (Barthlott, 1981). Sterols, principally cholesterol, are the common cyclic lipid in the surface waxes of animals. The presence of cholesterol in arthropod cuticle is particularly noteworthy in that neither insects nor arachnids are able to synthesize the steroid nucleus, and thus must obtain cholesterol or its precursors from their diet. Nevertheless, cholesterol is almost always present in the epicuticle, often in substantial quantities, a fact that strongly implicates its involvement in the proper function of this cuticular layer (Toolson and Hadley, 1977). Cholesterol is also present in the skin surface lipids of all mammals that have been studied where it is normally accompanied by other sterols (e.g., cholestanol, desmosterol), which are intermediates or by-products of its biosynthesis (Downing, 1976).

WAX COMPOSITION/PERMEABILITY RELATIONSHIPS

Many of the lipid features previously considered to be essential for surface function also likely influence the effectiveness of the waterproofing barrier. This point can be illustrated by examining surface hydrocarbons, which are a prime contributor to the water barrier. Our knowledge of the relationships between molecular structure and permeability is based largely on data obtained from experiments performed on artificial bilayers and plasma membranes. These experiments have shown that permeability to water is decreased by the incorporation of long-chain molecules and increased by the presence of unsaturated molecules and molecules containing methyl branches (de Gier et al., 1968; Silbert et al., 1973; Taylor et al., 1975). Assuming these results can be applied to hydrocarbons spread on the surface of a plant or animal (an assumption that awaits experimental verification), then long-chain saturated n-alkanes should provide maximum impermeability as well as thermal stability. As we shall see in future sections of this chapter, the data obtained thus far on hydrocarbons extracted from the cuticles of plants and arthropods indicate that this relationship indeed exists.

The situation regarding methyl branching is more difficult to explain. Branched hydrocarbons are common surface constituents and, in arthropods, often exceed n-alkanes in percent composition. This feature would appear to be maladaptive, for methyl branches are believed to increase permeability because of the inability of molecules to pack closely together. Moreover, branched alkanes would melt at a lower temperature than n-alkanes of corresponding chain length. Because arthropods are apparently incapable of synthesizing n-alkanes containing more than 35 carbon atoms, long-chain branched molecules may be a compromise for satisfying the environmentally imposed need for a more impermeable integument in the face of biochemical limitations. That is, although the permeability is somewhat greater than it would be if the cuticle were composed only of long-chain n-alkanes (> 35 carbon atoms), the presence of branched components of comparable size might provide superior waterproofing compared with a cuticle that features only C_{29} through C_{32} n-alkanes (Toolson and Hadley, 1977). Branched alkanes may also facilitate the secretion and transport of lipid to the surface and help orient surface lipid molecules for optimal waterproofing properties.

PLANTS

Cuticle Structure with Reference to Lipid Location

The plant cuticle is a multilayered structure that covers the epidermal cells (Fig. 5.1a and b). The outermost layer, or the cuticle proper, is composed of polymers of cross-esterified hydroxy C_{16} and C_{18} fatty acids called cutin. This substance, which is responsible for the structural integrity of the cuticle, appears to be deposited in the form of ultramicroscopic lamellae oriented parallel to the surface. The underlying layer is called the cuticular layer. It forms the bulk of the leaf cuticle and contains cutin along with embedded wax and pectin, the latter a cement-like substance (Holloway, 1977). A consistent ultrastructural feature of this layer is a network of fibrillae that permeate through the matrix, giving the matrix a globular or marbled appearance (Wattendorff and Holloway, 1980). A layer of more-or-less pure pectin usually fastens the cuticle to the epidermal cell wall. Histochemical tests have shown that each of these layers is rich in lipids; in fact, the plant cuticle is almost entirely lipoidal in composition. In addition, surface or epicuticular waxes are deposited on top of the cuticle membrane. The fingerlike projections of these surface waxes with respect to the cuticle are illustrated in Figure 5.2.

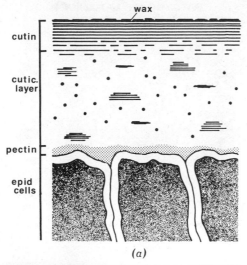

(a)

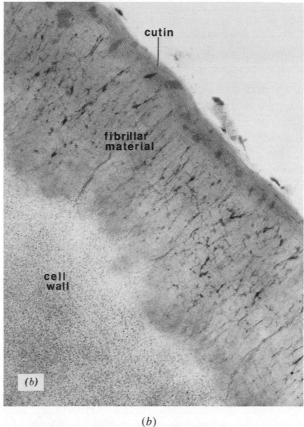

(b)

FIGURE 5.1. (*a*) Diagrammatic representation of the plant cuticle. Epicuticular wax overlies the outermost laminate cutin layer. The cuticular layer contains embedded wax (black) and cutin (banded). Relative thickness of layers not drawn to scale. Reprinted from Hadley (1981b). (*b*) A transverse section through the epidermis of the petiole of celery, *Apium sativum,* showing the lamellae of the cutin layer and fibrillar material in cuticular layer. ×3920. Micrograph with permission from N. D. Hallam and B. E. Juniper. The anatomy of the leaf surface. In: *Ecology of Leaf Surface Micro-organisms* (T. F. Preece and C. H. Dickinson, Eds.), 1971. Copyright: Academic Press Inc. (London) Ltd.

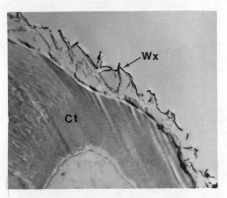

FIGURE 5.2. Micrograph of leaflet cuticle from velvet mesquite, *Prosopis velutina*, showing wax projections (Wx) that overlie the cuticle (Ct). ×5000. Reprinted from Bleckmann et al. (1980).

Surface Wax Morphology

The development of the scanning electron microscope (SEM) and various techniques associated with its use have greatly increased our knowledge of the morphology and arrangement of surface waxes. This has been a particularly active area of research among botanists, who have often combined morphological descriptions with experiments designed to determine those factors responsible for the formation and/or alteration of surface waxes. Micromorphologically, wax accumulations on the surface may be grouped into the following classes (Amelunxen et al., 1967): (1) granules, (2) rods and filaments, (3) plates and scales, (4) layers and crusts, (5) aggregate coatings, and (6) liquid or viscous coatings. Waxes that tend to lie flat on the cuticle (e.g., liquid or viscous films) and exhibit a well-defined orientation are difficult to discern with SEM. Leaves having this wax structure are likely to appear "green" (i.e., nonglaucous). In contrast, leaves with surface waxes that tend to project outward from the surface (e.g., rods, filaments, crusts) and that abundantly cover the entire surface can give the leaf a blue, gray, or even white appearance. This phenomenon is termed a "wax bloom," and leaves exhibiting such are considered to be glaucous. Two or more of the above wax forms may be present on a given leaf, either on different portions of the same surface or when the upper and lower surfaces are compared. Some examples of surface wax architecture on leaves of selected desert plants are shown in Figures 5.3 and 5.4.

The physical structure and arrangement of surface waxes are determined by a complex suite of interacting factors. Of primary importance is the chemical composition of the wax itself, which is under the control of genes located on different chromosomes (von Wettstein-Knowles, 1976; Bianchi et al., 1980). In barley, the formation of long, thin wax tubes depends on the presence of β-diketones in the wax (von Wettstein-Knowles, 1974). In pea leaves, the ribbonlike wax structures on the lower surface consist mainly

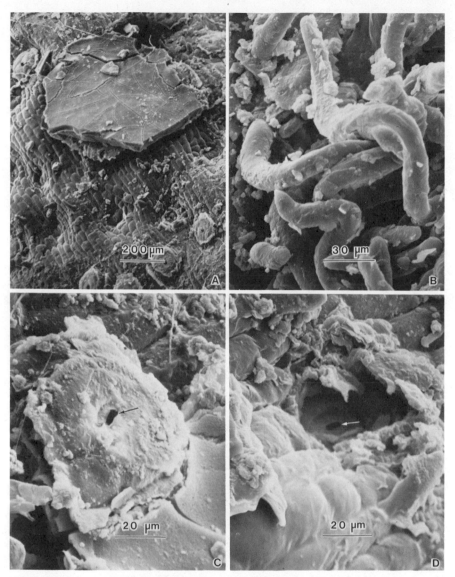

FIGURE 5.3. The candellia, *Euphorbia antisyphilitica*, whose wax is used in making candles, exhibits a variety of wax morphologies on the surface including (*A*) "plates," (*B*) tubular filaments, and (*C* and *D*) "encrustations." The stomates (arrows) in this species are either enclosed in a mound of wax (*C*) or sunken in a cavity surrounded by wax (*D*), both arrangements that effectively reduce water loss associated with gas exchange.

of alkanes, while the upper surface features interlocking wax plates composed mainly of primary alcohols (Holloway et al., 1977). Terpenoids and flavenoids have the tendency to crystallize into compound threads or into threads with complex microcrystalline surfaces (Barthlott, 1981). The micromorphology of epicuticular waxes is also influenced to some extent by

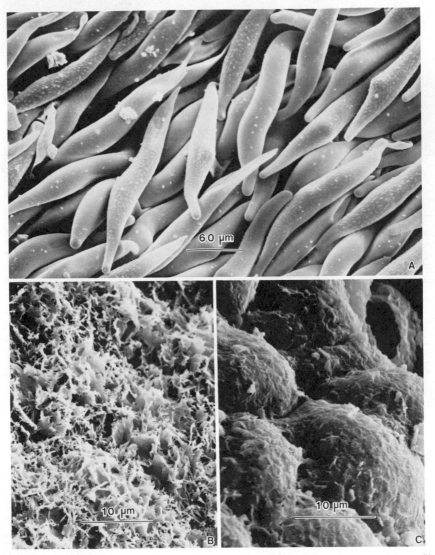

FIGURE 5.4. (A) The trichomes of guayule, *Parthenium argentatum,* are coated with a thick wax film and scattered wax encrustations. Large fields of guayule were cultivated in the western United States as a source of rubber during World War II. (B) Wax deposits on the surface of jojoba, *Simmondsia californica,* leaves impart a lettucelike appearance when viewed with SEM. These wax deposits are essentially removed following a brief wash in chloroform:methanol (C).

environmental factors such as temperature, humidity, light intensity, and wind. For example, Baker (1974) found that lower temperatures favored the formation of wax rods, filaments, and ribbons on the leaves of brussels sprout, whereas large plate waxes were formed at higher temperatures. A similar tendency for waxes to develop over rather than project from the cuticle with increased temperature was noted by Armstrong and Whitecross (1976) and Hunt et al. (1976). Temperature can also alter the chemical composition of epicuticular waxes, and thus, influence wax morphology. Brussels sprout grown at higher temperatures exhibited a lower content of alkanes and a higher content of aldehydes. The pattern of leaf cuticular wax deposits, on the other hand, does not seem to be dependent on the distribution or size of pores from which they might be excreted. Epicuticular waxes extracted from plants of several species having a variety of different wax morphologies recrystallized with form and dimensions similar to those present on the intact surfaces. Furthermore, wax from a single plant species crystallized to a similar form in a number of different solvents on porous surfaces having different pore sizes (Jeffree et al., 1975).

Cuticular Waxes and Transpiration

The importance of lipids in waterproofing the leaf cuticle was shown as early as 1896 by Kerner von Marilaun and Oliver, who described experiments in which water loss increased by almost one-third when the waxy bloom of leaves was rubbed off. Since this time numerous investigators have shown that lipids, deposited on the leaf surface or embedded within the cuticle, are essential for controlling water movement across the cuticle. Much of this supporting evidence, albeit strong, is indirect; that is, changes in the amount, composition, and physical structure of plant waxes, which are known or believed to enhance the waterproofing barrier (see earlier discussion), occur during times when transpiration potential is high. Probably the most frequently reported correlation involves the concentration of surface wax. Surface wax densities, which range from 0.01 to 0.50 mg per cm^2 (Holloway, 1977), typically increase in leaves exposed to the sun or in leaves of plants and seedlings grown at high temperatures (Baker, 1974; Hunt et al., 1976). Maximum deposits of wax also occur under high radiant energy flux and low ambient humidity (Daly, 1964; Hallam, 1970; Giese, 1975). In view of these findings, it is not surprising that many investigators have also found maximum surface concentration of leaf wax in plants during summer months (Freeman et al., 1979; Bukovac et al., 1979; Mayeux et al., 1981). Direct evidence for a positive relationship between the amount of surface wax and permeability is provided by Bengston and his colleagues (1978), who found that when oat seedlings were subjected to water stress, the amount of epicuticular wax increased while cuticular transpiration decreased in all six varieties tested.

The chemical composition and the architecture of epicuticular waxes are also believed to be important factors in the capability of the wax to reduce cuticular transpiration. The chemical composition of the wax is important for two reasons. First, as previously mentioned, the chemical composition of wax is a principal determinant of wax conformation. Many nonpolar lipid constituents (e.g., β-diketones) present in plant waxes produce rodlike or crystalline structures that project from the surface. A leaf surface with projecting wax rods and tubes would exhibit a decreased conductance to vapor diffusion because of the increased length and twisting of the vapor pathway. Moreover, the wax projections would reflect a greater portion of the incident radiation (Schulze et al., 1980). This would decrease the rise in leaf surface temperature, thereby reducing the diffusion gradient. Finally, surface "roughness" caused by wax projections may cause turbulence in laminar air flow, which also would increase the thermodynamic exchange, resulting in cooler leaf surface temperatures and decreased evaporation.

A second important consideration regarding chemical composition is that different wax components are not equally efficient in reducing evaporation. Grncarevic and Radler (1967) demonstrated that hydrocarbons, long-chain alcohols, and long-chain aldehydes extracted from grape leaves caused the greatest reduction in transpiration through an artificial membrane. Baker and Bukovac (1971) and Wilson (1975) also reported that nonpolar compounds were most effective in retarding evaporation when applied to artificial membranes. Unfortunately, only a few studies have investigated the possibility that plants increase their synthesis of these more hydrophobic wax constituents during times of drought or moisture deficiency. In peach leaves, the decreased permeability to water during summer is probably related to both an increase in surface wax and higher concentration of long-chain esters and alkanes that are present in extracts of leaf wax at this time (Bukovac et al., 1979). In a study by Bengston et al. (1978) on oat seedlings, however, alkanes that have been shown to be very effective in retarding evaporation were reduced in density in three of the six varieties tested, even though all stressed seedlings showed a reduction in water loss compared with controls (Table 5.2). The remarkably lower alkane content in Stormogul II, the most drought-resistant oat variety, may be due to a stimulated biosynthesis of β-diketones which, in turn, might lead to a changed epicuticular wax structure, and therefore increased impermeability for reasons described above. In all varieties, the major primary alcohol (hexadocosanol, C_{26}), which is also an effective deterent to water efflux, was slightly higher in stressed seedlings than in controls (Table 5.2).

Some of the best quantitative data demonstrating the importance of lipids in controlling the water permeability of plant cuticular membranes as well as the effects of temperature on the diffusion of water across the lipid barrier are found in Schönherr's studies of isolated cuticle segments. Schönherr (1976) considered the cuticular membrane to be composed of two components: the extractable waxes (embedded in the cutin matrix and deposited superficially) and nonextractable components in the matrix such as cellulose, polyuronic acids, proteins, and phenolic compounds. He assumed initially that the permeability coefficient of diffusion (P_d) of the cuticular membrane (CM) could be estimated by the equation

$$\frac{1}{P_d(\text{CM})} = \frac{1}{P_d(\text{MX})} + \frac{1}{P_d(\text{WAX})}$$

Table 5.2

The cuticular transpiration rate (CT) and the content of alkanes and hexadocosanol (C_{26}) in six oat varieties exposed to water stress. The varieties are arranged in decreasing drought resistance.

Variety	CT(mg cm^{-2} h^{-1}) Control	CT(mg cm^{-2} h^{-1}) Stress	Alkanes (μg cm^{-2}) Control	Alkanes (μg cm^{-2}) Stress	Hexadocosanol (μg cm^{-2}) Control	Hexadocosanol (μg cm^{-2}) Stress
Stormogul II	0.705 ± 0.016	0.409 ± 0.092[a]	0.96	0.03	8.1	8.8
Risto	0.392 ± 0.062	0.312 ± 0.041[b]	0.33	0.04	8.4	8.6
Sol II	0.366 ± 0.085	0.325 ± 0.069	0.06	0.09	6.9	7.0
Selma	0.416 ± 0.115	0.317 ± 0.025[b]	0.10	0.17	9.2	9.3
Sang	0.304 ± 0.064	0.288 ± 0.031	0.68	0.33	9.2	10.1
Pendek	0.408 ± 0.066	0.391 ± 0.028	0.29	0.50	5.3	5.8

SOURCE: Modified from Bengston et al. (1978).

[a]Denotes significant difference in CT between control and stressed seedlings for each variety at 0.01 confidence level.

[b]Denotes significant difference in CT between control and stressed seedlings for each variety at 0.05 confidence level.

where MX refers to the cutin matrix (waxes extracted), and WAX refers to combined epicuticular and embedded waxes. Using tritiated water to measure rates of water diffusion across isolated astomatous cuticle from the upper leaf surfaces of *Citrus aurantium* and the pear, *Pyrus communis,* he found that extraction of waxes from both species increased their water permeability 350- to 500-fold, whereas the contribution of the polymer matrix (MX) to overall conductance was negligible. Thus, the total resistance to water transport of cuticular membranes, at least in these species, can be attributed solely to extractable cuticular lipids. These results explain why water permeability of cuticular membranes, which consist mainly of polymer matrix, is independent of membrane thickness, a relationship observed in many earlier studies.

Subsequent studies on isolated cuticle segments of *Citrus* showed that, between 5 and 65°C, the water permeability of the cuticular membrane is completely determined by the soluble cuticular lipids. An Arrhenius plot showing the dependence of water permeability on temperature of cuticular membranes exhibited two linear portions, the first between 5 and 35°C and the second between 50 and 65°C, with the two lines intersecting at approximately 44°C (Schönherr et al., 1979). The effects of temperature were reversible below 44°C, but were irreversible if membranes were heated above this value. The sudden change in water permeability at 44°C suggests a change in membrane structure, specifically a phase transition of the constituent cuticular lipids to a relatively less ordered state. The latter explanation is supported by the electron spin resonance (esr) spectra of a fatty acid (18:0) spin label incorporated into the cuticular membrane. From these data, the authors concluded that water permeability depends strongly on the state and distribution of soluble lipids and that all soluble cuticular lipids in the cuticle are present as a homogeneous mixture rather than as individual layers differing in composition.

The discussion thus far has centered on the role of lipids in controlling permeability of water across the plant cuticle. Lipids, however, also play an important role in reducing transpiration associated with gas exchange through the stomata. This is an extremely important function, as the majority of plant water loss occurs through stomatal pathways when stomata are open. In the desert ironwood tree, *Olneya tesota,* the wax accumulation is especially thick in the vicinity of the stomata, with the encrustations and crystalline projections penetrating deep inside the outer stomatal cavity (Figure 5.5). The thick wax coating on leaves of the desert broom, *Baccharis sarothroides,* form ridges surrounding stomatal openings and also coat the walls of the stomatal cavity. Conifers often exhibit the greatest concentration of wax crystals near the stomata (Thair and Lister, 1975; Reicosky and Hanover, 1976). In Sitka spruce, *Picea sitchensis,* the stomatal antechambers are filled with a fine wax that extends from just above the epidermal surface down to the upper surfaces of the guard cells (Jeffree et al., 1971). Waxes associated with stomata help reduce moisture loss by increasing the thickness of the boundary layer, reducing the cross-sectional area available

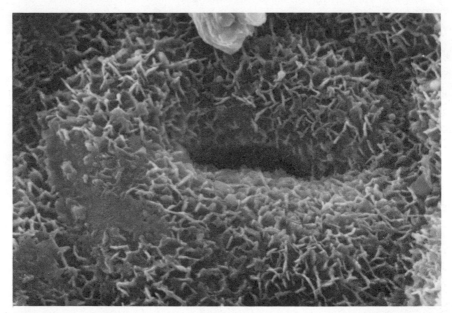

FIGURE 5.5.   Scanning electron micrograph of surface wax in the vicinity of a stomate on a leaf of the desert ironwood tree, *Olneya tesota.* Reprinted from Hadley (1981a).

for diffusion, and increasing diffusion resistance by increasing the tortuosity of the diffusion pathway. Jeffree et al. (1971) estimated that occlusion of the epistomal chambers by structural waxes reduces transpiration by two-thirds, but reduces photosynthesis by only one-third.

Other Functions of Surface Wax

Although the major role played by surface waxes involves the maintenance of water equilibrium, other important functions are recognized, many of which are a consequence of the hydrophobic barrier constructed to reduce transpiratory water loss. Since some of these functions are discussed in subsequent chapters, only brief coverage is presented here. Surface waxes likely provide some protection against plant disease by preventing the penetration of microorganisms such as spores and fungi. This protection may be a result of the physical barrier provided by the waxes, antifungal components in the wax, or because the water film required by an invading fungus is absent (see Hamilton and Hamilton, 1972). Still, the protection provided by the wax cuticle is limited because plant diseases can easily enter via the stomata or through other openings. Similarly, a heavy wax deposit can provide defense against insect attack. Again, protection may involve a physical barrier or the presence of specific chemical agents associated with the surface wax that act as repellents. Surface waxes also provide thermoregulatory benefits to the plant. The increased reflectance of solar insolation may help prevent leaf temperatures from reaching levels that are disruptive to normal protoplasmic functioning as well as protect them from a lethal rate of desiccation. At the other extreme, it has been suggested that surface waxes protect against frost damage, as more glaucous populations of *Eucalyptus* occur in regions of the species' range that are more susceptible to freezing temperatures. Finally, the decreased wettability of the surface caused by wax deposits may prevent inundation by water. The latter may be of special importance to mechanically unstable surfaces, such as petals in rain, and to seeds that must float in water for prolonged periods of time (Barthlott, 1981). Because of their unwettable nature, plant surface waxes also repel foliar applied herbicides. Surface waxes, of course, were not synthesized for this purpose, but this fact does require that special attention be given the chemical and physical properties of the herbicide that is to be applied.

ARTHROPODS

General Cuticle Structure and Chemical Composition

Despite the great number and diversity of arthropod species, the cuticular plan is basically similar in all groups. As in plants, the cuticle is a noncellular, multilayered membrane that covers a single layer of epidermal cells

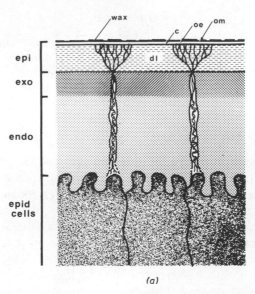

(a)

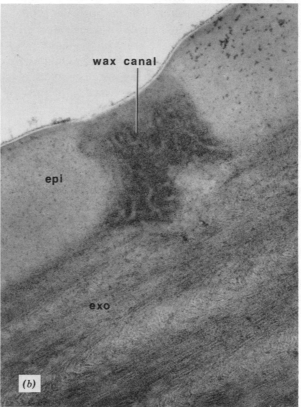

(b)

FIGURE 5.6. (*a*) Diagrammatic representation of insect cuticle. Major divisions include the epi-, exo-, and endocuticle. Sublayers of the epicuticle are the dense homogeneous layer (dl), cuticulin (c), outer epicuticle (oe), and outer membrane (om). Two pore canals divide into smaller wax canals as they approach the epicuticle. Relative thickness of layers not drawn to scale. (*b*) Transverse section of the sclerite cuticle of the scorpion, *Hadrurus arizonensis,* showing the epicuticle, the outer portion of the exocuticle, and a wax canal. ×125,000. Reprinted from Hadley (1980b).

(Figure 5.6a). Two major divisions are usually recognized: the procuticle and the epicuticle. The procuticle comprises the bulk of the cuticle and contains a chitin-protein complex that provides structural integrity (rather than lipid as in plant cuticle). The procuticle is usually divided into an outer, sclerotized exocuticle and an inner, soft endocuticle. The epicuticle is a thin (0.5 to 2.0 μm), nonchitinous layer that covers the procuticle. Ultrastructural studies have clearly demonstrated the presence of distinct sublayers within the epicuticle. Following the terminology of Filshie (1970), these are the dense homogeneous layer or protein epicuticle, the cuticulin layer, the outer epicuticle, and the outer membrane. Two of these, the homogeneous layer and the cuticulin layer, are secreted well before a molt, and both are rich in lipids. The cuticulin layer is the first layer deposited when a new cuticle is formed. When viewed with the electron microscope, it appears as a thin, extremely electron dense band of fairly uniform thickness (Figure 5.6b). Covering the cuticulin layer are the outer epicuticle (= wax layer; Weis-Fogh, 1970) and the outer membrane (= cement layer; Weis-Fogh, 1970). The outer epicuticle is a thin (ca. 3 to 10 nm) hydrophobic layer, which was considered by Locke (1965) to be the surface monolayer or oriented lipid layer. The outer membrane or cement layer, which is not always present, is believed to be secreted by dermal glands and may be impregnated with wax. The surface of the epicuticle is molded into polygonal areas, the margins of which correspond to the perimeters of the epidermal cells that secrete the cuticle (Neville, 1975). Surface detail is often masked by superficially deposited waxes, which in some species are sufficiently abundant to produce blooms (see later) similar in appearance to that described for plant leaves.

Although plant and arthropod cuticle exhibit many striking morphological similarities, there are some differences that warrant mention. The layers comprising the arthropod cuticle are rather distinct from one another whereas in plant cuticle each layer tends to merge gradually into the other. Perhaps of more functional significance is the presence of a complex, transverse canal system in arthropods that connects the epidermal cells with the outer region of the epicuticle (Figure 5.6). The system is composed of larger diameter pore canals, which connect with numerous smaller diameter wax canals at approximately the junction of the exo- and epicuticle. This pore/wax canal system is believed to be the route by which lipids travel from their site(s) of synthesis (epidermal cells, fat bodies, oenocytes) to locations on and within the cuticle. Despite the presence of the structural pathways, no satisfactory explanation has been put forth to explain the mechanisms underlying this transport. How synthesized lipids are transported and deposited on the surface of plant cuticle is even more of a mystery, as plant cuticle lacks a well-defined vertical canal system. There are scattered reports in the literature of the occurrence of pore and wax microchannels (see Hadley, 1981b); however, if they are present they do not routinely show up in electron micrographs. Their absence may indicate that the pores and channels are dynamic rather than stable structures; that is, they do not exist in dry cuticular membranes but develop upon hydration. Differential diffusion rates of polar molecules across isolated plant cuticle strongly support the existence of pores (Schönherr and Schmidt, 1979), but they may not be associated with lipid transport. There is also some evidence suggesting that surface wax components or their precursors move more-or-less randomly through the cuticle before being deposited on the surface (Armstrong and Whitecross, 1976). Whatever the nature of the conducting pathway, it is generally agreed that the lipid is carried to the surface in a solution that subsequently evaporates, causing the dissolved lipids to crystallize on the surface. The conducting solvent(s) responsible are unknown.

Surface Waxes

Wax deposits occur on the cuticle surface of many insects and arachnids. SEM of these superficial waxes provides some of the best evidence for the presence of cuticular lipids in terrestrial arthropods and the means by which they contribute to the waterproofing process. As in plants, the surface waxes of arthropods exhibit a variety of shapes and sizes that range from liquid or viscous coatings to plates, rods, and filaments (Hadley, 1981a). Surface lipids that cover the cuticle as a thin film are not readily discernible with SEM; in fact, their presence may go unnoticed unless cuticle segments treated with chloroform:methanol, which extracts the surface lipid film, are compared with untreated cuticle. An example of a liquid coating is found on the cuticle surface of the black widow spider, *Latrodectus hesperus* (Hadley, 1981b). SEM of untreated cuticle reveals a series of repeating polygonal shapes with slightly raised edges; small amorphous droplets are present within the boundaries of the polygons (Figure 5.7*a*). Considerably more surface detail can be seen in the same region of the cuticle after washing it in chloroform:methanol. The droplets are removed and the perimeters of the polygons are more pronounced, with sharper small ridges clearly visible within each area (Figure 5.7*b*).

In some arthropods the accumulation of superficial waxes is so substantial that they obscure the actual cuticle surface and give the species a powdery or matted appearance. Although Holdgate and Seal (1956) provided some of the earliest electron micrographs of wax blooms in insects, only recently have these deposits received the ultrastructural attention given similar blooms present on leaf surfaces (Hadley, 1982). The waxy threads of some coccinellid larvae (Coleoptera) are clearly visible to the naked eye, even though a full-grown individual may be no more than about 5 mm long (Pope, 1979). The shape and diameter of the wax threads appear to be dictated by the physical form of the pores from which they are secreted. The ribbon-like

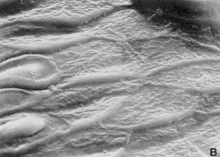

FIGURE 5.7. (A) Untreated dorsal cephalothorax cuticle of the black widow spider, *Latrodectus hesperus*, ×4000. (B) Same region of dorsal cephalothorax following a 15-minute wash in chloroform:methanol. Removal of the droplets and surface film reveals considerable fine surface detail. ×4250. From Hadley (1981b).

wax that covers the cuticle surface of the balsam woolly aphid, *Adelges piceae* (Homoptera), is believed to be formed from secretions from a row of cells that fuse together (Retnakaran et al., 1979).

Elaborate, multicolored wax blooms are often present in desert arthropods. These thick wax accumulations no doubt complement the role played by lipids associated with specific epicuticular layers in restricting water loss in this extreme environment. The banded grasshopper, *Trimerotropis pallidipennis,* which experiences air temperatures exceeding 50°C, has its thorax and abdomen uniformly covered by wax encrustations, crystals, and filaments. The elytral surface of the buprestid beetle, *Hippomelas planicosta,* which also inhabits hot, dry creosote flats, is covered with twisting tubular secretions. Perhaps the most unique and morphologically intricate wax blooms are found in certain desert tenebrionid beetles. *Cryptoglossa verrucosa,* a common beetle in the Sonoran Desert, exhibits distinct color phases that range from light blue to jet black when subjected to extremes of low and high humidity, respectively (Figure 5.8a) (Hadley, 1979). The basis for the humidity-induced color phases are waxes that are secreted from numerous miniature tubercles located on the dorsal thorax and abdominal elytra. At high humidities the wax appears as an amorphous secretion exuding from the tip of each tubercle (Figure 5.8b); however, when the beetles are transferred to low humidity the amorphous secretion organizes into numerous slender (0.14 μm) filaments, which radiate out and interconnect with other filaments, eventually creating a dense meshwork that covers the surface (Figure 5.8c). The individual wax filaments scatter incident light, producing a Tyndall blue effect and, hence, the characteristic whitish-blue color. In the Namib Desert, tenebrionid beetles produce surface wax that responds to changes in humidity in a similar manner to that described for *C. verrucosa.* The phenomenon, however, appears to be much more widespread in Namib

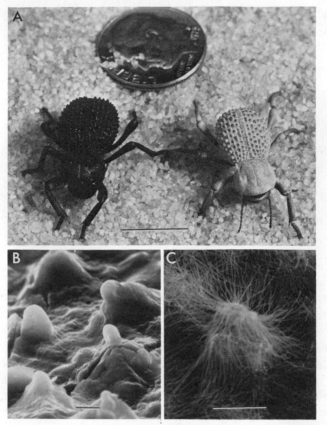

FIGURE 5.8. (A) Light blue (low humidity) and black (high humidity) color phases of the beetle *Cryptoglossa verrucosa*. (B) Miniature wax-secreting tubercles on the elytral surface of a black phase beetle. (C) Spreading of wax filaments from tip of single tubercle in response to low humidity. Reference bars: (A) 1 cm; (B) and (C) 10 μm.

tenebrionids, as wax blooms are found in at least 26 different species, with many of the species exhibiting specific colors and/or patterns (McClain et al., 1985).

Epicuticular Waxes

In addition to waxes deposited on the surface of the arthropod cuticle, there are also lipids associated with specific sublayers of the epicuticle. Separation of the discussion into "surface" and "epicuticular" lipids, however, is primarily designed to distinguish between SEM and TEM studies, and is not intended to imply that two distinct lipid entities exist. Although much of the surface wax previously described lies outside of the recognized epicuticular sublayers (e.g., wax blooms), a portion of these wax deposits likely impregnate or are continuous with the outermost layers of the epi-cuticle.

When appropriately stained arthropod cuticle is examined with the light microscope, all epicuticular layers as well as portions of the procuticle appear to contain lipids. Unfortunately, ultrastructural confirmation of the above has been difficult. For one thing, it is often not possible to relate with certainty layers observed with the light microscope with those seen under the electron microscope. Also, most conventional lipid stains designed for use in conjunction with the light microscope cannot be used to detect lipid in cuticle sections prepared for EM. New techniques for improving the visualization of lipids with the electron microscope are being developed, including the use of the unsaturated hydrocarbon myrcene ($C_{10}H_{16}$), which is incorporated into existing lipids in order to increase the amount of osmium they will bind (Wigglesworth, 1975a). Using this technique, Wigglesworth (1975b) and Walker and Crawford (1980) were able to show the presence of lipids in the epicuticle of the blood-sucking bug, *Rhodnius,* and the desert millipede, *Orthroporus ornatus,* respectively. A third and quite perplexing problem has been the failure to demonstrate changes in either the thickness or morphology of epicuticular layers in cuticle subjected to lipid solvents or acids. This was first shown by Filshie (1970), who found little difference between extracted and unextracted cuticle of *Calpodes ethlius* larvae (Lep-idoptera). These observations prompted Filshie to propose an epicuticular layer terminology that implies a physical position rather than a particular chemical composition. Similar results were obtained more recently by Had-ley and Filshie (1979) on a desert scorpion and by Hadley (1981b) on the black widow spider. Resistance of epicuticular layers to extraction with lipid solvents, however, should not be interpreted to mean that lipid is not present. Since application of the same organic solvents yields lipid for chemical analysis, it may be that much of the lipid is simply lost during the dehy-dration-embedding steps that precede TEM. It may also be that extraction procedures commonly employed for chemical analysis do not remove all the lipid associated with a specific sublayer or leave behind other chemical constituents (e.g., protein) that maintain the structural integrity of the layers.

Cuticular Lipids and Transpiration

The first experiments that showed that lipids associated with the epicuticle are primarily responsible for the low rates of water loss exhibited by most terrestrial arthropods were performed in the 1930s and 1940s by Ramsay, Beament, Wigglesworth, and others. They based their conclusions on the fact that cuticular water loss increased significantly when lipids were re-moved or disrupted by solvents or mechanical abrasion, adsorbed by dusts, or were structurally altered by exposure to high temperature (see Ebeling,

1974; Edney, 1977). Although more recent studies have indicated that the cuticular water barrier is perhaps more complex in that it likely involves lipids associated with other cuticular layers, nonlipid constituents, and possibly the apical membrane of the epidermal cells, none of the subsequent research conducted on arthropod cuticle has lessened the importance of lipids in this waterproofing function. In fact, current research efforts are beginning to establish strong correlations between the quantity, chemical composition, and morphology of epicuticular lipids and their effectiveness in reducing transpiratory water loss. Furthermore, it has now been shown that lipid composition can be modified by physical and biological factors in a manner that is adaptive when insects and arachnids require additional desiccation resistance.

Arthropods that exhibit low rates of cuticular transpiration often possess greater amounts of cuticular lipid in general or greater amounts of a specific lipid fraction (e.g., hydrocarbons). Moreover, during periods (seasonal or developmental) of high desiccation potential or moisture stress, many species increase their cuticular lipid content. Terrestrial adult stoneflies, *Pteronarcys californica,* have more surface lipid than do the immature aquatic naiad (Armold et al., 1969). Similarly, the early instar larvae of fleshflies, *Sarcophaga bullata,* which occupy moist habitats, have relatively little cuticular hydrocarbon compared with pupae and newly emerged adults, which are subjected to much drier conditions (Armold and Regnier, 1975). One example for which there is supporting ultrastructural evidence involves the tobacco hornworm, *Manduca sexta.* SEM of the cuticle surface of pupae in diapause demonstrated a much thicker wax layer than that observed in nondiapausing pupae (Bell et al., 1975). This increase is consistent with the need to conserve body water during the prolonged period (six to eight months) of arrested development. Both a thicker wax and a higher content of hydrocarbons were found in diapausing pupae versus nondiapausing pupae of the Bertha armyworm, *Mamestra configurata* (Hegdekar, 1979). Nymphs of the desert cicada, *Diceroprocta apache,* which live underground where temperature and moisture conditions are more moderate, contained only 10.9 μg of hydrocarbon per exuvium (cast skin), compared with 57.8 μg of hydrocarbon per adult (Hadley, 1980a). The latter is active at midday when air temperatures in the Sonoran Desert approach 50°C.

In many instances, changes in surface wax density are accompanied by changes in the chemical nature of the epicuticular lipids that are designed to maximize cuticular impermeability. The desert tenebrionid beetle, *Eleodes armata,* which exhibits lower cuticular transpiration rates in summer than winter, not only has a higher quantity of hydrocarbons during summer months, but also a higher percentage of long-chain hydrocarbon molecules when compared with winter-collected beetles (Figure 5.9a). A similar increase in the two parameters was observed when winter-collected beetles were acclimated to 35°C for 5- and 10-week periods, respectively (Figure 5.9b) (Hadley, 1977). Hydrocarbons in all instances were saturated. The

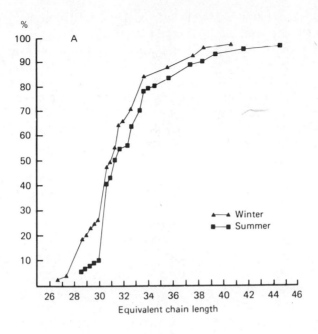

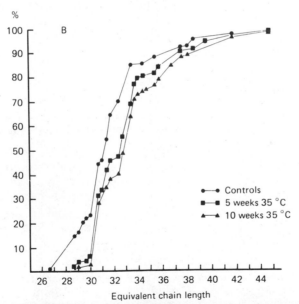

FIGURE 5.9. (*A*) Seasonal changes in the cumulative distribution functions of epicuticular hydrocarbons of the beetle *Eleodes armata*. The ordinate represents the percentage of hydrocarbon molecules that have an equivalent chain length less than or equal to the indicated number of carbon atoms. (*B*) Cumulative distribution functions of epicuticular hydrocarbons of winter-active beetles versus winter-active beetles acclimated to 35°C for 5 and 10 weeks. Reprinted from Hadley (1977).

desert-adapted scorpion *Hadrurus arizonensis* has a significantly lower cuticular permeability than its montane counterpart, *Uroctonus apacheanus*. Analysis of their respective cuticular lipid compositions revealed that the hydrocarbon density of *H. arizonensis* is twice that of *U. apacheanus,* and that the former species has higher proportions of long-chain, branched hydrocarbons and long-chain free fatty acids (Toolson and Hadley, 1977). Comparison of the free fatty acid composition of the two species is especially revealing. The longest free fatty acid in *U. apacheanus* contains 22 carbon atoms, whereas 67.5% of the free fatty acid molecules in *H. arizonensis* have 24 or more carbon atoms. Also, 84.7% of the free fatty acids in *H. arizonensis* are saturated, compared with only 50.6% in *U. apacheanus*. Transcuticular water loss in another scorpion, *Centruroides sculpturatus,* is also significantly lower in summer, during which time its cuticular lipids are characterized by a higher proportion of long-chain, branched alkanes (Toolson and Hadley, 1979). Longer chain, saturated molecules (summer condition) would increase the intensity of van der Waals interactions between hydrocarbon molecules, thus creating a more stable membrane that is less readily penetrated by water.

The hydrophobic nature of the arthropod cuticle is further enhanced by the arrangement and morphology of wax deposits. This is especially true if the surface wax consists of intertwining filaments and/or projecting rods and tubes, such as found in the wax blooms of tenebrionid beetles. These structures decrease the conductance to vapor diffusion by increasing the length and twisting of the vapor pathway. In *Cryptoglossa verrucosa,* there is also a boundary layer of air between the cuticle surface and the lower edge of the wax bloom, which further restricts the diffusion of water vapor. The architectural arrangement of its surface wax combined with the quantity and high percentage of long-chain saturated alkanes are no doubt essential factors in *C. verrucosa* having one of the lowest water loss rates recorded for any arthropod (Hadley, 1978b).

No discussion of the role played by lipids in waterproofing arthropod cuticle would be complete without some mention of transition temperatures and the monolayer hypothesis, two related concepts that for over 25 years have greatly influenced thinking concerning epicuticular structure and function. Transition temperature (t_c) refers to an abrupt, marked increase in transcuticular water loss at a specific temperature. It is often observed when water loss, corrected for saturation deficit, is measured in dead insects and arachnids whose spiracles have been occluded. This relationship between temperature and cuticular permeability was used initially as evidence to support the role of lipids in the waterproofing process and eventually led Beament (1958, 1961, 1964) to propose that the principal barrier to water was a monolayer of polar lipids located between the epicuticle proper and a thicker layer of wax randomly arranged on the epicuticle surface. To explain the functioning of the monolayer in relation to transition temperature phenomena, Beament further postulated that the polar lipids comprising the monolayer are normally positioned on the epicuticle with their hydrocarbon chains oriented at an angle of 65° to the cuticle surface, and that at t_c sufficient thermal energy exists to displace the polar molecules, causing them to assume a vertical position and thus producing abrupt changes in permeability. Although other models were subsequently proposed to account for permeability changes at t_c (Locke, 1965; Davis, 1974), the monolayer hypothesis with few exceptions gained wide acceptance among insect physiologists.

Modern analytical instrumentation and innovative research techniques have enabled investigators to examine the chemical and biophysical properties of the epicuticle in considerable detail. Thus far much of the data derived from these studies are in direct contrast with concepts implicit in the monolayer hypothesis. Major arguments against its existence, which are discussed in length by Gilby (1980), Machin (1980), and Hadley (1981a), can be briefly summarized as follows:

1. Chemical analyses of most arthropod cuticular lipids indicate that nonpolar constituents such as hydrocarbon often predominate; these compounds are not capable of forming monolayers and would interfere with or disrupt monolayers formed by other lipid classes.

2. Long-chain alcohols, which Beament assumed were responsible for the monolayer, are not present in the cuticular lipids of many species; nor are phospholipids, which are primarily responsible for lipid orientation in plasma membranes.

3. Fatty acids, which are usually present in at least trace amounts, can form monolayers only if artificially maintained in a high state of compression (e.g., on a film balance).

4. Freshly extracted cockroach lipid spread on the surface of water failed to form a tightly packed monolayer; instead the film readily collapsed and the constituent molecules were only weakly attracted to the water surface (Lockey, 1976).

5. Electron paramagnetic resonance (EPR) spectroscopic studies on the cuticular lipids of a scorpion indicated that the spin-labeled molecules do not have a preferred orientation, and that at 55°C, where water loss was increasing very rapidly, no changes in mobility were detected by EPR (Toolson et al., 1979).

6. The abrupt changes in permeability that characterize t_c may be an artifact created by incorrectly dividing the observed rate of water loss by the saturation deficit to correct for thermally induced changes in the diffusion gradient (Toolson, 1978).

If a lipid monolayer or oriented lipid layer is not a valid explanation for the operation of the lipid waterproofing barrier, what then is the correct working model? At present there is no satisfactory answer to this question.

The barrier to water efflux may simply be represented by a heterogeneous mixture of primarily nonpolar lipid molecules deposited on and/or near the surface of the epicuticle. Machin's (1980) calculation that it requires only 0.04 μm more of unoriented lipid to equal the barrier properties of an oriented layer (a 13% increase in lipid thickness) is certainly in harmony with the above possibility. Phase changes with concomitant increases in water permeability would still persist; however, they would be characterized by broad endothermic transitions expected of a heterogeneous mixture of cuticular lipids having different molecular sizes and chemical structures. A definitive answer must await further experimentation.

Other Functions of Surface Wax

Accumulations of wax on the cuticle surface can serve an arthropod in many ways in addition to reducing water loss. Moreover, many of these functions have their parallel in plants (see the earlier discussion of plants). Surface waxes are often an effective deterrent against predation. This protection may result from the physical barrier created by the wax, for example, the waxy threads of some coccinellid larvae, which entangle potential predators, or it may be a consequence of camouflage and/or mimicry produced by elaborate patterns of surface wax deposits. The blending with the underlying sand substrate caused by deposits of surface wax is well illustrated by the Namib tenebrionid beetle, *Onymacris rugatipennis albotessallata,* when individuals with and without a wax coating are compared (Figure 5.10). The wax accumulation also provides some degree of thermal protection. The increased percentage of reflected solar radiation from the lighter surface delays the rise of the beetle's body temperature, which in turn increases the time it can actively forage on the hot dry surface of the Namib Desert. Surface waxes also increase the reflectance of potentially harmful ultraviolet radiation, a fact that explains in part the usually high surface density of cuticular waxes found on insects and arachnids living in the alpine tundra. In mealybugs (Homoptera), the curled filaments of wax produced by trilocular and quinquelocular pores function primarily to protect the mealybugs from contamination by their own honeydew and defensive exudates (Cox and Pearce, 1983). There is also some evidence that the wax coating of aphids and house flies may have an inhibitory effect on the absorption of insecticides. Finally, many surface waxes provide chemical signals as well as visual cues that aid in species recognition and location. The latter are discussed in greater detail in the chapter on chemical communication.

VERTEBRATES

As it is in plants and arthropods, the framework and composition of the integument are important factors in controlling water flux to and from ver-

FIGURE 5.10. The Namib Desert tenebrionid beetle, *Onymacris rugatipennis albotessallata,* with and without a wax bloom.

tebrates. The rate of water loss through the skin of vertebrates is directly dependent on the difference in concentration of water vapor within the animal and in the free air beyond the adhering boundary layer next to the animal's surface (i.e., the driving force), and is inversely related to the combined resistance provided by the skin and boundary layer. These relationships are described by the following equation:

$$E_c = \frac{\text{driving force}}{\text{resistance}} = \frac{(\overset{*}{e}_{T_s} - e_{T_a})}{r_i + r_b}$$

where E_c = cutaneous water loss from the skin (g cm^{-2} sec^{-1}), $\overset{*}{e}_{T_s}$ = saturation vapor density of water at skin temperature (g cm^{-3}), e_{T_a} = vapor density of water at air temperature (g cm^{-3}), r_i = skin resistance (sec cm^{-1}), and r_b = boundary layer resistance (sec cm^{-1}) (Davis et al., 1980). Experimental studies have shown that, of these factors, skin resistance is primarily responsible for determining the rate of cutaneous water loss.

In view of the importance of skin resistance in controlling the water permeability of skin, it is not surprising that the evolution of vertebrates from an aquatic to a primarily terrestrial existence was accompanied by changes in the integument designed to increase skin resistance and thus reduce evaporation. Paramount among these changes was the development of a horny layer (the stratum corneum of the epidermis) comprised mainly of the tough, fibrous protein keratin. Keratin, like lipids, is insoluble in water and has a low chemical reactivity. The degree of keratinization, or hardening, observed depends to some extent on a species' terrestriality and functions of the integument other than preventing water loss. For example, adult amphibians, which rely on the integument for a portion of their gas exchange, normally develop a horny layer only one cell thick. Reptiles, on the other hand, which are more highly terrestrial and lack a dermal respiratory function, possess a thick, horny layer. The stratum corneum of birds consists of separate, loosely arranged lamellae, which are designed more for reducing weight in flight than for flexibility. Its structure is quite unlike the hollow keratinized cells that form the stratum corneum of mammals (Spearman, 1966).

Whereas keratinization in vertebrates is strongly correlated with terrestrial existence and the need to reduce cutaneous evaporation, recent experimental findings have altered some of the earlier concepts and ideas pertaining to integumental permeability. It has been shown that heavily cornified integument such as exhibited by reptiles does not provide an absolute barrier to water movement, nor is keratin solely responsible for restricting water flux. Data are accumulating that show that lipids, either in conjunction with keratin, or as a single entity deposited within special epidermal layers, or in some unique cases superficially deposited on the integument surface, are intimately involved in the waterproofing process. Moreover, as illustrated for plants and arthropods, the quantity and composition of lipids are important features for determining the effectiveness of the water barrier. The remainder of this chapter examines the water permeability of the vertebrate integument, with special emphasis on the role of lipids and their chemical and structural interactions with other integumental constituents that participate in the functioning of the water barrier.

Amphibians

The amphibian integument is characteristically highly vascular, flexible, and extremely permeable to the bidirectional movement of water (Lillywhite and Licht, 1974). With few exceptions, representative species exhibit rates of cutaneous water loss that often approach rates at which water evaporates from a free surface. Because of their high integumental permeability, amphibians are poorly adapted for a strict terrestrial existence. Only a few species are found in desert regions, and these spend most of the year burrowed deep within the soil to avoid desiccating climatic conditions. Surface activity is restricted to brief periods following summer rains, during which time the animal replenishes lost body water and fat reserves and reproduces. At this time a permeable integument is actually advantageous in that it facilitates the rapid uptake of water from pools or moist substrates.

Some amphibians that burrow during long periods of drought form cocoons composed of single or multiple layers of keratinized stratum corneum. Such structures have been discovered in xeric-adapted frogs in Australia (Lee and Mercer, 1967), Africa (Loveridge and Crayé, 1979; Loveridge and Withers, 1981), North America (Ruibal and Hillman, 1981), and South America (McClanahan et al., 1976). The cocoon typically covers the whole body dorsally and ventrally, the legs, and head including the eyes, but is not continuous over the nares. Electron micrographs of cocoon fragments in *Pternohyla fodiens* and *Lepidobatrachus llanensis* reveal a characteristic pattern of multiple squamous cell layers separated by less dense granular intercellular material (see Ruibal and Hillman, 1981). The anuran cocoon is an effective barrier to the outward diffusion of water, as is evidenced by reductions in evaporative water loss in cocooned individuals which range from 80 to over 90%. There have been no chemical analyses of anuran cocoons; hence, it is not possible to speculate if barrier function is the sole property of keratin associated with the stratum corneum or if lipids in some context contribute to the waterproofing process.

Although there have been numerous measurements of amphibian water loss, few studies have examined the possible effects of lipids on integumental water flux. Schmid and Barden (1965) found that the skin of the aquatic frog *Rana septentrionalis* was less permeable to water and had a higher lipid content than that of the terrestrial toad *Bufo hemiophrys*. The authors concluded that the lipid was likely an important factor in preventing body fluid dilution by acting as a barrier to the inward flux of water. Differences in evaporative water loss rates between four species of anuran amphibians could not be accounted for by differences in the thickness or lipid content of the skin (Bentley and Yorio, 1979). Certainly the most striking and dramatic evidence that an integumental lipid barrier is responsible for significantly lower cutaneous evaporative water loss rates in amphibians is found in an unique group of South American phyllomedusine frogs (Blaylock et al., 1976). Four species belonging to the genus *Phyllomedusa* possess integumental alveolar glands, which secrete a lipid material onto the surface of the skin. Each species, using a complex and stereotypic wiping behavior (Figure 5.11), spreads the lipid over its integument. The secretion, when dry, helps prevent evaporative water loss and allows the frogs to bask in the sun, during which time they are in a quiescent state. Analysis of the lipid secretion in one species, *P. sauvagei*, showed wax esters to be dominant (68%), with smaller quantities of triacylglycerols, hydrocarbons, cholesterol, cholesterol esters, and free fatty acids also present (McClanahan et al., 1978). The average chain length of the wax esters is about 46 carbons with each molecule containing an average of 0.8 double bonds. The lipid secretion is so effective in reducing evaporation that water loss rates for these frogs are comparable to rates observed for desert reptiles. Moreover, the molecular composition of the wax esters appears to be functionally significant. A strong correlation was noted between the temperature at which the wax melts (ca. 35 to 38°C) and the temperature at which the species exhibits a marked increase in evaporation. Thus, the surface lipid secretion apparently effectively retards water loss up to a point where thermal considerations outweigh those of water conservation. Transcuticular water loss rates similar to those

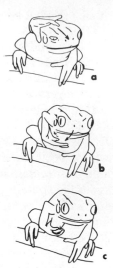

FIGURE 5.11.   Sequence by which the South American frog *Phyllomedusa sauvagei* uses its forelimbs to spread a lipid secretion over the body surface. Reprinted, with permission, from Blaylock et al. (1976). Copyright by the American Society of Ichthyologists and Herpetologists.

found for the phyllomedusine frogs have been recorded for African frogs belonging to the genus *Chiromantis* (Drewes et al., 1977) and *Hyperolius* (Withers et al., 1982), but thus far the waterproofing mechanism has not been elucidated.

Reptiles

Cutaneous water loss in squamate reptiles (snakes and lizards) has attracted much interest because of the unique structure and morphology of the squamate integument, plus the belief expressed earlier by some that the integument was actually impermeable to water. The success of lizards and snakes in extreme desert habitats throughout the world likely contributed to the development of the latter notion. While it is true that cutaneous water loss rates of xeric-adapted reptiles are typically lower than those for related species living in more mesic environments, the reptilian integument is not impervious to water. In fact, Bentley and Schmidt-Nielsen (1966) and Dawson et al. (1966) demonstrated that integumentary transpiration, although significantly lower than for most terrestrial vertebrates, accounts for the majority of total water loss in snakes and lizards, with rates of cutaneous water loss often two or more times higher than respiratory transpiration.

Of the several structures and components implicated as being responsible for the high degree of impermeability of reptilian integument, most investigations of barrier function have focused on the scales and/or the heavily keratinized epidermis. The structural attributes of reptilian scales are such that one is tempted to speculate as to their contribution to the integumental waterproofing process. Certainly the sculpturing created by their presence increases the surface area from which the diffusion of water vapor occurs, and may also enhance boundary layer resistance by creating unstirred air layers at the scalar surface (Lillywhite and Maderson, 1982). Despite these features, the presence of scales, at least in snakes, does not appear to be a major factor influencing cutaneous water loss. Studies by Licht and Bennett (1972) and Bennett and Licht (1975) showed that cutaneous evaporative water loss rates in mutant scaleless gopher and water snakes were not significantly different from rates in normal snakes. Moreover, several characteristic features of the scale itself (i.e., thick β-keratin, compact epidermis, superficial dermis) (see below) could not be regarded as adaptations to restrict water loss.

The epidermis of squamate reptiles consists of an outer epidermal generation above a basal stratum germinativum (Figure 5.12) and covers the outer and inner surface of the scales and the hinge region. The epidermis covering the hinge region may be thinner than that over the rest of the scale, but has the same fundamental layers and protein constituents. During the resting stage (shedding just completed), the outer epidermal generation can be further divided into an outermost serrated Oberhäutchen, a beta layer, a mesos layer, and an alpha layer. In addition, two or more layers of immature living cells lie between the alpha layer and the stratum germinativum (Figure 5.12). At the end of the resting phase, the onset of rapid proliferation in the stratum germinativum completes the outer generation and produces an inner epidermal generation that replaces the outer generation, which is shed approximately 14 days later. The system is unique in that an entire epidermal generation is shed either as a single unit (snakes) or in patches (lizards); in other vertebrates desquamation is continuous and involves only a few cells at a time.

The squamate epidermis contains two different conformational forms of keratin: alpha (hairlike) and beta (featherlike) (Baden and Maderson, 1970). α-Keratin exhibits a helical pattern, which confers flexibility to the layers in which it is found, namely the alpha and mesos layers. β-Keratin is relatively inflexible, and thus confers durability and protection to the skin

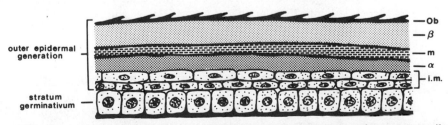

FIGURE 5.12. Diagrammatic drawing of the epidermis of the outer scale surface of a squamate reptile during the resting condition. Ob, Oberhäutchen; β, beta layer; m, meso layer; α, alpha layer; i.m., immature cells. Modified, by permission, from Maderson (1965).

where it occurs. The beta layer is quite thick in the outer scale region, but is represented by only a single layer of cells (the Oberhäutchen) on the inner surface and in the hinge.

If scales do not significantly contribute to the impermeable nature of the squamate integument, the main barrier must then rest with one or more of the keratinized epidermal layers. The observations on mutant scaleless snakes, which discredit the importance of scales in the waterproofing process, also provide evidence that β-keratin *per se* and the beta layer do not provide for integumental impermeability, as rates of water loss in mutants that lack the beta layer were comparable to those of normal scaled snakes. Furthermore, partial denaturation of epidermal keratin in the black rat snake (*Elaphe obsoleta*) increased cutaneous water loss only twofold (Roberts and Lillywhite, 1980). It appears that the primary function of β-keratin is to protect underlying layers from abrasion and injury. Maderson and co-workers (1978) suggested that the barrier to water is localized within the alpha layer. The removal of epidermal corneous materials with cellophane tape resulted in a sharp increase in rates of cutaneous water loss. These rates returned to normal levels following the reestablishment of the alpha layer. While this finding is consonant with the fact that an alpha layer of normal proportions was present in the scaleless snakes, Landmann et al. (1981) argue that cellophane stripping likely affects the mesos layer more than the alpha layer due to the more superficial position of the mesos, and that barrier function is more correctly a function of the mesos layer, with lipids associated with this layer having an integral part in the waterproofing mechanism.

Recent ultrastructural, histochemical, and physiological studies of squamate epidermis strongly support the existence of a mesos/lipid barrier. The mesos layer in the grass snake *Natrix natrix* consists of flat cornified cells similar in structure to the mammalian stratum corneum. Lamellar sheets that run parallel to the surface are found in the intercellular space between the cells (Figure 5.13) (Landmann, 1979). The lamellar sheets originate from mesos granules present in immature mesos cells and are rich in lipids (Landmann, 1980). The water-soluble tracer lanthanum percolates up freely from the dermis following its subcutaneous injection, but it does not diffuse past the level of transition between the alpha and mesos layers. When the lipids are removed with organic solvents, however, the tracer penetrates the mesos layer and labels the intercellular spaces completely (Landmann et al., 1981). Studies on the terrestrial snake *Elaphe obsoleta* by Roberts and Lillywhite (1980) strengthen these findings. Frozen sections of *E. obsoleta* epidermis stained with Oil Red O reveal the presence of lipids in all keratinized layers, with the most pronounced staining occurring in the mesos layer. Lipid extraction with chloroform:methanol resulted in roughly a fifteenfold increase in cutaneous water loss. Increased water loss was also noted following lipid extraction from the shed epidermis of three other snake species (including normal and scaleless individuals of one species) and the lizard *Iguana iguana*. Two recent studies have also demonstrated increased skin perme-

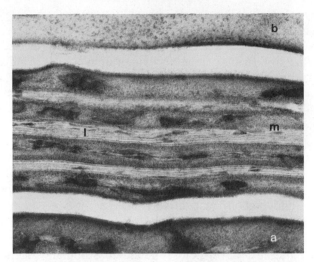

FIGURE 5.13. Electron micrograph of mesos layer (m) in the epidermis of the grass snake *Natrix natrix*. Lamellar sheets (l) composed of hydrophobic lipids fill the intercellular spaces. The mesos layer is sandwiched between the superficial beta layer (b) and alpha layer (a). × 177,000. Reprinted, by permission, from Landmann et al. (1981).

ability following lipid extraction in the water snakes *Regina septemvittata* and *Nerodia sipedon* (Stokes and Dunson, 1982) and in an arboreal snake, *Opheodrys aestivus* (Baeyens and Rountree, 1983).

Whereas lipid lamellar sheets in the mesos layer appear to pose the major barrier to transcutaneous water flux in the epidermis of squamate reptiles, other epidermal components likely contribute to total barrier function. Lipids associated with the beta and alpha layers probably help restrict the outward diffusion of water. Surface lipids may play a similar role, although wiping the surface of *E. obsoleta* with organic solvents did not alter their cutaneous water loss rates (Roberts and Lillywhite, 1980). The tracer and freeze-fracture techniques used by Landmann and colleagues in their study of the grass snake revealed a second permeability barrier associated with tight junctions that interconnect the apico-lateral plasma membranes of the uppermost stratum germinativum cells. Presently the tight junctions are thought to be more important in limiting the diffusion of electrolytes than water. Additional investigations that integrate epidermal morphology with physiology and a species' ecology are needed to further test these hypotheses. From such integrative approaches it may be possible to generate a working model applicable for reptilian integument in general.

Birds

There have been surprisingly few studies of cutaneous transpiration in birds. It was initially assumed that birds lose little water through their skin because they lack sweat glands. It was also thought that feathers might provide additional resistance to the outward diffusion of water. Studies in which cutaneous and respiratory water loss have been partitioned, however, clearly demonstrate that significant cutaneous evaporation occurs in birds. For example, in four species of birds native to xeric and mesic habitats, cutaneous evaporative water loss accounted for approximately 45 to 63% of the bird's total water loss (Bernstein, 1971). A comparable ratio of cutaneous to respiratory transpiration was found in the pigeon (Smith and Suthers, 1969). It was concluded that these rates of cutaneous evaporation were sufficient to have a major effect on avian thermoregulation. In support of the latter, Marder and Ben-Asher (1983) recently reported maximum cutaneous water loss rates of 6.8, 13.1, and 20.9 mg H_2O cm^{-2} hr^{-1} for the palm dove, collared dove, and pigeon, respectively. These rates are equivalent to a cooling capacity of 51.5, 86.1, and 96.5% of metabolic heat production during heat stress at an ambient temperature of 52°C.

The earlier findings, nevertheless, failed to stimulate much additional research on avian cutaneous water loss or on the nature of integumental barriers responsible for the rates observed. McNabb and McNabb (1977) measured skin changes during development of Japanese quail chicks and observed a direct relationship between the thickness of the cornified layer and cutaneous water loss. The data were interpreted as evidence that, as in mammals, the primary barrier to cutaneous water flux is the stratum corneum, despite the fact that avian epidermis is thinner and looser in structure than the corresponding layer of mammalian skin. As for lipids and their role in avian integumental water relations, studies for the most part have been restricted to secretions of the uropygial gland, which replaces the numerous sebaceous glands found in the skin of mammals. The uropygial gland is a large, exocrine structure located on the rump of the birds at the base of the tail feathers. Uropygial secretions (preen waxes) typically are complex mixtures of lipids that vary from order to order of birds. Monoester waxes usually predominate, with smaller quantities of saturated and unsaturated hydrocarbons and triacylglycerols present in some species (Jacob, 1976, 1978). The preen wax, when spread over the feathers, effectively prevents water penetration and facilitates the swimming ability of waterfowl. The secretion also likely protects against bacteria and fungi and may contain components that serve as chemical signals (Jacob and Ziswiler, 1982). Although such wax-impregnated feathers can prevent the skin from becoming wet, the resistance provided to the outward diffusion of water is apparently quite small when compared with the resistance provided by the bird's skin (Appleyard, 1979).

There is some recent ultrastructural and histochemical evidence that lipids in association with the stratum corneum might participate in barrier function in avian integument. Landmann (1980) described the presence of lamellar granules in the epidermis of the chicken that are believed to be homologous to the reptilian mesos granules previously discussed. The contents of these granules, which consist predominantly of polar lipids, are discharged into the intercellular space of the stratum corneum and to a lesser extent inside the horny cells themselves. After being discharged, the lamellar granules coalesce to form broad sheets oriented parallel to the surface of the keratinized cells. It has not been experimentally verified that these lipid-containing lamellar sheets effectively retard the efflux of water through the avian integument or, for that matter, what additional barrier mechanisms might be operative. An in-depth investigation of avian integumental water relations seems long overdue, and once conducted should provide much valuable comparative information.

Mammals

While the water relations of avian integument have received only token study, the permeability of mammalian integument, as well as lipids associated with the skin, has been actively investigated for many decades (see reviews by Nicolaides, 1963, 1974; Scheuplein and Blank, 1971, Downing, 1976). Much of this research has been conducted on human skin, with emphasis often on medical problems associated with the skin or its cosmetic function. The structural plan of mammalian integument basically follows that of a generalized vertebrate. The epidermis is composed of four or five cell layers of stratified squamous epithelium including a well-developed stratum corneum, which typically has been linked with barrier function as well as providing mechanical strength and protection. The situation in mammals, however, is complicated by the presence of sudoriferous, or sweat, glands, which are not found in any other vertebrate class. These epidermally derived glands secrete a thin, watery substance containing some salts and urea. Although their distribution is quite variable, most mammals have them at least on the soles of the feet and around the lips, genitalia, mammary glands, and anus (Webster and Webster, 1974). In humans they are profusely distributed over the entire body, with densities in some areas reaching 3000 per in.^{2}. During times of strenuous exercise or high ambient temperature, fluids from the sweat glands are literally poured onto the surface to evaporatively cool the organism and help maintain normothermic temperatures. Thus, you have an integumental barrier that greatly reduces the bidirectional flux of water interspersed with shunts designed to maximize the transport of water to the skin surface.

The mammalian integument also contains numerous sebaceous or oil glands, whose secretions have a direct bearing on the quantity and composition of skin lipids, and perhaps to a limited extent on the skin's permeability to water. These are simple branched acinar glands, which usually are connected to hair follicles (Figure 5.14). The sebaceous glands secrete an oily substance called sebum, which passes through the hair canal en route to the skin surface. The density of sebaceous glands varies from 50 to 900 per cm^2 in humans; they are absent from the palms and soles (Nicolaides,

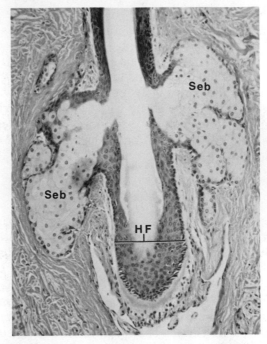

FIGURE 5.14. Photomicrograph of two sebaceous glands (Seb) opening into a hair follicle (HF). ×160. Reprinted, by permission, from Reith and Ross (1977).

1974), two regions where sweat glands are most numerous. Sebum is composed primarily of sterols and sterol esters, wax mono- and diesters, and, in humans, triacylglycerols. Squalene, an intermediate in the synthesis of cholesterol, is also abundant in human sebum, as are free fatty acids, which Downing (1976) suspects result from the bacterial action on triacylglycerol molecules. This complex mixture of lipids is believed to help keep hair from becoming dry and brittle, keep skin soft and pliable, help prevent the entry of certain pathogens, and possibly serve in chemical communication. Its role in integumental waterproofing remains controversial. Although such lipids spread on the surface would likely have some effect on the diffusion of water, sebum itself probably contributes little to the major integumental water barrier. Kligman (1963) noted that it would require at least 10 times the normal amount of sebum on the skin surface before the effect on permeability would be noticeable, and at this thickness the sebum would tend to flow into valleys, minimizing its effectiveness.

Mammalian skin lipids are also produced by epidermal cells. Although the amount of these lipids is considerably less than the quantity secreted in the sebum, nonetheless it is the epidermal lipids that are believed to be essential for epidermal barrier function. Because the barrier to water flux in mammals has historically been strongly linked with the stratum corneum, specifically the keratin fibrillar matrix inside the cell itself, it was generally thought that the epidermal lipids were also located inside the cornified cells, where they possibly surround the keratin filaments. Studies of neonatal mice (hairless and ductless), however, clearly demonstrate that these lipids are associated with intercellular spaces between adjacent layers of cornified cells and not with the cell interior (Elias and Friend, 1975; Elias et al., 1977). The origin and formation of the lipid barrier follows a sequence of events that are similar to those described for reptiles and birds. The barrier originates in the stratum granulosum, which contains the lipid-rich lamellar bodies. These bodies are secreted into the intercellular spaces where the lamellar material greatly expands and, following cornification, appears to form broad sheets that constitute the primary barrier to water efflux. Similar observations have been reported by Landmann (1980).

Studies by Elias and co-workers on the neonatal mice also provide information on the composition of epidermal lipids and the location of specific lipid groups within the epidermal layers. Extraction of isolated epidermal sheets with chloroform : methanol : water yielded 3 to 4% lipid per wet weight with neutral and polar lipids present in approximately equal amounts (Table 5.3) (Elias et al., 1977). A significant finding was the presence of high

TABLE 5.3

Distribution of polar and nonpolar lipids in newborn mouse epidermal barrier layers. Numbers represent average of three separate preparations for each lipid class.

	Total lipids (wt. %)
Neutral lipids	
Hydrocarbons	3.7 ± 1.9
Sterol esters	3.2 ± 1.6
Triacylglycerols	8.1 ± 3.6
Free fatty acids	11.7 ± 3.6
Free sterols	18.4 ± 4.5
Polar lipids	
Phosphatidylethanolamine (PE)	11.6 ± 4.7
Phosphatidylcholine (PC)	11.6 ± 1.4
Lysolecithin	1.5 ± 0.6
Glycolipid-rich	27.0 ± 10.6

SOURCE: Reprinted, by permission, from Elias et al. (1977). © 1977 The Williams & Wilkins Co., Baltimore.

quantities of hydrocarbons, free fatty acids, and glycolipids, with all polar lipid fractions and free fatty acids containing large amounts of long-chain fatty acids (24:0 and 26:0). Histochemical stains indicated that the neutral lipids were localized in the intercellular spaces of the stratum corneum, while phospholipids and possibly the glycolipids were associated with stratum granulosum cells. The combined histochemical and biochemical evidence confirms earlier reports that there is a shift from polar to neutral lipids during cornification, with perhaps the polar lipids of the granular layer being transformed into the neutral lipids of the cornified layer (Elias et al., 1977). A mixture of hydrocarbons and long-chain free fatty acids should be well suited for providing an effective water barrier.

Elias and Brown (1978) provided further evidence that intercellular lipids derived from epidermal lamellar bodies are primarily responsible for barrier function by comparing transepidermal water loss in mice deficient in essential fatty acids with control mice. Mice fed a diet deficient in essential fatty acids exhibited water loss rates of between 4 and 10 mg cm^{-2} h^{-1}, compared with a mean of 0.4 mg cm^{-2} h^{-1} in controls (Figure 5.15). Essential fatty acid deficiency was confirmed by showing altered ratios of arachidonic acid (20:4) to eicosatrienoic acid (20:3) in control versus deficient animals. The increased transepidermal water loss in deficient mice is caused by the progressive development of an abnormal epidermal water barrier. Although normal numbers of epidermal lamellar bodies were pro-

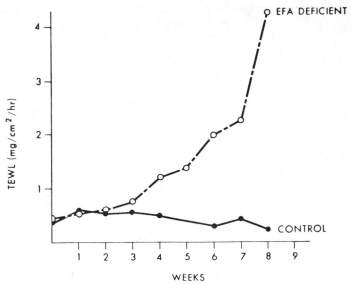

FIGURE 5.15. Transepidermal water loss (TEWL) in control and essential fatty acid–deficient mice. Reprinted, by permission, from Elias and Brown (1978). © 1978, US-Canadian Division of the International Academy of Pathology.

duced by deficient animals, the internal lamellae decreased in comparison to controls and, after secretion, failed to fill the intercellular spaces and coalesce into the broad, neutral lipid-rich sheets that fill stratum corneum interstices (Elias and Brown, 1978). Hartop and Prottey (1976) were able to restore skin permeability in rats deficient in essential fatty acids by adding linoleic (18:2) and linolenic (18:3) (both essential) to their diet, a finding that complements the observations cited above on deficient mice. In view of these results, the increased proportion of linoleic acid in the unesterified fraction of skin lipids (predominantly sebum-derived) of cattle during prolonged heat exposure (Poon et al., 1978) may reflect overall alterations in lipid synthesis designed to reduce transepidermal water loss during these stressful periods.

CONCLUDING REMARKS

Despite the advances made in our knowledge of lipid function in integumental waterproofing processes, much remains to be learned about lipid/water barrier relationships in plants and animals. Future research is likely to take several directions. In plants, a central problem continues to be the identification of a transport system and a transport medium that can account for the movement of waxes or their precursors from the site of synthesis to the cuticle surface. In both plants and arthropods, combined ultrastructural and histochemical studies are needed to identify the location of specific lipid classes in the various layers of the cuticle. Experiments using isolated cuticle segments or artificial membranes should be conducted to establish the effect of a specific lipid class, combination of classes, or structural features of specific lipid molecules on diffusion resistance. In vertebrates, the contribution of lipids to barrier function of the stratum corneum should be examined in a variety of species, especially those that inhabit areas with high desiccation potential yet exhibit extremely low transcutaneous water loss rates. In all groups more information is needed on the extent to which integumental lipid profiles can be facultatively changed in response to environmental stimuli. Investigators should also be alert to the presence and possible role of unique lipid compounds in the formation and function of the water barrier. Examples here include alkylresorcinols and alkylpyrones, two families of amphiphilic lipids found in the membranes of the soil bacterium *Azotobacter vinelandii,* which are believed to provide resistance to desiccation during cyst formation (Reusch and Sadoff, 1983), and acylglucosylceramides, which may be responsible for aggregating the disks of membrane that are stacked in the lamellar granules in mammalian epidermis (Wertz and Downing, 1982). The results of these efforts should enhance our appreciation for the widespread adoption of hydrophobic lipids by evolutionarily divergent organisms to combat potential dehydration.

REFERENCES

Ahern, D. G. and D. T. Downing (1974). Skin lipids of the Florida indigo snake. *Lipids* 9:814.

Amelunxen, F., K. Morgenroth, and T. Picksak (1967). Untersuchungen an der Epidermis mit dem Stereoscan-Elektronemikroskop. *Z. Pflanzenphysiol.* 57:79-95.

Appleyard, R. F. (1979). Cutaneous and respiratory water losses in the ring dove, *Streptopelia risoria.* M.S. Thesis, Washington State Univ., Pullman.

Armold, M. T. and F. E. Regnier (1975). A developmental study of the cuticular hydrocarbons of *Sarcophaga bullata. J. Insect Physiol.* 21:1827-1833.

Armold, M. T., G. J. Blomquist, and L. L. Jackson (1969). Cuticular lipids of insects. III. The surface lipids of the aquatic and terrestrial life forms of the big stonefly, *Pteronarcys californica* Newport. *Comp. Biochem. Physiol.* 31:685-692.

Armstrong, D. J. and M. I. Whitecross (1976). Temperature effects on formation and fine structure of *Brassica napus* leaf waxes. *Aust. J. Bot.* 24:309-318.

Baden, H. P. and P. F. A. Maderson (1970). Morphological and biophysical identification of fibrous proteins in the amniote epidermis. *J. Exp. Zool.* 174:225-232.

Baeyens, D. A. and R. L. Rountree (1983). A comparative study of evaporative water loss and epidermal permeability in an arboreal snake, *Opheodrys aestivus,* and a semi-aquatic snake, *Nerodia rhombifera. Comp. Biochem. Physiol.* 76A:301-304.

Baker, E. A. (1974). The influence of environment on leaf wax development in *Brassica oleracea* var. gemmifera. *New Phytol.* 73:955-966.

Baker, E. A. and M. J. Bukovac (1971). Characterization of the components of plant cuticles in relation to the penetration of 2,4-D. *Ann. Appl. Biol.* 67:243-253.

Barthlott, W. (1981). Epidermal and seed surface characters of plants: Systematic applicability and some evolutionary aspects. *Nord. J. Bot.* 1:345-355.

Barthlott, W. and E. Wollenweber (1981). Zur Feinstruktur, Chemie und taxonomischen Signifikanz epicuticularer Wachse and ähnlicher Sekrete. *Trop. Subtrop. Pflanzenwelt* (Akad. Wiss. Lit. Mainz). 32:1-67.

Beament, J. W. L. (1958). The effect of temperature on the waterproofing mechanism of an insect. *J. Exp. Biol.* 35:494-519.

Beament, J. W. L. (1961). The water relations of insect cuticle. *Biol. Rev.* 36:281-320.

Beament, J. W. L. (1964). Active transport and passive movement of water in insects. *Adv. Insect Physiol.* 2:67-129.

Beament, J. W. L. (1976). The ecology of cuticle. In: *The Insect Integument* (Hepburn, H. R., Ed.), pp. 359-374. Elsevier, Amsterdam.

Bell, R. A., P. R. Nelson, T. K. Borg, and D. L. Cardwell (1975). Wax secretion in nondiapausing pupae of the tobacco hornworm, *Manduca sexta. J. Insect Physiol.* 21:1725-1729.

Bengston, C. S., S. Larsson, and C. Liljenberg (1978). Effects of water stress on cuticular transpiration rate and amount and composition of epicuticular wax in seedlings of six oat varieties. *Physiol. Plant.* 44:319-324.

Bennett, A. F. and P. Licht (1975). Evaporative water loss in scaleless snakes. *Comp. Biochem. Physiol.* 52A:213-215.

Bentley, P. J. and K. Schmidt-Nielsen (1966). Cutaneous water loss in reptiles. *Science* 151:1547-1549.

Bentley, P. J. and T. Yorio (1979). Evaporative water loss in anuran amphibia: A comparative study. *Comp. Biochem. Physiol.* 62A:1005-1009.

Bernstein, M. H. (1971). Cutaneous water loss in small birds. *Condor* **73**:468–469.

Bianchi, G., E. Lupotto, B. Borghi, and M. Corbellini (1980). The effects of chromosomal deficiencies on the biosynthesis of wax components. *Planta* **148**:328–331.

Blaylock, L. A., R. Ruibal, and K. Platt-Aloia (1976). Skin structure and wiping behavior of phyllomedusine frogs. *Copeia* **1976**:283–295.

Bleckmann, C. A., H. M. Hull, and R. W. Hoshaw (1980). Cuticular ultrastructure of *Prosopis velutina* and *Acacia Greggii* leaflets. *Bot. Gaz.* **141**:1–8.

Blomquist, G. J. and J. W. Dillwith (1983). Pheromones: biochemistry and physiology. In: *Invertebrate Endocrinology,* Vol. 1. Endocrinology of Insects (Downer, R. G. H. and H. Laufer, Eds.) pp. 527–542. Alan R. Liss, New York.

Blomquist, G. J. and L. L. Jackson (1979). Chemistry and biochemistry of insect waxes. *Prog. Lipid Res.* **17**:319–345.

Bukovac, M. J., J. A. Flore, and E. A. Baker (1979). Peach leaf surfaces: Changes in wettability, retention, cuticular permeability, and epicuticular wax chemistry during expansion with special reference to spray application. *J. Am. Soc. Hort. Sci.* **104**:493–496.

Cox, J. M. and M. J. Pearce (1983). Wax produced by dermal pores in three species of mealybug (Homoptera: Pseudococcidae). *Int. J. Insect Morphol. Embryol.* **12**: 235–248.

Daly, G. T. (1964). Leaf-surface wax in *Poa colensoi. J. Exp. Bot.* **15**:160–165.

Davis, J. E., J. R. Spotila, and W. C. Schefler (1980). Evaporative water loss from the American alligator, *Alligator mississippiensis:* The relative importance of respiratory and cutaneous components and the regulatory role of the skin. *Comp. Biochem. Physiol.* **67A**:439–446.

Davis, M. T. B. (1974). Critical temperature and changes in cuticular lipids in the rabbit tick, *Haemaphysalis leporispalustris. J. Insect Physiol.* **20**:1087–1100.

Dawson, W. R., V. H. Shoemaker, and P. Licht (1966). Evaporative water losses in some small Australian lizards. *Ecology* **47**:589–594.

de Gier, J., J. G. Mandersloot, and L. L. M. Van Deenen (1968). Lipid composition and permeability of liposomes. *Biochim. Biophys. Acta* **150**:666–675.

Downing, D. T. (1976). Mammalian waxes. In: *Chemistry and Biochemistry of Natural Waxes* (Kolattukudy, P. E., Ed.), pp. 17–48. Elsevier, Amsterdam.

Drewes, R. C., S. S. Hillman, R. W. Putnam, and O. M. Sokol (1977). Water, nitrogen and ion balance in the African treefrog *Chiromantis petersi* Boulenger (Anura: Rhacophoridae), with comments on the structure of the integument. *J. Comp. Physiol.* **116**:257–267.

Ebeling, W. (1974). Permeability of insect cuticle. In: *The Physiology of Insecta,* 2nd ed. (Rockstein, M., Ed.), pp. 271–343. Academic Press, New York.

Edney, E. B. (1977). *Water Balance in Land Arthropods.* Springer-Verlag, New York.

Elias, P. M. and B. E. Brown (1978). The mammalian cutaneous permeability barrier. *Lab. Invest.* **39**:574–583.

Elias, P. M. and D. S. Friend (1975). The permeability barrier in mammalian epidermis. *J. Cell Biol.* **65**:180–191.

Elias, P. M., J. Goerke, and D. S. Friend (1977). Mammalian epidermal barrier layer lipids: Composition and influence on structure. *J. Invest. Dermatol.* **69**:535–546.

Filshie, B. K. (1970). The resistance of epicuticular components of an insect to extraction with lipid solvents. *Tissue and Cell* **2**:181–190.

Freeman, B., L. G. Albrigo, and R. H. Biggs (1979). Ultrastructure and chemistry of cuticular waxes of developing *Citrus* leaves and fruits. *J. Am. Soc. Hort. Sci.* **104**:801–808.

Giese, B. N. (1975). Effects of light and temperature on the composition of epicuticular wax of barley leaves. *Phytochemistry* **14**:921–929.

Gilby, A. R. (1980). Transpiration, temperature and lipids in insect cuticle. *Adv. Insect. Physiol.* **15**:1–32.

Grncarevic, M. and R. Radler (1967). The effect of wax components on cuticular transpiration: Model experiments. *Planta* **75**:23–27.

Hadley, N. F. (1977). Epicuticular lipids of the desert tenebrionid beetle, *Eleodes armata:* Seasonal and acclimatory effects on composition. *Insect Biochem.* **7**:277–283.

Hadley, N. F. (1978a). Cuticular permeability and lipid composition of the black widow spider, *Latrodectus hesperus.* In: *Symp. Zool. Soc. London,* No. 42 (Merrett, P., Ed.), pp. 429–438. Academic Press, London.

Hadley, N. F. (1978b). Cuticular permeability of desert tenebrionid beetles: correlations with epicuticular hydrocarbon composition. *Insect Biochem.* **8**:17–22.

Hadley, N. F. (1979). Wax secretion and color phases of the desert tenebrionid beetle *Cryptoglossa verrucosa* (LeConte). *Science* **203**:367–369.

Hadley, N. F. (1980a). Cuticular lipids of adults and nymphal exuviae of the desert cicada, *Diceroprocta apache* (Homoptera, Cicadidae). *Comp. Biochem. Physiol.* **65B**:549–553.

Hadley, N. F. (1980b). Surface waxes and integumental permeability. *Am. Sci.* **68**:546–553.

Hadley, N. F. (1981a). Cuticular lipids of terrestrial plants and arthropods: A comparison of their structure, composition, and waterproofing function. *Biol. Rev.* **56**:23–47.

Hadley, N. F. (1981b). Fine structure of the cuticle of the black widow spider with reference to surface lipids. *Tissue and Cell* **13**:805–817.

Hadley, N. F. (1982). Cuticle ultrastructure with respect to the lipid waterproofing barrier. *J. Exp. Zool.* **222**:239–248.

Hadley, N. F. and B. K. Filshie (1979). Fine structure of the epicuticle of the desert scorpion, *Hadrurus arizonesis,* with reference to location of lipids. *Tissue and Cell* **11**:263–275.

Hadley, N. F. and L. L. Jackson (1977). Chemical composition of the epicuticular lipids of the scorpion, *Paruroctonus mesaensis. Insect Biochem.* **7**:85–89.

Hallam, N. D. (1970). Growth and regeneration of waxes on the leaves of *Eucalyptus. Planta* **93**:257–268.

Hallam, N. D. and B. E. Juniper (1971). The anatomy of the leaf surface. In: *Ecology of Leaf Surface Micro-organisms* (Preece, T. F. and C. H. Dickinson, Eds.), pp. 3–37. Academic Press, New York.

Hamilton, S. and R. J. Hamilton (1972). Plant waxes. In: *Topics in Lipid Chemistry,* Vol. 3 (Gunstone, F. D., Ed.), pp. 199–266. Elek Science, London.

Hartop, P. J. and C. Prottey (1976). Changes in transepidermal water loss and the composition of epidermal lecithin after applications of pure fatty acid triglycerides to the skin of essential fatty acid-deficient rats. *Br. J. Dermatol.* **95**:255–264.

Hegdekar, B. M. (1979). Epicuticular wax secretion in diapause and non-diapause pupae of the Bertha armyworm. *Ann. Entomol. Soc. Am.* **72**:13–15.

Holdgate, M. W. and M. Seal (1956). The epicuticular wax layers of the pupa of *Tenebrio molitor* L. *J. Exp. Biol.* **33**:82–106.

Holloway, P. J. (1977). Aspects of cutin structure and formation. *Biochem. Soc. Trans.* **5**:1263–1266.

Holloway, P. J., G. A. Brown, E. A. Baker, and M. J. K. Macey (1977). Chemical composition and ultrastructure of the epicuticular wax in three lines of *Brassica napus* (L). *Chem. Phys. Lipids* **19**:114–127.

Hunt, G. M., P. J. Holloway, and E. A. Baker (1976). Ultrastructure and chemistry of *Clarkia elegans* leaf wax: A comparative study with *Brassica* leaf waxes. *Plant Sci. Lett.* **6**:353–360.

Jackson, L. L. and G. J. Blomquist (1976). Insect waxes. In: *Chemistry and Biochemistry of Natural Waxes* (Kolattukudy, P. E., Ed.), pp. 201–233. Elsevier, Amsterdam.

Jacob, J. (1976). Bird waxes. In: *Chemistry and Biochemistry of Natural Waxes* (Kolattukudy, P. E., Ed.), pp. 93–146. Elsevier, Amsterdam.

Jacob, J. (1978). Hydrocarbon and multibranched ester waxes from the uropygial gland secretion of grebes (Podicipediformes). *J. Lipid Res.* **19**:148–153.

Jacob, J. and V. Ziswiler (1982). The uropygial gland. In: *Avian Biology,* Vol. 6 (Farner, D. S., J. R. King, and K. C. Parkes, Eds.), pp. 199–324. Academic Press, New York.

Jeffree, C. E., R. P. C. Johnson, and P. G. Jarvis (1971). Epicuticular wax in the stomatal antechamber of sitka spruce and its effects on the diffusion of water vapor and carbon dioxide. *Planta* **98**:1–10.

Jeffree, C. E., E. A. Baker, and P. J. Holloway (1975). Ultrastructure and recrystallization of plant epicuticular waxes. *New Phytol.* **75**:539–549.

Kligman, A. M. (1963). The uses of sebum? *Br. J. Dermatol.* **75**:307–319.

Kolattukudy, P. E. (1975). Biochemistry of cutin, suberin and waxes, the lipid barriers on plants. In: *Recent Advances in the Chemistry and Biochemistry of Plant Lipids* (Gallard, T. and E. I. Mercer, Eds.), pp. 203–246. Academic Press, New York.

Kolattukudy, P. E. (1976). Introduction to natural waxes. In: *Chemistry and Biochemistry of Natural Waxes* (Kolattukudy, P. E., Ed.), pp. 1–15. Elsevier, Amsterdam.

Kolattukudy, P. E. and T. J. Walton (1972). The biochemistry of plant cuticular lipids. *Prog. Chem. Fats Other Lipids* **13**:121–175.

Landmann, L. (1979). Keratin formation and barrier mechanisms in the epidermis of *Natrix natrix* (Reptilia: Serpentes): An ultrastructural study. *J. Morph.* **162**:93–126.

Landmann, L. (1980). Lamellar granules in mammalian, avian and reptilian epidermis. *J. Ultrastruct. Res.* **72**:245–263.

Landmann, L., C. Stolinski, and B. Martin (1981). The permeability barrier in the epidermis of the grass snake during the resting stage of the sloughing cycle. *Cell Tiss. Res.* **215**:369–382.

Lee, A. K. and E. H. Mercer (1967). Cocoon surrounding desert-dwelling frogs. *Science* **157**:87–88.

Licht, P. and A. F. Bennett (1972). A scaleless snake: Tests of the role of reptilian scales in water loss and heat transfer. *Copeia* **1972**:702–707.

Lillywhite, H. B. and P. Licht (1974). Movement of water over toad skin: Functional role of epidermal sculpturing. *Copeia* **1974**:165–171.

Lillywhite, H. B. and P. F. A. Maderson (1982). Skin structure and permeability. In: *Biology of the Reptilia,* Vol. 12 (Gans, C. and F. H. Pough, Eds.), pp. 397–442. Academic Press, New York.

Locke, M. (1965). Permeability of insect cuticle to water and lipids. *Science* **147**:295.

Lockey, K. H. (1976). Cuticular hydrocarbons of *Locusta, Schistocerca* and *Periplaneta,* and their role in waterproofing. *Insect Biochem.* **6**:457–472.

Lockey, K. H. (1980). Insect cuticular hydrocarbons. *Comp. Biochem. Physiol.* **65B**:457–462.

Loveridge, J. P. and G. Crayé (1979). Cocoon formation in two species of southern African frogs. *South Africa J. Sci.* **75**:18–20.

Loveridge, J. P. and P. C. Withers (1981). Metabolism and water balance of active and cocooned African bullfrogs *Pyxicephalus adspersus. Physiol. Zool.* **54**:203–214.

Machin, J. (1980). Cuticle water relations: Towards a new cuticle waterproofing model. In: *Insect Biology in the Future* (Locke, M. and D. S. Smith, Eds.), pp. 79–103. Academic Press, London.

Maderson, P. F. A. (1965). Histological changes in the epidermis of snakes during the sloughing cycle. *J. Zool.* **146**:98–113.

Maderson, P. F. A., A. H. Zucker, and S. I. Roth (1978). Epidermal regeneration and percutaneous water loss following cellophane stripping of reptile epidermis. *J. Exp. Zool.* **204**:11–32.

Manheim, B. S. and T. W. Mulroy (1978). Triterpenoids in epicuticular waxes of *Dudleya* species. *Phytochemistry* **17**:1799–1800.

Marder, J. and J. Ben-Asher (1983). Cutaneous water evaporation—I. Its significance in heat-stressed birds. *Comp. Biochem. Physiol.* **75A**:425–431.

Martin, J. T. and B. E. Juniper (1970). *The Cuticles of Plants.* St. Martin's Press, New York.

Mayeux, H. S., Jr. W. R. Jordan, R. E. Meyer, and S. M. Meola (1981). Epicuticular wax on goldenweed (*Isocoma* spp.) leaves: Variation with species and season. *Weed Sci.* **29**:389–393.

McClain, E., M. K. Seely, N. F. Hadley, and V. Gray (1985). Wax blooms in tenebrionid beetles of the Namib Desert: Correlations with environment. *Ecology* (in press).

McClanahan, L. L., V. H. Shoemaker, and R. Ruibal (1976). Structure and function of the cocoon of a ceratophryd frog. *Copeia* **1976**:179–185.

McClanahan, L. L., J. N. Stinner, and V. H. Shoemaker (1978). Skin lipids, water loss, and energy metabolism in a South American tree frog (*Phyllomedusa sauvagei*). *Physiol. Zool.* **51**:179–187.

McNabb, F. M. A. and R. A. McNabb (1977). Skin and plumage changes during the development of thermoregulatory ability in Japanese quail chicks. *Comp. Biochem. Physiol.* **58A**:163–166.

Neville, A. C. (1975). *Biology of the Arthropod Cuticle.* Springer-Verlag, Berlin.

Nicolaides, N. (1963). Human skin surface lipids—Origin, composition and possible function. *Adv. Biol. Skin* **4**:167–187.

Nicolaides, N. (1974). Skin lipids: Their biochemical uniqueness. *Science* **186**:19–26.

Poon, W. Y., D. M. Jenkinson, R. C. Noble, and J. H. Moore (1978). *In vivo* changes in the composition of cattle skin surface lipid with time. *Res. Vet. Sci.* **25**:234–240.

Pope, R. D. (1979). Wax production by coccinellid larvae (Coleoptera). *Syst. Entomol.* **4**:171–196.

Reicosky, D. A. and J. W. Hanover (1976). Seasonal changes in leaf surface waxes of *Picea pungens. Am. J. Bot.* **63**:449–456.

Reith, E. J. and M. H. Ross (1977). *Atlas of Descriptive Histology,* 3rd ed. Harper and Row, New York.

Retnakaran, A., T. Ennis, L. Jobin, and J. Garnett (1979). Scanning electron microscopic study of wax distribution on the balsam woolly aphid, *Adelges piceae* (Homoptera: Adelgidae). *Can. Entomol.* **111**:67–72.

Reusch, R. N. and H. L. Sadoff (1983). Novel lipid components of the *Azotobacter vinelandii* cyst membrane. *Nature* **302**:268–270.

Roberts, J. B. and H. B. Lillywhite (1980). Lipid barrier to water exchange in reptile epidermis. *Science* **207**:1077–1079.

Ruibal, R. and S. Hillman (1981). Cocoon structure and function in the burrowing hylid frog, *Pternohyla fodiens. J. Herpetol.* **15**:403–408.

Scheuplein, R. J. and I. H. Blank (1971). Permeability of the skin. *Physiol Rev.* **51**:702–747.

Schmid, W. D. and R. E. Barden (1965). Water permeability and lipid content of amphibian skin. *Comp. Biochem. Physiol.* **15**:423–427.

Schönherr, J. (1976). Water permeability of isolated cuticular membranes: The effect of cuticular waxes on diffusion of water. *Planta* **131**:159–164.

Schönherr, J. and H. W. Schmidt (1979). Water permeability of plant cuticles: Dependence of permeability coefficients of cuticular transpiration on vapor pressure saturation deficit. *Planta* **144**:391–400.

Schönherr, J., K. Eckl, and H. Gruler (1979). Water permeability of plant cuticles: The effect of temperature on diffusion of water. *Planta* **147**:21–26.

Schulze, E.-D., B. M. Eller, D. A. Thomas, D. J. v. Willert, and E. Brinckmann (1980). Leaf temperatures and energy balance of *Weltwitschia mirabilis*. *Oecologia* **44**: 258–262.

Silbert, D. F., R. B. Ladenson, and J. L. Honegger (1973). The unsaturated fatty acid requirement in *Escherichia coli:* Temperature dependence and total replacement by branched-chain fatty acids. *Biochim. Biophys. Acta* **311**:349–361.

Smith, R. M. and R. Suthers (1969). Cutaneous water loss as a significant contribution to temperature regulation in heat stressed pigeons. *Physiologist* **12**:358.

Soliday, C. L., G. J. Blomquist, and L. L. Jackson (1974). Cuticular lipids of insects. VI. Cuticular lipids of the grasshoppers *Melanoplus sanguinipes* and *Melanoplus packardii*. *J. Lipid Res.* **15**:399–405.

Spearman, R. I. C. (1966). The keratinization of epidermal scales, feathers and hair. *Biol. Rev.* **41**:59–96.

Stokes, G. D. and W. A. Dunson (1982). Permeability and channel structure of reptilian skin. *Am. J. Physiol.* **242**:F681–689.

Taylor, A. R., W. T. Roubal, and U. Varanasi (1975). Effects of structural variation in β-monoglycerides and other lipids on ordering in synthetic membranes. *Lipids* **10**:535–541.

Thair, B. W. and G. R. Lister (1975). The distribution and fine structure of the epicuticular leaf wax of *Pseudotsuga menziezii*. *Can. J. Bot.* **53**:1063–1071.

Toolson, E. C. (1978). Diffusion of water through the arthropod cuticle: Thermodynamic consideration of the transition temperature phenomenon. *J. Therm. Biol.* **3**:69–73.

Toolson, E. C. and N. F. Hadley (1977). Cuticular permeability and epicuticular lipid composition in two Arizona vejovid scorpions. *Physiol. Zool.* **50**:323–330.

Toolson, E. C. and N. F. Hadley (1979). Seasonal effects on cuticular permeability and epicuticular lipid composition in *Centruroides sculpturatus* Ewing 1928 (Scorpiones: Buthidae). *J. Comp. Physiol.* **129**:319–325.

Toolson, E. C., T. R. White, and W. S. Glaunsinger (1979). Electron paramagnetic resonance spectroscopy of spin-labelled cuticle of *Centruroides sculpturatus* (Scorpiones: Buthidae): Correlation with thermal effects on cuticular permeability. *J. Insect Physiol.* **25**:271–275.

Tulloch, A. P. (1976). Chemistry of waxes of higher plants. In: *Chemistry and Biochemistry of Natural Waxes* (Kolattukudy, P. E., Ed.), pp. 235–287. Elsevier, Amsterdam.

Tulloch, A. P. (1981). Composition of epicuticular waxes from 28 genera of Gramineae: differences between subfamilies. *Can. J. Bot.* **59**:1213–1221.

von Wettstein-Knowles, P. (1974). Ultrastructure and origin of epicuticular wax tubes. *J. Ultrastructure Res.* **46**:483–498.

von Wettstein-Knowles, P. (1976). Biosynthetic relationships between β-diketones and esterified alkan-2-ols deduced from epicuticular wax of barley mutants, *Molec. Gen. Genet.* **144**:43–48.

Wattendorff, J. and P. J. Holloway (1980). Studies on the ultrastructure and histochemistry of plant cuticles: The cuticular membrane of *Agave americana* L. in situ. *Ann. Bot.* **46**:13–28.

Walker, L. J. and C. S. Crawford (1980). Integumental ultrastructure of the desert millipede, *Orthoporus ornatus* (Girard) (Diplopoda: Spirostreptidae). *Int. J. Insect Morphol. Embryol.* **9**:231–249.

Webster, D. and M. Webster (1974). *Comparative Vertebrate Physiology.* Academic Press, New York.

Weis-Fogh, T. (1970). Structure and function of insect cuticle. In: *Insect Ultrastructure* (Neville, A. C., Ed.), pp. 165–185. Blackwell, London.

Wertz, P. W. and D. T. Downing (1982). Glycolipids in mammalian epidermis: Structure and function in the water barrier. *Science* **217**:1261–1262.

Wigglesworth, V. B. (1975a). Lipid staining for the electron microscope: A new method. *J. Cell Sci.* **19**:425–437.

Wigglesworth, V. B. (1975b). Incorporation of lipid into the epicuticle of *Rhodnius* (Hemiptera). *J. Cell Sci.* **19**:459–485.

Wilson, L. (1975). Wax components as a barrier to aqueous solutions. Ph.D. Dissertation, University of California, Davis.

Withers, P., G. Louw, and S. Nicolson (1982). Water loss, oxygen consumption and colour change in "waterproof" reed frogs (*Hyperolius*). *South African J. Sci.* **78**:30–32.

6

ENERGY
AND FUEL STORAGE

All living organisms must obtain from their environment the energy needed for maintenance, growth, movement, and other processes that are characteristic of life. The ultimate source of energy for nearly all organisms is the sun. Autotrophic organisms, which include most plants, utilize carbon dioxide as their sole source of carbon to construct complex biomolecules directly or indirectly through the process of photosynthesis. Heterotrophic organisms, which include all multicellular animals, cannot use carbon dioxide directly from their environment, but instead obtain their energy in a rather complex chemical form from plants or from other heterotrophs occupying one or more trophic levels. The energy utilized is thus derived from specific types of chemical substances that together comprise the food of an organism.

The chemical forms of energy degraded by organisms usually are grouped into three categories: proteins, carbohydrates, and lipids. Nucleic acids, the other major organic group in biological systems, make up a high percentage of the cell weight but are not typically utilized as a source of energy. In most instances, the energy required by an organism is provided by the immediate metabolism of ingested foodstuffs or synthesized compounds. During times of starvation or intense activity, however, energy expenditure exceeds energy input and the organism must rely on the catabolism of previously stored food reserves. The latter are usually in the form of lipids; in fact, the function of lipids as energy reserves is one of the most general and significant functions of these compounds in plants and animals. This chapter examines the properties of lipids that make them ideally suited for fuel storage, interactions between lipids and other foodstuffs during times of energy production, the sites of lipid storage in different phylogenetic groups, and some of the biological activities of organisms that require adequate lipid depots. Much of the emphasis is on adipose tissue, which in vertebrates serves as both a fat depot and as a facility for synthesizing new fat from carbohydrates (Dole, 1965). Other important functions of adipose tissue in addition to energy storage (e.g., thermal insulation, heat production, mechanical buffering, lubrication) are covered in the appropriate chapters that follow.

ADVANTAGES OF LIPIDS AS A FUEL SOURCE

There are many advantages of storing and utilizing lipids as a source of fuel. Many of these advantages relate directly to the physical and chemical features of lipids, which were discussed in Chapter 1. Certainly a major consideration has to be differences in both the heat energy and potential biologically useful energy of lipid relative to carbohydrate or protein. In most organisms, lipid is stored primarily in the form of triacylglycerols, or "fats." Calorimetric determinations show that the energy derived from the combustion (oxidation) of 1 g of fat is more than twice as high as from the oxidation of carbohydrate or protein (Table 6.1). The potential biological energy available in lipid can be seen by comparing the oxidation of an individual fatty acid such as palmitate, a common constituent of triacylglycerol molecules, with that of glucose. The complete oxidation of 1 molecule of glucose to carbon dioxide and water yields 38 ATP, or approximately 277 kcal (38 × 7.3 kcal per mole ATP). In contrast, the complete oxidation of palmitic acid results in the production of 129 ATP, or about 942 kcal (Stryer, 1975). The complete balanced equation for the latter reactions is

$$\text{palmitate} + 23\ O_2 + 129\ P_i + 129\ ADP \rightarrow$$
$$16\ CO_2 + 145\ H_2O + 129\ ATP$$

TABLE 6.1
Energy and water production from the metabolism of the basic foodstuffs

	Energy source		
	Carbohydrate	Fat	Protein
Average metabolizable energy (kcal g^{-1})	4.1	9.3	4.1
Oxygen consumed (ml) g^{-1} of foodstuff	829	2019	966
Caloric equivalent (kcal liter^{-1} O$_2$)	5.05	4.69	4.60
Metabolic water production (ml g^{-1})	0.56	1.07	0.40[a]

SOURCE: Data from Edney (1977) and Hochachka and Somero (1973).

[a]Metabolized to urea.

Thus, on a molar basis, fatty acid oxidation captures approximately 3.4 times more biologically useful energy than does glucose metabolism (Hochachka and Somero, 1973). The number of ATP molecules, and hence the potential chemical energy obtained from the oxidation of an unsaturated fatty acid having the same number of carbon atoms, is slightly lower because unsaturated fatty acids have fewer hydrogen atoms and thus fewer electrons to be transferred via the respiratory chain to oxygen (Lehninger, 1975). Triacylglycerols also have an advantage over glycogen, the storage form of carbohydrate, in terms of ATP generation. On an equivalent weight basis, oxidation of pure triacylglycerols yields nearly 2.5 times as many molecules of ATP as pure glycogen.

Triacylglycerols possess other physicochemical properties that enhance their suitability as an energy reservoir. Triacylglycerols are large, viscous molecules. Overall, they are less reactive, less polar, and less water soluble than either carbohydrates or protein or, for that matter, other lipid groups such as fatty acids and alcohols. In adipose tissue, triacylglycerols can be stored compactly as relatively pure single droplets within each adipocyte, or fat cell (Needham, 1965). Moreover, they can be stored in a nearly anhydrous form, whereas proteins and carbohydrates are much more polar and hence more hydrated. This difference is dramatically apparent when triacylglycerols are again compared with glycogen. It is estimated that 1 g of glycogen is accompanied by anywhere from 2 to as much as 5 g of water (Stryer, 1975; Schmidt-Nielsen, 1979). This water of hydration further reduces the kcal per g value of glycogen. As a result, the mass of glycogen and its associated water can be as much as 10 times that of a fat store containing the same potential amount of energy. If the amount of stored triacylglycerols in adult humans was replaced by glycogen in a quantity sufficient to yield the same amount of ATP on combustion, the total body weight would have to be increased by about 60 kg, or 130 lb (Lehninger, 1975). As we shall discuss later, the "lightness" quality of stored fat or oil is of particular importance for flying insects and birds, especially during long-range migratory flights.

The equation summarizing the ATP production from the combustion of palmitic acid also shows that considerable water is generated from the oxidation of fat. Triacylglycerols are an excellent source of metabolic water, yielding about 1.07 ml of water per gram of fat. This value is approximately twice the metabolic water produced from an equivalent weight of carbohydrate, and nearly 2.5 times the amount of water formed from the oxidation of protein to urea (Table 6.1). Thus, triacylglycerols provide for the simultaneous storage of both energy and water. Specific examples are given later in this chapter illustrating the importance of metabolic water for animals restricted to a diet of dry foodstuffs, fasting animals such as hibernators that rely on stored fat, and marine mammals and pelagic sea birds, which either never have access to osmotically free water or spend long periods away from a source of fresh water.

A strict interpretation of the values presented in Table 6.1 can be misleading, however, without consideration of oxygen requirements and resultant energy production. This is especially true when comparing fats with carbohydrates. Whereas combustion of fat produces over twice the calories that can be obtained from an equal weight of carbohydrate, fat metabolism also requires nearly 2.5 times more oxygen for its complete oxidation. As a result, the caloric equivalents (kcal per liter O_2) for fat and carbohydrate are very similar (Table 6.1). It has been suggested, especially by insect physiologists, that the additional water loss associated with increased respiratory activity might negate any potential net gain of metabolic water from the oxidation of fat. (I return to this argument later in this chapter.) The net production of water from fat metabolism must also be weighed against the fact that water input is required for the synthesis of fat; hence, apparent differences between fat and carbohydrate metabolic water production may not be as significant as values would suggest. Presumably the same argument can be used when comparing the ATP produced from fat versus carbohydrate metabolism. As Edney (1977) suggests, important considerations in evaluating comparative values include the amount of fat contained in the ingested food plus the nutritional and hydration state of the animal during the time that fat stores are synthesized.

The above discussion leads to another important point when comparing the benefits of fat versus carbohydrate metabolism. Whereas the oxidation of fat yields more calories and water per unit weight of foodstuff, a different picture emerges when metabolic water production is expressed on a per kcal basis. For example, 0.133 g of water per kcal are produced by oxidizing carbohydrates, while the same energy output using fat produces only 0.112 g of water (Edney, 1977). This relationship was noted by Schmidt-Nielsen (1964) in examining the adaptations of desert vertebrates and later elaborated by Loveridge and Bursell (1975) in their study of locust water relations. A figure from the latter paper illustrates the relationships between oxygen consumed, carbon dioxide expired, and metabolic water produced, all expressed per unit of ATP formed (Figure 6.1). The graph shows that the main effect of a shift from carbohydrate to fat metabolism is a reduction in the quantity of reserve required to maintain a specific level of energy expenditure. Because of the higher energy content of fat, less fat must be metabolized in order to generate the energy required to maintain biological activity. This in turn results in reduced water production, even though fat per unit weight is a richer source of metabolic water. Actual values obtained by Loveridge and Bursell (1975) show that the switch toward increased utilization of fat in starving locusts results in a 10% decrease in the amount of metabolic water produced. Thus, on an equal potential energy basis, the use of fat rather than carbohydrate as a respiratory substrate may actually be counterproductive so far as water is concerned. This finding, however, does not minimize the importance of fat storage from the standpoint of weight economy for these locusts.

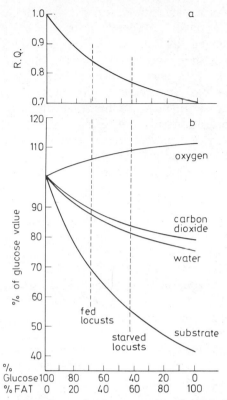

FIGURE 6.1. The relation between the use of various proportions of carbohydrate and fat by the locust and (*a*) respiratory quotient, R.Q., and (*b*) oxygen consumed, substrate utilized, and carbon dioxide evolved, all expressed per unit of ATP formed. Consumption and production are shown as percentages of the respective values if glucose had been the sole substrate. Reproduced from Loveridge and Bursell (1975), Bull. Ent. Res. **65,** by permission of the Commonwealth Agricultural Bureaux.

NONLIPID FUEL STORAGE

In some organisms nonlipid compounds serve as the principal source of fuel storage. In these cases the lipid is invariably replaced by carbohydrates, specifically starch (plants) and glycogen (animals). These high molecular weight polymers of glucose residues are usually deposited in the form of starch grains in plastids (e.g., chloroplasts, amyloplasts) or as glycogen granules in the cytoplasm of cells. In certain plant species, large amounts of starch are found in seeds, roots, and tubers, whereas in animals glycogen is especially abundant in the liver and muscle. Other polysaccharide fuel stores (e.g., fructans, xylans, and galactomannans in plants and galactan in animals) are rather rare in their occurrence (Candy, 1980).

Carbohydrate molecules, like fats, have unique physical and chemical properties that also make them suitable compounds for fuel storage, and in

some cases, superior to fats. Starches and glycogen can be rapidly synthesized during periods when carbohydrate supplies are plentiful. Moreover,
in the case of glycogen, its conversion from glucose-6-phosphate requires
only one ATP. Since the complete oxidation of glucose-6-phosphate yields
37 ATP, the overall efficiency of storage is nearly 97% (Stryer, 1975).
Storage polysaccharides can also be rapidly mobilized and degraded to meet
the energy needs of an organism. In this respect they are somewhat superior
to fats, whose mobilization for energy usage is slower. The rapid metabolism
of glycogen and amylopectin is believed to be due in part to their chainlike
structure, which offers many loose ends for sites of enzyme action, plus the
fact that the enyzmes involved in the degradation of polysaccharides are
actually loosely bound to the polysaccharide molecule, and hence readily
available (Candy, 1980).

Perhaps a more important consideration is the fact that glycogen can
provide energy under anoxic conditions, whereas fat metabolism requires
free oxygen. Certain organisms such as intestinal parasites occupy environments in which oxygen is virtually absent. These parasites must obtain their
energy through the anaerobic degradation of glycogen to carbon dioxide and
various organic acids. Other animals, such as intertidal molluscs or benthic
invertebrates, spend long periods during which time oxygen availability is
insufficient to permit aerobic metabolism. During this time they acquire
their energy through the anaerobic metabolism of stored glycogen. Even
strict aerobes (e.g., vertebrates) compensate for the temporary depletion or
unavailability of oxygen during vigorous activity by maintaining high capacities for anaerobic generation of ATP for supporting muscle work (Hochachka and Somero, 1973). In the case of parasites or sessile invertebrates,
where mobility is not a significant factor, the additional weight of glycogen
is not a serious problem (Schmidt-Nielsen, 1979). In fact the high affinity
for water exhibited by saccharides may provide an organism with some
degree of antifreeze protection. This may be one reason why temperate and
polar plant species tend to store carbohydrates, whereas tropical members
store lipids, whose hydrophobic properties have little effect on the aqueous
phase.

LIPID METABOLISM AND MOBILIZATION

In Chapter 2, background information was provided on how organisms
acquire lipids from their diet or synthesize these *de novo,* and how these
lipids are transported and stored in fuel depots such as seeds or adipose
tissue. In this section we examine how stored lipids are enzymatically degraded to capture energy in the form of ATP or, in the case of oil seeds,
converted into carbohydrates (mainly sucrose) before being transported to
the growing seedling. Emphasis is on the metabolism of triacylglycerols,
the major storage form of lipids, and specifically the oxidation of the fatty

acid moieties of the triacylglycerol molecule. Coverage is intended to provide a basic understanding of the mechanisms and pathways involved in the degradation of acyl lipids; more detailed information on enzymatic steps involved in the activation and oxidation of fatty acids can be found in many of the references cited.

Lipases

Earlier it was shown that dietary lipid in the form of triacylglycerol had to be digested to produce glycerol and free fatty acids before it could be absorbed and resynthesized into triacylglycerols. Similarly, stored triacylglycerols must first be hydrolyzed to glycerol and fatty acids before they are fed into energy-yielding enzymatic pathways. The enzymes that attack the triacylglycerol molecule are termed lipases; they act only at an oil-water interface (Galliard, 1980). Lipase activity has been reported in a wide variety of plant and animal tissues, especially in those structures that serve as fat or oil storage depots. Thus, lipases have been identified in several oil-rich seeds (e.g., castor bean, peanut, cotton) and also in cereal seeds (e.g., wheat, oats, rice) that contain significant amounts of triacylglycerols in the embryo and/or outer layers of the grain. Lipases normally exhibit a marked increase in activity that is coincident with the onset of germination, often in response to increased levels of gibberellic acid, a plant hormone (growth regulator). Similarly, animal triacylglycerols stored in adipose tissue or analogous structures are hydrolyzed by intracellular lipases whose activities are also controlled by various hormones including epinephrine, adrenocorticotrophic hormone, and glucagon (Figure 6.2). The free fatty acids and glycerol diffuse into the blood and are delivered to the various tissues by the cardiovascular system. Because glycerol is water-soluble, it readily dissolves in the plasma and is carried primarily to the liver and kidney, where it is either used for energy, as a source of α-glycerophosphate for glyceride synthesis, or converted into glucose (Masoro, 1968). Free fatty acids, in contrast, are water-insoluble and must be transported bound to serum albumins. The majority of these bound fatty acids are delivered to the liver, heart, muscles, and kidneys, where they are subsequently used for energy. In all tissues of vertebrates, with the exception of the brain, fatty acids delivered in this manner can be completely oxidized to carbon dioxide and water (Lehninger, 1975).

β-Oxidation

The sequence of events by which fatty acids are degraded in a stepwise fashion to acetyl-CoA coupled with the synthesis of ATP from ADP is termed β-oxidation. The process is so named because the second, or β-, carbon from the carboxyl group on the fatty acid is oxidized each time the process is repeated. The basic concept of β-oxidation was proposed first by

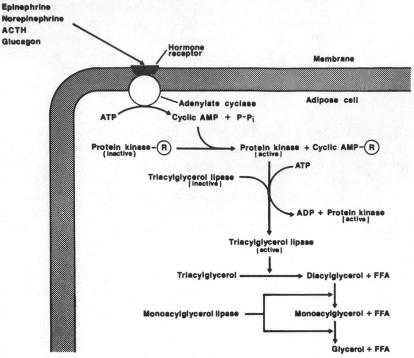

FIGURE 6.2. Endocrine control of fat mobilization from adipose tissue of mammals. Modified from Rex Montgomery, Robert L. Dryer, Thomas W. Conway, and Arthur A. Spector, *Biochemistry: A Case-Oriented Approach,* 4th ed., St. Louis, 1983, The C. V. Mosby Co.

Knoop at the turn of the century, using fatty acids labeled with phenyl groups; however, over 50 years passed before the intermediate compounds in Knoop's scheme were identified using modern enzymological techniques (Gurr and James, 1975).

The basic steps in this cyclic process in mitochondria are presented in Figure 6.3. Before the long-chain fatty acids can be oxidized, they must first be converted into a thioester derivative, namely fatty acyl-CoA. This process, referred to as the activation step, occurs only once in the degradation of a fatty acid to acetyl-CoA, and is the only step that requires energy derived from ATP. The subsequent oxidation takes place entirely inside the mitochondrial matrix. The fatty acyl-CoA formed in the initial step next loses two hydrogen atoms from the α- and β-carbons, forming an enoyl-CoA compound. This reaction requires an FAD-linked dehydrogenase. The enoyl-CoA is next enzymatically hydrated, forming β-hydroxyl acyl-CoA. In the next step, the β-carbon of β-hydroxyl acyl-CoA is oxidized further to β-oxyacyl-CoA; this reaction requires an NAD-linked dehydrogenase. In the final step of the first sequence, β-oxyacyl-CoA is cleaved at the β-carbon and another molecule of CoA added, such that a fatty acyl-CoA molecule

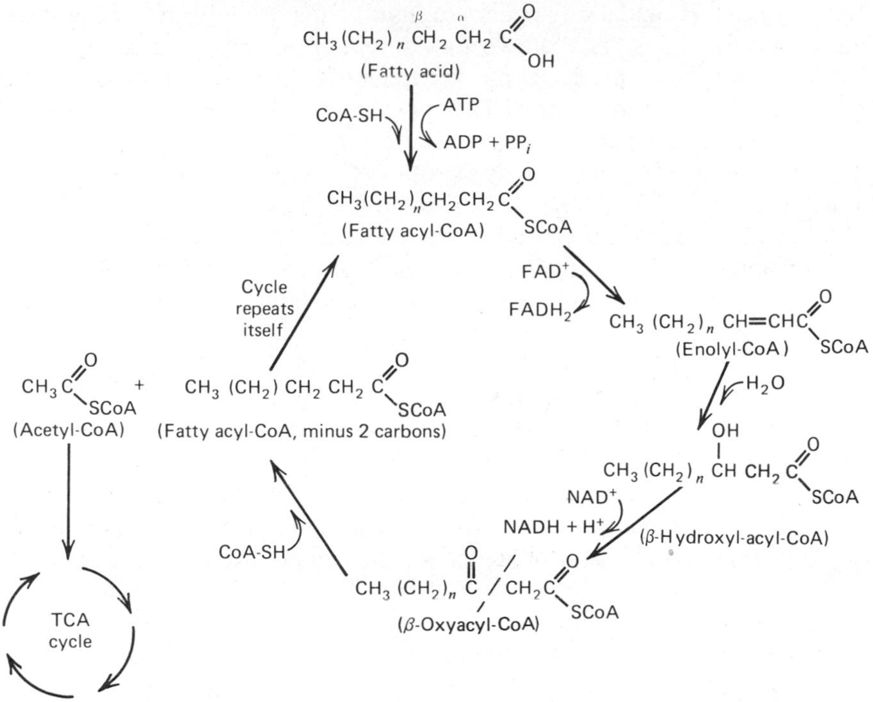

FIGURE 6.3. Pathway for the β-oxidation of fatty acids. The enzymes that catalyze these reactions are not shown. After Jensen (1976).

having two fewer carbon atoms than the original and one molecule of acetyl-CoA are formed as products. The former now becomes the substrate for another round of reactions until the original long-chain fatty acid is ultimately degraded into acetyl-CoA fragments. The acetyl-CoA fragments can now enter the reactions of the tricarboxylic cycle and be oxidized to carbon dioxide and water, or they can be used to form other compounds.

For many years β-oxidation was thought to occur only in the mitochondrial matrix in animal tissues. In 1969, however, it was described in glyoxysomes (see Figure 2.8) from germinating castor bean endosperm (Cooper and Beevers, 1969), and subsequently in rat liver peroxisomes (Lazarow and de Duve, 1976). Since 1976, peroxisomal β-oxidation has been demonstrated in many different types of mammalian cells. Although peroxisomal β-oxidation, like mitochondrial β-oxidation, generates acetyl-CoA through successive steps of dehydrogenation, hydration, dehydrogenation, and thiolytic cleavage (see Figure 6.3), the first dehydrogenation step of peroxisomal β-oxidation is catalyzed by an FAD-linked acyl-CoA oxidase rather than the FAD-linked acyl-CoA dehydrogenase of mitochondria. The acyl-CoA oxidase transfers electrons directly to oxygen to form hydrogen peroxide, which is decomposed in turn by a catalase. The physiological significance of peroxisomal β-oxidation appears to involve the catabolism of long-chain fatty acids, for it has been shown that diets rich in very long-chain mono-unsaturated fatty acids produce a marked induction of peroxisomal β-oxidation (Neat et al., 1981). Berry et al. (1983) have recently shown that mitochondria and peroxisomes cooperate in the oxidation of long-chain fatty acids, with reducing equivalents produced by peroxisomes subsequently transferred to mitochondria for completion of the oxidative process.

Earlier in this chapter, a complete balanced equation for the oxidation of palmitic acid (16:0) was given, indicating that a net total of 129 ATP were produced. The basis for this total can now be examined by considering the energetics of β-oxidation as it would occur in mitochondria. For each of the first 7 acetyl-CoA molecules generated by β-oxidation, a minimum of 5 ATP will be produced as a result of the transport of electrons from the coenzymes FAD and NAD to the terminal respiratory chain. Thus, for these 7 acetyl-CoA molecules, a sum of 35 ATP will be produced. The complete oxidation of 1 molecule of acetyl-CoA by the TCA cycle will produce 12 ATP. Since 1 palmitic acid will ultimately produce 8 molecules of acetyl-CoA, the high energy phosphate bonds generated from the oxidation of acetyl-CoA will amount to 96 (8 × 12) ATP. Adding the two sources of potential ATP production together, the total is 131 ATP. However, since 2 molecules of ATP are utilized to form palmitoyl-CoA from palmitic acid (Figure 6.3), the net yield is 129 ATP, or approximately 40% of the standard free energy of oxidation (Lehninger, 1975).

The sequence of enzymatic reactions presented in Figure 6.3 is for the oxidation of an even-numbered, saturated fatty acid. Unsaturated fatty acids, such as oleic acid (18:1), are oxidized by the same general pathway as saturated fatty acids, but the pathway has to be modified somewhat because the double bonds of naturally occurring unsaturated fatty acids are in the *cis* configuration, whereas the unsaturated acyl-CoA intermediates formed during the β-oxidation sequence of saturated fatty acids are *trans*. The problem is resolved by the participation of an additional enzyme, enoyl-CoA isomerase, which catalyzes a reversible shift of the double bond from the *cis* to *trans* configuration (Lehninger, 1975). In the case of polyunsaturated fatty acids, a second auxiliary enzyme is necessary to complete their oxidation. These modifications have potentially important metabolic and energetic consequences. Not only do unsaturated fatty acids of a given chain length yield correspondingly less free energy upon their complete oxidation because there are less hydrogen atoms available to enter the terminal respiratory chain, but the additional enzymatic steps necessary to convert the substrates into a form accepted by the β-oxidation system may also slow the rate at which unsaturated fatty acids are metabolized.

When they occur, odd-numbered chain fatty acids and even-numbered branched fatty acids are also metabolically degraded by β-oxidation. In these cases one of the products produced will be the three-carbon proprionyl-CoA rather than the acetyl-CoA. In animals, proprionyl-CoA is ultimately converted by most tissues into succinyl-CoA and eventually into succinic acid, both intermediates of the tricarboxylic acid cycle. In the liver this conversion is handled readily, while the heart is not equipped enzymatically to perform proprionate oxidation (Gurr and James, 1975).

The β-oxidation pathway for oxidizing fatty acids in animals also occurs in plants. In fact, evidence for a similar enzymatic pathway in plants was obtained using experiments analogous to the classic work of Knoop. In plants, however, much of the β-oxidation process is extramitochondrial, being associated with specialized cell organelles called glyoxysomes (Figures 2.8, 6.4). Acetyl-CoA is converted through the glyoxylate bypass into carbohydrate rather than being oxidized in the tricarboxylic acid cycle for energy conservation in ATP. In the case of unsaturated fatty acids, the auxiliary enzymes involved in the β-oxidation process in animal tissues are presumed to operate in plants as well (Galliard, 1980). Plants also contain an efficient process for the oxidation of proprionate obtained from odd-chain-length fatty acids, but this differs from the succinate pathway used by animals, and instead is superficially similar to β-oxidation processes exhibited by bacteria (Galliard, 1980).

Before considering other mechanisms by which fatty acids are metabolically degraded, some further discussion of the fate of acetyl-CoA is warranted. Normally this end-product of β-oxidation is directed into the energy-releasing reactions of the tricarboxylic acid cycle. However, during times when the production of acetyl-CoA via β-oxidation in the liver exceeds the rate at which acetyl-CoA can condense with oxaloacetate to form citrate in the tricarboxylic acid cycle, two molecules of acetyl-CoA will instead condense to form acetoacetyl-CoA, and ultimately acetoacetic acid. The latter can then be reduced to β-hydroxybutyric acid or be decarboxylated to form acetone. The last three chemical compounds are collectively termed "ketone

acetyl-CoA
   +   ⟶ acetoacetyl-CoA ⟶
acetyl-CoA

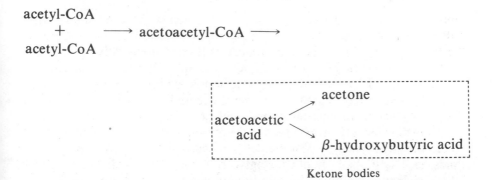

Ketone bodies

bodies." These compounds can be transported via the blood to the peripheral tissues, where they can be used as a source of energy. For example, the brain, which normally utilizes only glucose for energy, can, under periods of prolonged fasting, use β-hydroxybutyric acid as its major oxidative fuel. A recent report also indicates that ketone bodies are a potential source of energy during hibernation in arctic ground squirrels (Rauch and Behrisch, 1981).

α-Oxidation

Whereas β-oxidation is the principal fatty acid oxidation route in organisms, other so-called "minor" pathways for oxidizing fatty acids exist. One of these, α-oxidation, is found in both plants and animals and offers some interesting contrasts when compared with β-oxidation. In α-oxidation, the number two or α-carbon (counting from the carboxyl group) becomes oxidized and only one carbon, the carboxyl carbon, is cleaved at each step. α-Oxidation also differs from β-oxidation in that enzymes involved in the former are located in the microsomal rather than the mitochondrial fraction, and that the α-oxidation process does not require that fatty acids first be activated to acyl thioesters. Perhaps the most significant difference in the context of the present discussion is the fact that α-oxidation is not apparently used for energy production. What then is its physiological role? Presently this has not been fully defined. It many simply serve as a mechanism for degrading very long-chain fatty acids, which are not readily transported or utilized by certain tissues via β-oxidation (e.g., brain tissue) (Gurr and James, 1975). In plants, fatty aldehydes formed from the oxidation are often components of volatile products that serve the plant in a protective or communicative capacity. Alternatively, the aldehydes may be reduced to long-chain fatty alcohols, which are components of wax esters, an abundant constituent of the surface lipids of plants (Chapter 5).

Glyoxylate Cycle

Unlike α- and β-oxidation, which are concerned with the metabolic degradation of fatty acids, the glyoxylate cycle (Figure 6.4) is concerned with the fate of acetyl-CoA, the end-product of β-oxidation. Although the glyoxylate cycle was first described in bacteria as the pathway by which acetate or ethanol could be utilized as the sole source of carbon for growth, subsequent research has demonstrated that it occurs in many organisms where it operates under various nutritional and physiological conditions. It has been most intensively investigated in the germinating seeds of higher plants, where the glyoxylate cycle provides a mechanism for converting reserve lipids (usually in the form of triacylglycerols) into the carbohydrates needed by the growing seedlings. Its role in germinating seeds forms the basis of discussion in this chapter. The glyoxylate cycle, however, is also found in lower plants such as algae and fungi, and in protozoans and certain free-living nematodes when grown heterotrophically on C_2 compounds such as acetate or ethanol. It also occurs in parasitic nematodes such as *Ascaris*,

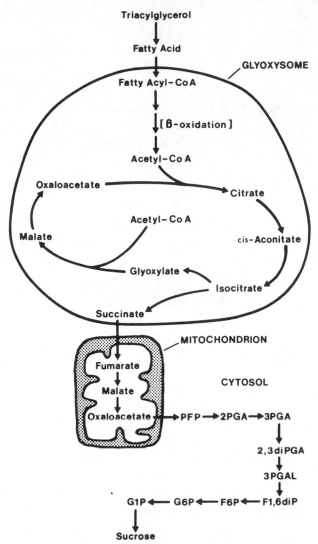

FIGURE 6.4. Summary of pathways involved in the conversion of triacylglycerols (stored in lipid bodies) to sucrose as it occurs in germinating oil seeds. The pathways are housed in several different subcellular compartments to effect the overall lipid-to-carbohydrate conversion. Glyoxysomes are the main organelles involved, containing enzymes of β-oxidation and the glyoxylate cycle. Intermediates derived from part of the TCA cycle in the mitochondria are converted to sucrose in the cytosol through reversed glycolysis.

where it functions in the conversion of reserve lipid to glycogen during embryogenesis (Barrett et al., 1970). A list of organisms in which the glyoxylate cycle operates, as well as a comprehensive review of its comparative biochemistry, are provided by Cioni et al. (1981).

Triacylglycerols constitute the primary energy storage compound in the seeds of many plants. Following germination there is a marked increase in lipase activity, which results in the rapid breakdown of triacylglycerols into glycerol and fatty acyl moieties. The latter are then catabolized to acetyl-CoA by the β-oxidation sequence (Huang et al., 1983). However, unlike in mammalian cells where the acetyl-CoA is metabolized to carbon dioxide and water, the bulk of the acetyl-CoA formed in germinating oil seeds is instead converted to sucrose, which is water-soluble and thus can be easily transported to growing parts of the plant (shoot and root), where it provides a source of carbon for biosynthesis of other compounds and/or an energy source. The conversion is extremely efficient, for it has been shown in castor bean endosperm that 1 g of carbohydrate accumulates for each gram of oil lost (see Beevers, 1980) for a 70% carbon conversion, taking into account the hydration of the carbohydrate. This conversion takes place until all of the reserve lipid has been utilized and the growing plant begins autotrophic growth (photosynthesis), thus generating its own carbon compounds via solar energy conversions.

LIPID UTILIZATION

Organisms use stored lipids for a variety of needs. Certain of these (e.g., growth, maintenance, reproduction) are essentially universal in their occurrence. Other biological activities that require an adequate amount of stored lipid for their successful completion are either unique to special groups of organisms or are activities shared by widely differing taxa. Examples here include flight, migration, hibernation (or its equivalent), and increased metabolic demands associated with endothermy. In this final section of Chapter 6, discussion centers on how lipids support the above activities from the standpoint of energy utilization, using examples from plants and animals including humans. Special attention is given to how patterns of lipid storage and utilization are influenced by environmental parameters such as food availability, precipitation, and temperature.

Plants

The energy needed by plants for morphological and metabolic changes associated with reproduction, development, and growth is invariably accompanied by changes in the content and metabolism of stored lipids. We have already mentioned the rapid mobilization of reserve lipids during the germination process and their conversion via the glyoxylate cycle into sucrose, which serves as a source of both carbon and energy for the growing seedling. There have been a few studies of the changes in lipid composition and metabolism that accompany the germination process. Notable among these is Zimmerman and Klosterman's (1965) investigation of changes in lipid composition of flax seed. They found that, up until 18 hours after

germination, triacylglycerol levels declined only slightly, but thereafter triacylglycerols were metabolized at a much higher rate so that by 90 hours
the lipid content of the seeds had decreased by approximately 50%. Also
noted was a concomitant increase in the level of free fatty acids, especially
the number of acids with odd carbon chains (e.g., 15:0, 17:0) and long-
chain fatty acids (e.g., 22:0, 24:0, 26:0). The authors attributed the increase
in odd-chain fatty acids to α-oxidation, which, as previously discussed,
would remove a single carbon atom from the even-chain fatty acids that
were dominant prior to the onset of germination, and the presence of long-
chain saturated acids to the development of an active fatty acid synthetase
system (Chapter 2) in the seedling tissues. Except for these relatively minor
changes, no specific class of fatty acid or triacylglycerol was preferentially
metabolized for the generation of energy. Studies of germinating watermelon
and oil palm seeds have led to basically the same conclusions (see Hitchcock
and Nichols, 1971).

Although triacylglycerols are the most common seed storage lipid, there
are a few plant species that contain little or no triacylglycerol, but instead
obtain the energy needed for germination and growth from other lipid storage
compounds. A unique species in this regard is the jojoba plant, *Simmondsia
chinensis,* which is native to arid regions of southwestern North America.
The cotyledons of jojoba seeds contain 50 to 60% of their weight as liquid
wax esters (Moreau and Huang, 1977), which are stored in lipid bodies, as
are triacylglycerols in other oil seeds. During germination, there is a gradual
decline in wax content with a corresponding increase in carbohydrates,
indicating a highly efficient gluconeogenic process in this species. Thin
layer chromatography showed that both hydrolysis products of the wax
esters, fatty acids and fatty alcohols, were metabolized. Both components
enter glyoxysomes, where they are modified and activated before being
β oxidized and processed in the glyoxylate pathway (Moreau and Huang,
1977). Other lipid compounds (e.g., sulpholipids, galactosyl diglycerides)
have been suggested as possible metabolic energy reserves, but experimental
data supporting this function, or for that matter any specific function of
these two lipid groups, are sparse (Douce and Joyard, 1980; Harwood,
1980).

The high percentage of stored lipids in most seeds can contribute to the
reproductive success of a plant in ways other than providing a source of
chemical energy for germination and growth. The low thermal conductivity
of lipids, specifically triacylglycerols, may provide the seed with some
degree of protection against excessive heat or cold. The high caloric value
of lipids might also aid seed germination for plants growing in temperate
and polar climates. Finally, the specific gravity of lipids should be a factor
in the dispersal of seeds. The lightness of "oils" (relative to starches and
proteins) should facilitate seed transport by streams or wind. Also, the anhydrous nature of lipids permits more compact storage, and hence a smaller
seed. The size of the seed is likely an important factor in determining the

extent to which seeds are disseminated by birds. These possibilities, along with supporting evidence, are examined in more detail in chapters to follow.

Marine Invertebrates

Lipids are a prominent storage compound in most marine invertebrates, although among this group no members exhibit a structure that can be strictly defined as adipose tissue (Vague and Fenasse, 1965). In some primitive phyla (e.g., Porifera), there is no visible trace of lipid stores. In echinoderms such as starfish and sea urchins, lipids are often stored in cells that line the digestive tract or are found in accessory gastrointestinal structures. An abundant fatty reserve is usually found in association with the hepatopancreas system of molluscs, crustaceans, and arachnids.

There appear to be a number of conditions under which marine invertebrates will utilize stored lipids as a source of fuel. These include periods when food availability is reduced, times of the year when the species is inactive, during periods of reproduction which require increased rates of synthesis, and possibly to provide additional energy needed during periods of increased activity (Lawrence, 1976). Utilization of stored lipids during starvation has been demonstrated for a variety of marine invertebrates representing many phyla. Whether marine invertebrates utilize lipid stores during seasons of reduced food availability is more difficult to ascertain. For example, barnacles may demonstrate lipid utilization when starved during winter, but it is not always possible to establish that a decrease in lipid levels corresponds to the nonavailability of food in the water. Lawrence (1976), based on his survey of the existing literature, concluded that benthic marine invertebrates overall do not rely on lipid stores during periods of low food availability, but may simply lower their energetic requirements to such a degree that lipid reserves are unnecessary. Complicating this issue is the fact that benthic species do utilize lipids for the production of gametes, and that changes in reproductive condition often occur simultaneously with periods of reduced food availability.

In pelagic marine invertebrates, the evidence for utilization of stored lipid for both maintenance energy and gametogenic activity is more conclusive. Moreover, several studies have noted increased lipid utilization during periods of low food availability, and also during times of increased activity (see references in Lawrence, 1976 and Sargent et al., 1976). Crustaceans have received the most study, especially copepods, which are the dominant zooplankton in most marine waters, and krill (euphausids), which are abundant in arctic and antarctic waters. In a recent study on krill, it was found that lipid accumulated in spring and summer was depleted during the following winter (Falk-Petersen et al., 1981). Although it was felt that a portion of the stored lipid was used for maintenance of basal metabolism during overwintering, the authors concluded that the primary function of the lipid reserve was to provide energy necessary for gonad formation. These data are, however, inconsistent with the supposition that most of the lipid accumulated during spring and summer is ultimately transferred into eggs since lipid in most of the krill investigated was heavily depleted before spawning.

Copepods have generated special interest with the discovery that in addition to fat storage they also store considerable amounts of wax esters. In fact, wax esters can account for 90% of the stored lipid, and up to 70% of the animal's dry weight (Lee, 1974). Copepods with this amount of wax esters are typically species that occur in deep water or in cold surface water. Although both fat and wax serve as metabolic fuels, the fats are generally distributed throughout the body and provide the copepod with energy for short-term purposes, whereas waxes are found in more specific locations such as oil sacs (see Figure 8.2) and are utilized during times of starvation or are transferred to the eggs, where they provide energy for the developing nauplii larvae (Benson and Lee, 1975). The ability to accumulate wax and retain it until all of the fat reserve has been used is apparently under enzymatic control. Wax esterase, which hydrolyzes the wax into fatty acids and fatty alcohols, is inhibited by some unknown mechanism until the time that the fat reserve has been essentially depleted. As a result, wax esters are a relatively stable lipid reserve. Wax esters are also an important dietary fuel source for numerous smaller fish that feed on calanoid copepods and thus ultimately provide energy for the higher trophic levels. A simplified food chain involving shallow-occurring copepods and energy transfer is depicted in Figure 6.5.

In addition to serving as a metabolic energy reserve, other functions have been assigned to lipids in marine invertebrates. One of these, buoyancy, has been firmly established, and is examined in detail in Chapter 8. Possible roles in thermal insulation and biosonar are at best speculative for invertebrates, but are of functional significance in marine vertebrates, and are discussed in Chapters 7 and 9, respectively. The extent to which metabolic water produced from the oxidation of lipids by marine invertebrates is involved in osmoregulatory processes remains largely unknown. Most marine invertebrates are osmoconformers; the osmotic concentration of their body fluids is in passive equilibrium with that of the ocean, although some species show varying degrees of volume and/or ionic regulation. In strict osmoconformers, any excess metabolic water produced would likely diffuse rapidly to the external medium. Oxidation water, however, may provide an osmotic buffer to some estuarine or brackish-water species, who are subject to daily fluctuations in salinity. Here any advantage derived from metabolic water would have to be weighed against the energetic cost incurred in obtaining osmotically free water from the sea (Sargent et al., 1976). These are interesting physiological and biochemical problems which deserve further study.

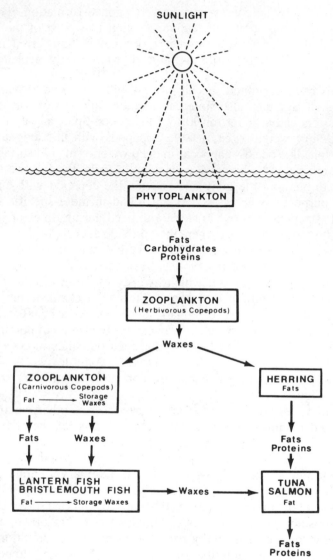

FIGURE 6.5. Portion of food chain in temperate zone oceans depicting role of wax in energy transfer. Foodstuffs manufactured by phytoplankton living near the surface are consumed by herbivorous copepods, which store large amounts of wax. Copepods (and other zooplankton) are fed upon by deep-water carnivorous copepods or by fish such as sardines and herring. These fish, which can metabolize the wax but do not store it, are in turn consumed by larger fish such as tuna and salmon. Carnivorous copepods also store large amounts of wax, which is consumed and stored by deep-water fish such as the lantern fish and bristlemouth fish. The stored waxes in the latter two species are transferred directly to tuna and salmon, which also prey on these deep-water fish. After A. A. Benson and R. F. Lee. The role of wax in oceanic food chains. Copyright © 1975 by Scientific American, Inc. All rights reserved.

166

Insects

Lipids play an important role in several morphogenetic and physiological functions in insects. In Chapter 5 we examined the contribution of various classes of lipids to the waterproofing mechanism of the cuticle. Here we shall see how lipids provide energy for basic biological activities such as maintenance and reproduction as well as for more dramatic events such as metamorphosis, migration, and diapause. The lipid required for these diverse functions is largely derived from storage compounds deposited in adipose tissue or, as more commonly termed, the insect fat body. The fat body consists of a loosely arranged meshwork of lobes invested in delicate connective tissue membranes. In many species, it is present as a peripheral layer beneath the integument and a central layer around the gut (Wigglesworth, 1965). The extent of the fat body may be substantial, filling three-fourths of the insect's body and accounting for 90% of the weight of some larvae (Vague and Fenasse, 1965). The term fat body, however, is misleading as it suggests that the only function of this tissue is the synthesis and storage of fat, whereas it can be an equally important storage site for proteins and glycogen as well. Furthermore, the fat body may assume some of the biochemical functions associated with the liver and kidney of vertebrates (Kilby, 1963). Thus, no true homology exists between the insect fat body and the adipose tissue of vertebrates.

The physiological and biochemical processes associated with insect development have received extensive study. The role of lipids in development is especially significant in view of the complexity of the developmental process and the energetic demands imposed by various stages of an insect's life history. The following section examines the importance of lipids in providing an efficient source of metabolic energy for embryogenesis, post-embryonic development, metamorphosis, and diapause in insects. More complete coverage, along with extensive bibliographies, can be found in reviews by Agrell and Lundquist (1973), Downer and Matthews (1976), and Beenakkers et al. (1981).

With few exceptions, insect eggs are rich in lipids (especially triacylglycerols), which provide an energy source for the developing embryo as well as a source of phospholipids needed for membrane synthesis. Evidence for the utilization of lipid during embryogenesis is based on changes in lipid weight during the embryonic period and respiratory data that typically show a decline in the respiratory quotient (RQ) from a brief initial value near 1.0 to an RQ of approximately 0.7, indicating a switch from carbohydrate to fat catabolism. The advantages of lipid metabolism during embryogenesis are basically the same as those described for organisms in general. Lipids contain more potential energy and yield more metabolic water than an equal weight of either carbohydrate or protein. The high metabolic water production, combined with the weight economy of stored triacylglycerols, may have special adaptive significance to some species. For example, female cockroaches, *Leucophaea madera,* carry the embryos within their blood sac for two months, during which time the desiccation potential is quite high (Gilbert, 1967). In the closed system of a cleiodic egg, lipid metabolism has an additional advantage when compared with protein metabolism. Lipids are completely oxidized to carbon dioxide and water; no end-products containing unutilized energy are produced and excretion is not a problem. Proteins, in contrast, are incompletely metabolized; toxic nitrogenous end-products accumulate, which require both energy and water for their elimination. By relying solely on lipid oxidation, these problems are avoided, and the protein stores can be used for synthesis of enzymes and structural materials.

Whereas during embryogenesis there is a transformation of stored triacylglycerols into active protoplasm, true growth during insect development occurs only in the larval stage (Agrell and Lundquist, 1973). During this period, insects typically accumulate large amounts of lipid to be used as energy reserves for later nonfeeding stages and metamorphosis. This is particularly true for holometabolous insects, whose larvae differ considerably in morphology and often in mode of life from the final adult form. The profound reorganization of body tissues that occurs during metamorphosis requires that a pupal stage be interposed between the last larval instar and the adult insect. The general pattern is for lipid content to increase steadily until about the middle instar and then increase quite rapidly, so that the pupa contains from two to three times the quantity of lipid present in the early last larval instar (Beenakkers et al., 1981). The molts, or ecdyses, that produce the successive larval instars are also energy-requiring processes, which apparently draw upon stored lipid, as a decrease in lipid content is often noted during every molting period. In the pupa, which superficially appears to be a quiescent stage, lipids accumulated during larval development are metabolized along with some carbohydrate to provide energy for breaking down larval structures and forming those of the adult. The extent to which lipid is the preferred fuel during metamorphosis and the periods during which its utilization are maximum vary among different insect species as well as between males and females of the same species. Changes in fatty acid composition during the larva-pupa-adult transformation suggest that saturated fatty acyl chains from the triacylglycerols are preferred over unsaturated moieties, but at present the data are too limited to assign any physiological significance to this trend.

In many insect species, development is not a continuous process. Instead there is a period of arrested activity, termed diapause, during which both growth and development are suspended. The delay insures that critical stages of morphogenesis take place when environmental conditions are suitable for the species. Diapause can occur at any developmental stage, but the stage is typically constant for a given species. Physiologically, diapause is characterized by a marked reduction in metabolism and brain electrical activity; in diapausing larvae and adults, there is also a cessation of feeding. Thus, in all diapausing stages, the insect must rely on stored nutrients (principally lipid) for meeting energy requirements. Lipid and/or triacylglycerol contents of diapausing individuals are often several times greater than those of non-diapausing individuals of the same species. For example, total lipid and triacylglycerol levels in diapausing face flies, *Musca autumnalis,* are seven and nine times higher than respective levels in nondiapausing flies (Pitts and Hopkins, 1965; Valder et al., 1969). Other species exhibiting a similar pattern are given in Downer and Matthews (1976). Lipid accumulation also occurs in adult insects prior to overwintering, although arrested activity during the latter should not be confused with true diapause (Patton, 1963).

Of all the activities of insects, none is more energy demanding than flight. This is especially true of species such as the monarch butterfly, which engages in long, continuous migratory flights, or bees and flies, which have small wings that operate at extremely high frequencies. Many insects can hover continuously within an enclosed space. For these individuals energetic costs are maximal, as all air movement must occur as a result of the beating wings. Bartholomew and Casey (1978) reported that a 1-g sphinx moth hovering at an ambient temperature of 23°C had a metabolic rate about 170 times greater than that of the same insect at rest. This increase in metabolic rate must be accompanied by comparable increases in enzyme activity, substrate flux, and ATP production (Kammer and Heinrich, 1978). In many cases, the primary source of the energy required to meet these additional metabolic demands is lipid. The experiments that have been conducted to establish the importance of lipids in flight metabolism, as well as the physiology and biochemistry of lipid utilization during flight, are comprehensively reviewed by Beenakkers et al. (1981).

Carbohydrates and lipids are the two principal storage fuels used to sustain muscle activity during flight. Which substrate is utilized depends largely on the duration of the flight, a species' feeding habits, and the nutritional value of the food, more so than on the previously emphasized phylogenetic relationships (Martin and Lieb, 1979). Diptera (flies) and Hymenoptera (bees, wasps), which feed frequently on nectar or other sugar solutions and are capable of only relatively short periods of continuous flight, obtain their energy from the oxidation of carbohydrates. Lepidoptera (butterflies, moths), which feed rarely if at all as adults, and thus must rely on fuel substrates accumulated during the larval stage, tend to oxidize primarily lipids during flight. Species, such as the monarch butterfly and silk moth, that make long-distance flights benefit from the weight economy that lipids offer plus the high production of metabolic water. Other insects, especially the Orthoptera (grasshoppers, locusts, cockroaches), utilize both substrates, oxidizing carbohydrates (glycogen, glucose, trehalose) during early flight, but turning to lipid for prolonged flight. Exceptions to these general patterns exist. Lepidoptera that do feed as adults will typically utilize both carbohydrates and lipids for flight. The blowfly, whose flight muscles contain large quantities of glycogen, continues to metabolize carbohydrates during longer flights (Sacktor, 1975). A variety of species representing six insect orders have been shown to utilize the amino acid proline to some extent (Bursell, 1981). Paramount among these is the tsetse fly, in which the mitochondria of the flight muscle oxidize proline some 76 times more rapidly than pyruvate.

Insect species that rely primarily on lipids for energy during flight mobilize lipid reserves in the fat body. Although the major storage form is triacylglycerol, these molecules are not the immediate source of energy. Instead triacylglycerols are first converted to diacylglycerols, which are released into the hemolymph and transported to the flight muscles in combination with the diacylglycerol-carrying lipoprotein (DGLP, see Chapter 2). Evidence that diacylglycerol is the principal fuel substrate is largely the fact that the hemolymph concentration of diacylglycerol increases significantly during flight. Little is known about the mechanisms of diacylglycerol uptake by flight muscle or how these mechanisms are regulated (Beenakkers et al., 1981). It is known, however, that substrate mobilization and transport during flight are under neuroendocrine control, specifically the adipokinetic hormone (AKH), which is stored and released from the corpus cardiacum. Oxidation of the fatty acid moieties of the diacylglycerols by flight muscle takes place by the reactions of β-oxidation described earlier.

Most insect species that utilize lipid as a substrate for flight have insufficient lipid present in muscle tissue to provide energy for both the initial stages and extended periods of flight. Thus, they usually oxidize carbohydrates in the muscle and hemolymph until a continuous supply of diacylglycerol can be provided from the fat body. An exception to this general pattern, however, was recently reported by Ward et al. (1982) for the triatomine bugs *Rhodnius prolixus* and *Triatoma infestans*. Light and electron microscopy showed that lipid droplets are abundantly present in the flight muscles of both species (Figure 6.6). This lipid, which is primarily in the form of triacylglycerol and may account for about one-half of the total dry weight of the muscle, exhibits a significant decrease during the first hour of flight. Moreover, chemical analysis showed that the glycogen content of the muscle and hemolymph carbohydrate levels are too low to contribute significantly to overall flight metabolism. The authors propose that, by storing large quantities of lipid within the flight musculature where it is available for immediate oxidation, any problems resulting from the delay in mobilizing and transporting lipid to the flight muscles are minimized or eliminated.

Fishes

The adipohepatic balance discussed for invertebrates also occurs among the vertebrates. The development and functional importance of the liver is typically inversely proportional to the development of adipose tissue (Vague and Fenasse, 1965). Fish illustrate this relationship well. Adipose tissue *per*

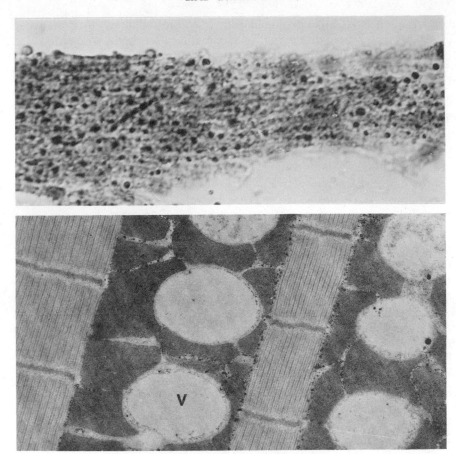

FIGURE 6.6. (Upper) Photomicrograph of longitudinal section of *Triatoma* flight muscle stained with Sudan IV. ×520. (Lower) Electron micrograph of longitudinal section of unflown *Rhodnius* flight muscle. ×22,230. V = vesicle. Reprinted, with permission, from J. P. Ward, D. J. Candy, and S. N. Smith, Lipid storage and changes during flight by triatomine bugs (*Rhodnius prolixus* and *Triatoma infestans*). *J. Insect Physiol.* **28.** Copyright 1982, Pergamon Press, Ltd.

se is present in only limited quantities, and its lipid content is low in comparison with adipose deposits in higher vertebrates. Depending on the species, adipose tissue may be found at the base of fins, along the lateral line, within the skin itself, between muscles, and surrounding the swimbladder. A far more important storage site of lipid in fish is the liver. Some sharks and many bottom-dwelling fish may deposit 75% of their total body fat in the liver, with the liver's fat content approaching 90% (Tashima and Cahill, 1965). The liver is also an important lipid storage site for more pelagic species, although in certain fast-swimming species appreciable lipid reserves are incorporated into muscles as well.

Evidence indicates that lipids are the preferred energy source in fish. In fact, although fish utilize glycogen during swimming bursts, their ability to digest dietary carbohydrates and to metabolize absorbed carbohydrates is limited. Raw starches are only 30 to 40% digestible by salmonids, and digestibility appears to decrease markedly when carbohydrate levels exceed 25% of the ration (Brett and Groves, 1979). Two principal storage forms of neutral lipid are available to fish in their natural environment. Triacylglycerols predominate in fresh-water species, whereas in many marine species (e.g., mackerel, herring, young salmonids), which feed heavily on calanoid copepods (see Figure 6.5), wax esters become a major storage form, accounting for between one-third and two-thirds of their body weight (Cowey and Sargent, 1977). The digestion of wax esters by marine fish poses some special problems, as they are more hydrophobic than triacylglycerols and, hence, more difficult to emulsify. To compensate, the marine species apparently have elevated levels of lipase enzymes, but more importantly possess numerous ceca in their intestinal tract, which increase the retention time for the food and, in so doing, facilitate its digestion (Cowey and Sargent, 1979). As pointed out by the latter authors, this strategy, which relies on time rather than catalytic efficiency, is effective only in animals that feed rather infrequently. In addition to triacylglycerols and wax esters, certain species of sharks utilize the hydrocarbon squalene (Chapter 1) as an energy source. They also replace part of their triacylglycerols with glycerol ethers containing an aliphatic alcohol linked to the glycerol molecule in the alpha position (Chapter 1). The two most widespread glycerol ethers in elasmobranch tissues are batyl and selachyl alcohol.

Although the maintenance energy requirements of fish are quite low, locomotion is energetically very costly (Figure 6.7). A tenfold increase in metabolic rate has been measured in some streamlined species during sustained rapid swimming (Brett and Groves, 1979). Storage lipids are an important fuel source during these times. Their role is perhaps most vividly illustrated by migratory sockeye salmon, which travel hundreds of miles to spawning grounds without eating. Not only does lipid oxidation provide energy for the difficult ascent upriver, the previously stored lipids are also required for the maturation of gonads. Analysis of salmon tissue in individuals before and after a spawning run showed that 91 to 96% of the fat reserve is utilized (Idler and Clemens, 1959). A 50 to 60% depletion of total protein indicates that it too is an important fuel substrate during the migration.

Two distinct types of swimming or myotomal musculature are typically recognized in fish. Red muscle provides the propulsive force during slow- to medium-speed sustained swimming. Its predominantly aerobic nature is evidenced by the presence of numerous mitochondria, an abundance of oxidative enzymes, and a high degree of vascularization. White muscle, which comprises the bulk of the musculature, is primarily geared for anaerobic metabolism. Its major function is to provide energy for more vigorous short-term swimming bursts such as those used to capture prey. The fat content of red muscle is about twice that of white muscle. Moreover,

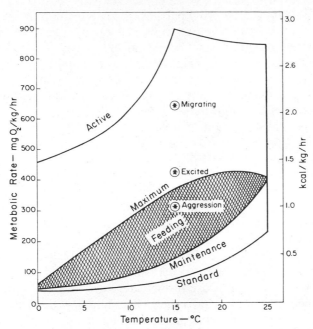

FIGURE 6.7. Rate of energy expenditure of fingerling sockeye salmon showing standard and active metabolic rates (lower and upper lines) in relation to temperature. Metabolic rates associated with feeding (from maintenance to maximum food ration) shown cross-hatched. Maximum oxygen consumption rate observed for aggression and excitement, and calculated rate for lake migration shown for a temperature of 15°C. Reprinted, with permission, from Brett and Groves (1979).

red muscle can oxidize free fatty acids some 10 times more rapidly than can white muscle (Driedzic and Hochachka, 1978). Much of the fat that is stored intracellularly in red muscle is surrounded by mitochondria so that it can be rapidly channeled into the energy-producing reactions of β-oxidation and the TCA cycle. The high degree of vascularization assures that, in addition, an adequate supply of fatty acids from extramusculature storage sites can be delivered to the red muscle fibers. Ultrastructural studies of tuna white muscle indicate that it is unique in having relatively substantial amounts of intracellular lipid as well as higher than usual capillarity and mitochondrial numbers (Hulbert et al., 1979). Although this implies the potential for aerobic metabolism, the ultrastructural and histochemical studies clearly demonstrate that glycogen is still the principal energy and carbon source for tuna white muscle. Heat production resulting from the oxidation of fats by red and white muscle fibers and its contribution to the elevated temperatures of tuna are discussed in Chapter 7.

Amphibians

Amphibians utilize lipids for meeting energy demands associated with the production of gametes, metamorphosis, and metabolic maintenance during periods of dormancy. The principal storage sites include the abdominal fat bodies (corpora adiposa), which in frogs are large lobulated yellowish organs located near the gonads of each sex, and the carcass. In salamanders fat stored in the tail can represent a significant fraction of their total lipid reserve (Fitzpatrick, 1973; Maiorana, 1975). Subcutaneous adipose tissue is poorly developed, if it occurs at all, and the fat content of amphibian liver is much reduced in comparison with fish.

The storage and utilization of lipids for energy by amphibians has been reviewed by Fitzpatrick (1976). The fact that abdominal fat bodies are often smallest immediately after the breeding season is used as evidence for their role in the maintenance and development of gonads. Direct evidence for this role has come from fat body extirpation experiments conducted on salamanders. When both the left and right fat body were removed from *Notophtalmus viridescens,* the testes and ovaries degenerated in fed and unfed individuals (Adams and Rae, 1929). There was some regeneration of the fat bodies in the fed newts. Rose (1967) also observed a significant reduction in ova development in the salamander, *Amphiuma means,* as a result of fat body excision. Rose further suggested that, in addition to stored lipid, fat-soluble vitamins and other micronutrients deposited in the fat bodies were essential for maintenance of the gonads.

The transition from tadpole to adult in anuran amphibians features dramatic changes in morphology, physiology, and biochemistry. The significance of alterations in lipids during amphibian metamorphosis has received little study except for a rather thorough investigation of the frog *Rana tigrina* (Sawant and Varute, 1973). Lipids formed an important constituent of tadpole bodies; the percentage of total and neutral lipids increased during the growth stages and prometamorphosis. During metamorphosis, however, lipid levels decreased significantly. Since this is a period when tadpoles do not feed, it is likely that they use the neutral lipids, especially triacylglycerols, as a source of energy for histolytic events such as degeneration of internal gills and tail regression. That triacylglycerols are likely the preferred fuel substrate is further suggested by the fact that phospholipid levels are not significantly altered during metamorphosis. Thus, the pattern and importance of lipid storage and utilization during amphibian metamorphosis is very similar to that described for insects.

A number of investigators have noted correlations between the variation in the size of fat bodies and winter dormancy or inactivity during unfavorable weather (see Fitzpatrick, 1976). It is generally inferred from these observations that lipid stores in the fat bodies are utilized for metabolic maintenance during these periods, although in amphibians in which vitellogenesis continues during dormancy it is not always possible to quantitatively differentiate between lipid use for gonadal development and for maintenance functions. Seymour (1973) partitioned the energy utilization for these functions in spadefoot toads. One species in his study, *Scaphiopus couchi,* has attracted the interest of many environmental physiologists because of its success in hot dry North American deserts. In Arizona, *S. couchi* becomes active on the surface following summer rains (July), during which time the frogs breed, feed, and rehydrate. After 2 or 3 months on the surface, they burrow and become continuously dormant for 9 to 10 months. When they emerge again in the following summer, the size of their fat bodies and their total body lipid content are greatly diminished from levels prior to burrowing (September) (Figure 6.8). The seasonal changes in lipid content are closely matched to the energy content of the metabolized lipid and the predicted metabolic rate over the dormancy period (Seymour, 1973). In a later study, Jones (1980) showed that there is a shift from lipid to protein metabolism and, hence, increased urea synthesis in *S. couchi* in the latter stages of dormancy when the species becomes osmotically stressed due to the increased drying of the soil microenvironment. By storing urea in the body fluids, a more favorable osmotic relationship is obtained between the toad and surrounding soil.

Reptiles

Reptiles, like amphibians, utilize energy from stored lipids for growth, maintenance, and reproduction. In most lizards and snakes, the bulk of this lipid is contained in the abdominal fat bodies, although some species may

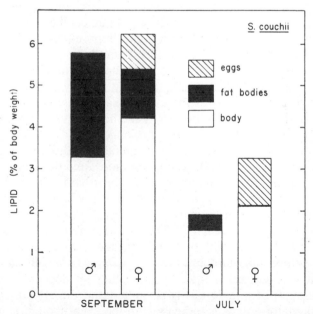

FIGURE 6.8. Lipid distribution in male and female *Scaphiopus couchi* before (September) and after (July) natural dormancy. Reprinted, with permission, from Seymour (1973). Copyright by the American Society of Ichthyologists and Herpetologists.

also store significant quantities of lipid in their tail, liver, muscle, and subcutaneously (Derickson, 1976). Turtles are unique among reptiles in that they do not have organized fat bodies located in close proximity to the gonads, but rather have adipose tissue that is dispersed in small "pads" throughout the carcass (McPherson and Marion, 1982). Triacylglycerols represent the principal lipid in these deposits as well as in lipid-rich reptilian eggs (Hadley and Christie, 1974).

Information on patterns of lipid storage and utilization in reptiles is based largely on the examination of lizard species and to a lesser extent snakes. In Derickson's (1976) review of reptilian life history literature related to lipid utilization, four general patterns were identified. The most common pattern was the use of lipids predominantly, and sometimes even exclusively, for reproduction. Included among the species in this category are the western diamondback and cottonmouth snakes, and the side-blotched lizard and desert iguana. Other species, such as the green anole and common garter snake, apparently utilize their lipid primarily for maintenance energy during winter dormancy. In species such as the ovoviviparous montane lizard *Sceloporus jarrovi,* the size of fat bodies decreased during vitellogenesis and after winter dormancy, indicating that stored lipids are used for both functions. Finally, fat bodies are absent in some tropical species that inhabit frost-free areas and undergo no seasonal cycle of reproductive activity. Derickson (1976) further showed that the factor ultimately responsible for these patterns of lipid storage and utilization is food availability, which in turn is strongly influenced by both precipitation and temperature. The more food available to an organism, the more lipid that can be stored and utilized for reproduction. This can result in a higher reproductive effort in a given season, and thus influence a species' life span (Tinkle and Hadley, 1975).

Many lizard species store significant amounts of lipid in the tail (Figure 6.9). In the gecko *Coleonyx variegatus,* the adaptive significance of tail lipid storage poses some interesting evolutionary questions in that this species readily sacrifices its tail to predators and hence must frequently contend with a substantial loss of stored energy (Congdon et al., 1974). Moreover, the regeneration of the tail requires considerable additional energy, as regenerated tails are often larger and contain more lipid than the original tail. The energy needed for regeneration must be diverted from growth and reproduction. One solution that might provide for both effective predator avoidance and energy conservation is for the gecko to ingest the lost tail (assuming it escapes the predator). This was not observed in experimental trials with *C. variegatus,* although tail ingestion has been found in other species that exhibit tail autotomy (see Vitt et al., 1977). In the case of *C. variegatus* it was suggested that, by allocating fat deposition to the tail, rapid regeneration of a large tail is permitted and the effectiveness of tail display for predator escape is increased. In the event the tail is not lost, an emergency energy store is still available (Congdon et al., 1974). In another tail autotomizing gecko, *Coleonyx brevis,* it was found that tailless females

FIGURE 6.9.   The gecko *Coleonyx variegatus* with thickened tail due to fat deposition. Photograph courtesy of L. J. Vitt.

increased their rate of food ingestion substantially during vitellogenesis to compensate for the loss of caudal reserves (Dial and Fitzpatrick, 1981).

Birds

No vertebrate class exceeds birds in the use and storage of lipid as an energy reserve. The energy derived from the oxidation of triacylglycerol fatty acids is used for migratory flights, for helping to maintain a high metabolic rate during cold winter nights, and for a variety of activities associated with reproduction. The role of lipids in these and other functions has been intensively investigated and is well documented in the literature, making adequate coverage within the scope of this chapter all but impossible. Blem (1976), however, has provided an excellent overview of the patterns of lipid storage and utilization in birds, and I have paralleled his treatment of this subject, updating specific areas wherever possible. Similar comprehensive reviews of more restricted topics concerning lipid storage and utilization in birds can be found in Odum (1965), King (1972), and Ricklefs (1974).

The functional importance of lipid stores is reflected in the first appearance among the vertebrates of a well-developed subcutaneous adipose tissue. Adipose cells form thick layers, initially in regions underlying feather tracts, and eventually around the neck, rump, and in the abdomen. In the fattest birds, all regions of the body, with the exception of the pericardium, exhibit some fat deposition. The greatest accumulation of fat occurs in migratory species, with especially large amounts stored in adipose tissue located beneath the skin and in the abdominal body. Odum (1965) refers to these two sites as the "migratory fat bodies." At peak deposition, long-range migrants may contain 3 g of dry fat for each gram of nonfat body weight, with about one-half of the lipid stored in the migratory fat bodies. Several investigators have reported that not only are migratory birds capable of increasing their body fat by as much as tenfold, the increase occurs without a corresponding increase in fat-free dry weight or relative water content (see references in Blem, 1976). It appears that migratory obesity involves the filling of preexisting adipocytes with dry fat, a process analogous to filling the tanks of an airplane with high-octane fuel (Odum, 1965). By simply adding lipid to preexisting cells, the high energetic cost of producing new cells prior to migratory flights is also avoided.

Pure depot fat is rich in triacylglycerols containing predominantly unsaturated fatty acids such as oleic (18:1), linoleic (18:2), and linolenic acid (18:3). The most common saturated fatty acids are palmitic (16:0) and stearic acid (18:0) (Caldwell, 1973). Distinct geographic trends in the degree of saturation are not apparent, nor are differences between migratory and nonmigratory birds, except for a tendency for migratory species to have a higher proportion of unsaturated fatty acids (Johnson, 1973). There is some evidence that shows that the shift toward greater unsaturation occurs prior to migration, and that this increase may be related to changes in diet, temperature, and photoperiod (Yom-Tov and Tietz, 1978). It has been suggested that a higher proportion of unsaturated fatty acids may permit a more rapid mobilization of the fat store, but experimental evidence for this is lacking. In view of the very narrow size range of acyl chains in stored triacylglycerols (predominantly 16 and 18 carbon acids), differences in energy or water production resulting from the oxidation of saturated versus unsaturated molecules would appear to be negligible, and hence of little adaptive significance to the species.

Migration is one of the most spectacular and energetically costly events in the life history of many birds. It is also an event that perhaps most dramatically illustrates the functional role played by lipids as a fuel storage compound among the vertebrates. Approximately one-third of all the world's species are migratory to some extent, with the number of individuals involved running into the tens of billions. Some long-distance migrants, such as the arctic tern, bobolink, and barn swallow, will cover distances of 7,000 to 10,000 miles in migrating between their summer and winter homes. The energy for these as well as shorter flights is derived almost exclusively from stored fat. It has been estimated that the amount of fat deposited in adipose tissue is sufficient to provide small land birds with enough energy to make nonstop flights of 1000 to 1500 miles at an average flight speed of 25 mph (Odum, 1965). Even the tiny ruby-throated hummingbird, with its very high metabolic rate during flight, is able to store enough fat for a nonstop flight across the Gulf of Mexico. During intracontinental flights, however, many species make periodic stops, during which time they feed and replenish their lipid fuel supplies (King, 1972; Berthold, 1975). In reference to this flight/ feeding pattern, Cherry (1982) found that the amount of fat deposited and the length of stopover for migrant white-crowned sparrows were both significantly related to the amount of fat upon arrival, with lean birds gaining more weight and staying longer than fat birds.

Not only are birds capable of depositing extensive fat stores prior to migration, they accumulate the fat quite quickly. In some small passerines, the entire fattening period may require only 6 to 9 days (King, 1972). The premigratory fat deposition appears to be due largely to increased food intake (hyperphagia), rather than to an adaptive shift in food selection. The amount of fat accumulated is often correlated with a species' migratory pattern. Birds that migrate slowly and forage en route typically carry smaller lipid reserves than species that must fly for extended periods in crossing oceanic and desert barriers. In addition to providing a compact, lightweight, high-energy fuel source, migratory fat deposits have a direct effect on flight performance and efficiency. Pennycuick (1969) analyzed the aerodynamic aspects of flight and fat storage and showed that birds with heavy fuel loads (i.e., stored fat) should fly faster than individuals with a lighter fuel load, assuming the fat-free weight is the same for both groups. The increased flight velocity for heavier birds, plus the reduction in frequency and time needed to replenish fuel supplies during migration, is offered as rationale for why fat birds tend to move faster in migration than lean ones (King, 1972). It is also consonant with the observation that species that tend to store more fat in spring than in fall often exhibit a more rapid vernal migration. The latter may be adaptively significant in that it provides for the earliest possible arrival on the breeding grounds (Blem, 1976).

Seasonal fattening also occurs in nonmigratory birds, especially species that are year-round residents at high latitudes where winter nights are long and cold. Here fat reserves, deposited daily, are indispensable in enabling a bird to survive a night of fasting and to initiate foraging on the following morning (King, 1972). Winter fat storage is much lower than the maximum capacity possible based on accumulation levels exhibited by migratory species. In fact, as indicated in Table 6.2, winter fat storage in small birds is only sufficient to permit survival overnight and for part of the following day. The magnitude of lipid storage in these species may be limited by increased susceptibility to predation because of the weight of the extra fat. Another important consideration here is the fact that other means for storing or conserving energy may modify the extent of fat deposition. Many winter acclimatized birds adapt to colder temperatures behaviorally (huddling, seeking shelters), physiologically (partial hypothermia, food storage in the crop), and morphologically (increased fat insulation), which combined reduce the energetic demands placed on stored fat. The thermogenic capacity and insulative benefits provided by fat depots are discussed in Chapter 7.

Birds, as do other animals, allocate large amounts of energy to reproductive activities such as defense of breeding territories, gonadal growth

TABLE 6.2
Overnight midwinter energetics of several species of birds

Species	Weight (g)	Depot fat (kcal)	Overnight energy requirement (kcal)
Black-capped chickadee	12.0	7.2	6.9
Tree sparrow	20.4	23.3	19.3
Bullfinch	24.8	11.3	8.2
House sparrow			
Florida sparrow	27.1	18.4	10.2
Illinois sparrow	31.0	28.6	15.1
Saskatchewan sparrow	33.5	36.5	22.2
Yellow bunting	29.2	22.3	17.4
Dickcissel	30.6	27.9	6.8
Starling	91.0	108.9	31.7
Willow ptarmigan	620.0	103.6	99.0

SOURCE: Reprinted, by permission, from Blem (1976).

and gametogenesis, incubation, nestling requirements, and parental care. The energy expenditure associated with these activities has been considered by Ricklefs (1974) and Walsberg (1983). Although valid quantitative data on the patterns of energy use during the reproductive cycle are still somewhat limited, there is strong evidence for the importance of lipids as a fuel substrate for most of these reproductive events. Paramount among these perhaps is the energy required for egg synthesis. This energy is a product of egg mass and energy density, clutch size, and the energetic efficiency of biosynthesis (Walsberg, 1983). The mass of a single egg can represent a significant proportion of a female's body mass. Furthermore, the yolk, which contains 99% of the egg lipid, comprises a significant fraction of the egg content. Precocial species tend to have a higher proportion of yolk (and hence lipid) in their eggs than do altricial species (Ricklefs, 1974). The increased fat deposition in the eggs is adaptive in that it helps provide the additional energy for the longer incubation period that is characteristic of precocial birds. The energy demands of egg production are also evident when changes in the mass of females are examined. In wood ducks, which produce energy-rich eggs, the depletion of fat stores in the female is equivalent to about 88% of the requirements for egg formation (Drobney, 1980). In mallards, the female loses approximately 25% of her body fat during egg laying and early incubation (Krapu, 1981). Besides providing a major source of energy for egg synthesis, utilization of fat stores by the female satisfies much of her own energetic requirements. The latter enables the female to concentrate her foraging on invertebrates, which are scarce and thus more energetically costly to obtain than waste grains on agricultural lands, but nevertheless are required to provide essential amino acids needed for protein components of the egg.

Lipid stores are also an important energy source for adult birds during incubation, especially when the incubation involves prolonged periods of fasting. Emperor penguins are a classic example in this regard. The males and females spend several months during the antarctic summer at sea feeding and accumulating extensive lipid deposits. The latter may account for 30% of their body weight by the time they return to sea ice in early April to pair and breed (Groscolas and Clément, 1976). The female lays a single egg about 15 days after copulation, briefly takes full charge of the egg's welfare, and then leaves it with the male while she goes to sea to feed (Le Maho, 1977). During the 45 days from pairing to egg laying the female fasts, relying on energy generated from the oxidation of stored fat. The situation is even more demanding for the male, who now takes over the egg brooding at temperatures approaching $-50°C$ accompanied by stiff winds that significantly add to the chill factor. Incubation lasts for about 65 days. After the egg hatches, the male feeds the young chick for about 10 days by means of an esophageal "curd" until the female returns to assume parental care. Thus, the male must endure some 115 days of fasting under bitter winter conditions. It is estimated that about 93% of his energy requirements during this period are provided by stored fat (Groscolas and Clément, 1976). The problem is compounded in that males must still have enough fat left after incubation to walk the approximately 200 km between the rookery and the open sea before energy reserves can be replenished (Pinshow et al., 1976).

Mammals

Stored lipids play much the same critical role in the life history of mammals as that previously described for birds and most lower vertebrates. The food and energy provided by fats deposited in adipose tissue are used by mammals to survive food shortages and stresses associated with severe weather, competition for mates, territorial defense, gestation and lactation, and in some species to accomplish migrations (Young, 1976). Certainly equally important are functions associated with heat production and insulation, two topics to be covered in Chapter 7.

As in birds, the importance of lipids is reflected in the degree of development of adipose tissue. In addition to often large accumulations of subcutaneous adipose tissue between the skin and superficial fascia, adipose tissue also accumulates in deeper areas of the limbs (i.e., perivascular and intermuscular), in the peritoneal and retroperitoneal areas, within synovial spaces, and in bone marrow (Vague and Fenasse, 1965). The development of adipose tissue, especially subcutaneous, varies with the temperature of the environment and an animal's insulation (see Chapter 7). Certain animals (e.g., ruminants) store lipid in structures such as the hump, rump, or tail, which is then drawn upon during times of starvation. These stores are typically subject to rather rapid fluctuations in amount relative to other lipid deposits within the body. In hibernating mammals there is usually a substantial development of brown adipose tissue, or "brown fat," in addition to a large accumulation of white adipose tissue. The anatomical distribution and functions of brown adipose tissue relative to the hibernation process are also discussed in detail in Chapter 7. Finally, fat distribution is strongly influenced by genetic factors in certain families or races. This is especially true in humans, the one group of mammals for which the anatomical distribution of adipose tissue has and continues to generate the most interest and study. A classic example of the hereditary influence is seen in the ethnic steatopygias of South African Bushmen and Hottentots (Figure 6.10). The protrusion of the buttocks, especially in females, is largely the result of extensive fat deposition which begins at puberty. These fat deposits, which likely evolved in response to frequent periods of inadequate food supplies experienced by these primitive wandering tribes, are thought to provide extra energy during pregnancy and lactation. It is interesting, however, that during weight loss this fat accumulation is resistant to mobilization (Vague and Fenasse, 1965).

Yearly cycles of lipid deposition and utilization associated with seasonal availability of food, reproductive stress, and migration are found in many mammals. The gray whale exhibits rapid weight gain and fat deposition

FIGURE 6.10.   Ethnic steatopygia in a Bushman woman. Photograph courtesy of Phillip V. Tobias.

during summer months prior to its migration from cold arctic waters to the sheltered lagoons along the coast of Baja California, where breeding and calving take place (Rice and Wolman, 1971). During the extensive migration, the whales lose much of their accumulated fat as feeding activity is greatly reduced and food is relatively scarce. Furthermore, reproduction itself creates a drain on energy reserves, for the female whale not only produces a massive calf but also generates tons of milk (Young, 1976). The species begins its northward return with the onset of warmer weather, and by summer is feeding heavily on the abundant plankton in the North Pacific and Arctic Oceans and once again accumulating fat reserves. Young (1976) cites several other studies that have documented the energy requirements of pregnancy and lactation in mammals with respect to lipid utilization. The tendency for females to have more body fat than males enables the former to give birth and nurse the offspring during times of food shortage or when the female is unable to feed. The importance of lipid storage and potential energy has also been applied to human females by Frisch and McArthur (1974) who estimated that 16 kg of stored fat (144,000 kcal) would be sufficient for a pregnancy and three month's lactation. The energetic demands placed upon males are also greater during the breeding period. Spermatogenesis, at least the final stages, often takes place in early spring when food supplies are limited and temperatures are still cold. In addition, males often put on extra weight (fatted male phenomenon) during the mating period, which apparently improves their fighting ability when it comes to establishing dominance as well as serving as a signal of their ability to breed (Young, 1976).

The role of lipids in the biochemistry and physiology of lactation warrants special mention. The milk secreted by female mammary glands is an important source of nutrients and energy for the newborn mammal until the time that it can fend for itself. Lipids are a variable but vital component of this milk. In human and cow's milk, lipid accounts for only 4% of the milk by weight, but this value approaches 50% in some marine mammals (Dils, 1983). The high energy content of seal milk enables the pups to put on a thick layer of subcutaneous fat, and hence cope with low environmental temperatures during their development (see Chapter 7). Triacylglycerols are the main lipid constituent in milk; they are secreted in the form of small globules (0.1 to 20 μm in diameter) that are surrounded by a unit membrane.

Most mammals produce a milk fat that contains substantial amounts of short- and medium-chain fatty acids. The *de novo* synthesis of these unique fatty acids was described in Chapter 2. Fatty acids obtained from the diet and/or hydrolyzed from body fat stores and transported to the mammary gland by the blood are also incorporated into milk fat. The influence of diet and other factors on the fatty acid composition of milk fat varies among the different groups of mammals. In nonruminants, the nature of the diet can have a marked effect on milk fat composition. The high content of palmitoleic acid (16:1) in the milk of marine mammals reflects the abundance of this compound in the plankton and fish upon which they feed. Ruminant milk, in contrast, appears to be little affected by dietary factors. The high content of butyric acid (4:0) and the predominantly saturated nature of long-chain fatty acids reflects the action of the intestinal flora in these animals (Smith and Abraham, 1975). Even the addition of unsaturated fatty acids to the diet of ruminants has comparatively little effect on the composition of the milk because of the extensive biohydrogenation that occurs in the rumen. For this reason commercial infant formulas based on cow's milk are usually supplemented with vegetable oils to insure that an adequate level of the essential fatty acid linoleic acid (18:2) is present. Interestingly, ruminant neonates have adjusted their metabolic processes so that they grow normally despite low levels of this essential acid (Noble, 1979).

The functional significance of short- and medium-chain acids in the metabolism of a newborn mammal is not fully understood. Dils (1983) suggests that it may be advantageous to the neonate that milk triacylglycerols containing such fatty acids are hydrolyzed more rapidly by lipases, and that the released fatty acids are transported directly to the liver via the portal vein rather than being incorporated into chylomicra. These short-chain fatty acids would also be readily absorbed by the young mammal. Another possibility is that short- and medium-chain fatty acids are used for ketogenesis by the liver of neonates to provide a source of energy during this period of rapid development. The presence of short- and medium-chain fatty acids also has important consequences for the physical state of the milk fat molecule. In ruminants, butyric and hexanoic acids are esterified exclusively to position 3 of glycerol, with high proportions of palmitic acid in positions 1 and 2 (Christie, 1979). The presence and asymmetric distribution of short-chain fatty acids on the glycerol moiety give the milk fat a sufficiently low melting point to be readily secreted as liquid droplets. Thus, these short-chain fatty acids function in a capacity normally served by polyunsaturated fatty acids, which are not readily available to the ruminant mammary gland (Moore and Christie, 1979).

A seasonal pattern of lipid cycling is also observed in mammals that overwinter in cold environments. Although some mammals that cache food in their winter hibernacula or in accessible places often also increase their fat reserves in late summer and autumn, the most pronounced changes in lipid deposition and utilization occur in the true hibernators, such as ground squirrels, hedgehogs, and dormice. These species exhibit a marked decrease in body temperature (usually to 4 to 5°C) and metabolic rate, reduced activity, usually a curled posture (except bats), plus the ability to periodically spontaneously rewarm to normal body temperatures. Heat production associated with the arousal periods is strongly linked to oxidization of brown fat, and is considered in the next chapter; coverage here is restricted to the utilization of fat reserves to provide sufficient energy to support those metabolic activities that continue at the low body temperatures.

An overview of the major patterns and events regarding lipid usage in mammalian hibernation is given in Mrosovsky (1976). The prehibernation phase of fat deposition is linked to a period of hyperphagia similar to that described for birds preparing for migratory flights. In mammals food intake apparently slackens well before hibernators reach their maximum weight, suggesting that other factors, such as decreased motor activity, lower body temperatures, and perhaps changes in feeding efficiency, also contribute to the prehibernation obesity. Other pertinent observations are that the rate of weight (fat) gain is not maximal, and that the extent of fat accumulation is apparently regulated at specific levels referred to as set points (Mrosovsky, 1976). While the adaptive value of prehibernation fattening is quite obvious, why hibernating species refuse to eat when food is readily available is not so clear. It has been proposed that anorexia prior to and during hibernation, even though food may be available, reduces the rate at which metabolic end-products accumulate, and thus lengthens the duration of hibernation bouts, as these end-products (including metabolic water) are usually eliminated during the arousal periods (Mrosovsky and Barnes, 1974). This likely requires that the hibernator have some mechanism for lowering its set point to prevent compensatory eating as some weight loss is inevitable. Nonetheless, despite the heavy reliance on stored fat during the hibernation period, most mammals still possess a substantial fat reserve at its termination. For example, squirrels and badgers deplete only about one-half to two-thirds of their stored fat during winter hibernation (Fisher and Manery, 1967; Harlow, 1981), and hence have a fuel source for early spring activities.

Although it is clearly evident that fat is the principal fuel source for hibernating mammals, temporal changes in intermediary metabolism related to fat usage have received relatively little study. For example, details on seasonal changes in the mechanisms of or the capacity for lipolysis and cellular fatty acid degradation are essentially lacking (Willis, 1982). There are data that show that the size and number of mitochondria of the heart of ground squirrels increase during the hibernation season (fat metabolism is an oxidative process), and that the number of fat droplets near mitochondria also increase, but these changes are not uniform among mammalian hibernators (see references in Willis, 1982). The dependence on body fat as an energy source by hibernators does pose an interesting biochemical problem for tissues such as the brain and renal medulla, which rely solely or primarily on glucose metabolism. Since the capacity to store glycogen is limited, mammalian hibernators must depend on gluconeogenesis for this substrate or have these glucose-dependent tissues adapt so that other substrates are acceptable. Of the three main substrates used in gluconeogenesis (amino acids, lactate, glycerol), only glycerol seems to be very important in the resynthesis of carbohydrates by hibernators. Galster and Morrison (1975) estimated that sufficient glycerol was produced from the catabolism of triacylglycerols to restore as much as two-thirds of the carbohydrate reserve depleted during a single bout of hibernation. During starvation associated with hibernation, glucose reserves may be spared by using ketones as a primary source of energy. In the arctic ground squirrel, all tissues showed an elevation of ketone bodies (acetoacetate and β-hydroxybutyrate) during the hibernating state (Table 6.3). There is some evidence that brain tissue in these ground squirrels and other hibernators may be capable of utilizing ketone bodies directly, thus further sparing limited glucose reserves.

No coverage of winter hibernation and fat utilization would be complete without at least brief mention of bears. The fact that their body temperature decreases only slightly below euthermic levels during dormancy does not qualify bears as "true" hibernators. However, they still have an extensive fattening phase in late summer and early autumn, and then rely on stored fats for energy during their "winter sleep." During the latter, which may last for over 100 days, bears do not eat, urinate, or defecate. [The belief held by some Indian tribes that they obtain nourishment by sucking their feet is in reality a response to tenderness caused by shedding of keratinous fat pads (Rogers, 1974).] Urea production is significantly reduced as a result of the exclusive metabolism of fat, which does not generate nitrogenous end-products. A bear in summer cannot survive starvation without using protein as well as fat as a source of energy. This results in extensive urine formation, ketosis, and dehydration (Nelson et al., 1975).

Seasonal variation in body fat reserves is by no means restricted to temperate mammals. Many tropical and subtropical desert species metabolize fat during times of food shortage to generate both energy and metabolic

TABLE 6.3

Acetoacetate and β-hydroxybutyrate concentrations (μmoles per g wet tissue) of selected organs of nonhibernating and hibernating arctic ground squirrels. Mean values for the two physiological states are statistically different at $P < 0.0001$ in all cases

Organ	Physiological state	Acetoacetate (μmol g^{-1})		β-Hydroxybutyrate (μmol g^{-1})	
		Mean ± SE	n	Mean ± SE	n
Blood	Nonhibernating	0.0293 ± 0.0027	9	0.0536 ± 0.0061	7
	Hibernating	0.2475 ± 0.0110	10	1.2317 ± 0.0585	10
Liver	Nonhibernating	0.0385 ± 0.0020	9	0.1164 ± 0.0103	9
	Hibernating	0.2465 ± 0.0098	12	0.9062 ± 0.0487	8
Cecum	Nonhibernating	0.0388 ± 0.0022	8	0.1506 ± 0.0075	8
	Hibernating	0.2186 ± 0.0192	10	0.8647 ± 0.0494	9
Heart	Nonhibernating	0.0335 ± 0.0022	8	0.0736 ± 0.0035	8
	Hibernating	0.1122 ± 0.0053	8	0.2795 ± 0.0310	6
Brown adipose tissue	Nonhibernating	0.0370 ± 0.0029	6	0.1033 ± 0.0079	6
	Hibernating	0.1827 ± 0.0101	17	0.4680 ± 0.0217	16

SOURCE: From Rauch and Behrisch (1981). Reproduced by permission of the National Research Council of Canada from the *Canadian Journal of Zoology*, Volume 59, 1981.

water. These species typically do not deposit extensive amounts of fat in their subcutaneous tissues as do polar and temperate species, as this would interfere with heat dissipation. Instead, they circumvent this problem by storing the fat in specialized structures such as the hump (camel) or tail (sheep). The camel's fat-laden hump may account for up to 20% of its body weight when food is plentiful (Macfarlene, 1964). Many small desert rodents also accumulate large amounts of depot fat, which they use during hot dry summer months when food intake is reduced (Hayward, 1965; McNab, 1968; Goyal et al., 1981). It is generally agreed by investigators that fat serves as an energy buffer against starvation and that it is only indirectly associated with the problem of water restriction. Seasonal variation in fat reserves even occurs in bats inhabiting tropical rain forests. McNab (1976) found that the annual amplitude in fat reserves was strongly influenced by the food habits of the bats, with insectivorous species showing the greatest amplitude because of the decline of flying insects during the dry season.

CONCLUDING REMARKS

The high energy content, compactness, weight economy, and potential yield of metabolic water make lipids, especially triacylglycerols, an ideal fuel storage compound for plants and animals. For lipids to function in this manner, they must be synthesized and deposited during times of plentiful food, and then metabolized at a rate that is consonant with the storage capacity. This requires rather stringent control mechanisms of which our understanding is far from complete. There is no question that hormones play an important role in controlling fat deposition and mobilization on both a daily and seasonal basis, and that these hormones may be released in conjunction with circadean rhythms. For example, several laboratories have demonstrated daily variations in fattening responses to prolactin in vertebrates, with the daily rhythm entrained by the daily photoperiod. Moreover, photoperiodic entrainment appears to be mediated by adrenal corticosteroids (Meier and Burns, 1976). How these neuroendocrine events organize the total animal so that its metabolism, reproduction, and behavior are fully integrated remains to be discovered.

An important underlying theme in this chapter, which relates to control mechanisms, is the ability of an organism to accumulate fat stores when they have an adaptive function and *eliminate* them at other times. In other words, heavy deposits of fat can be disadvantageous, both from an ecological and physiological standpoint. The extensive premigratory fat deposition in birds is accompanied by vigorous flight activity during which time these fuel reserves are largely depleted. A similar degree of fat deposition at other times in their life history would be detrimental in that it would likely impede their locomotion, and hence increase the risk of predation. As noted by Meier and Burns (1976), "the inverse relation between locomotor activity and amount of body fat that is often assumed for humans does not even superficially exist in migrants.''

Examination of the delicate balance between fat deposition and usage in humans uncovers some interesting and perplexing evolutionary and immediate problems. Human adipose tissue in an individual with a normal weight contains a reserve of energy equivalent to 40 days or more of active metabolism; in a moderately obese individual this energy reserve increases to a year or more of normal metabolism (Dole, 1965). While this reserve may have been important for the survival of humanity under primitive conditions (and still is in some cultures), in modern society, which has largely eliminated the seasonality associated with food supply, this large deposit seems to have little function in daily metabolism. Instead, adiposity (obesity) is now closely linked with physiological ills and undesirable cosmetic consequences. At no other time has the interest been so great in why some people get fat on the same amount of food that keeps others thin, and why it is so difficult for one to rid the body of excess fat once accumulated.

A partial answer to the above may be found when the situation in humans is compared with that in birds. Premigratory fat deposition in birds is accomplished by filling existing cells with fat without actually increasing the number of fat cells. In humans, not only does the size of fat cells increase when calorie input exceeds output, but at a certain point these cells begin to multiply, reaching levels five times the normal cell number in some parts of the body. The key factor in this problem is that the process has never been known to work in reverse; fat cells may shrink but they never disappear (Adler and Gosnell, 1982). Thus, it is easy to gain weight if you already have a lot of empty cells to put it in.

The chemical basis of lipid utilization and its complex controlling mechanisms in humans are probably very similar to those operational in higher vertebrates. A set point, proposed for migratory birds and hibernating mammals, likely functions in humans to stabilize weight for long periods within a fairly narrow range. Although it has not been established if a person's set point is inborn, it has been shown that it can be lowered by exercise and increased by consumption of carbohydrate-rich diets (i.e., sweets). The problem of weight gain and loss is obviously complex and is not likely to be unraveled in the immediate future; however, answers are needed if much of modern society is going to benefit from lipid fuel storage and energy production in a manner similar to that exhibited by most other organisms.

REFERENCES

Adams, A. D. and E. E. Rae (1929). An experimental study of the fat bodies in *Triturus (Diemyctylus) viridescens. Anat. Rec.* **41**:181–203.

Adler, J. and M. Gosnell (1982). What it means to be fat. *Newsweek* (December):84–90.

Agrell, I. P. S. and A. M. Lundquist (1973). Physiological and biochemical changes during insect development. In: *The Physiology of Insecta,* Vol. 1 (Rockstein, M., Ed.), pp. 159–247. Academic Press, New York.

Barrett, J., C. Ward, and D. Fairbairn (1970). The glyoxylate cycle and the conversion of triglycerides to carbohydrates in developing eggs of *Ascaris lumbricoides. Comp. Biochem. Physiol.* **35**:577–586.

Bartholomew, G. A. and T. M. Casey (1978). Oxygen consumption of moths during rest, pre-flight warm-up, and flight in relation to body size and wing morphology. *J. Exp. Biol.* **76**:11–25.

Beenakkers, A. M. T., D. J. Van der Horst, and W. J. A. Van Marrewijk (1981). Role of lipids in energy metabolism. In: *Energy Metabolism in Insects* (Downer, R. G. H., Ed.), pp. 53–100. Plenum, New York.

Beevers, H. (1980). The role of the glyoxylate cycle. In: *The Biochemistry of Plants,* Vol. 4 (Stumpf, P. K., Ed.), pp. 117–130. Academic Press, New York.

Benson, A. A. and R. F. Lee (1975). The role of wax in oceanic food chains. *Sci. Am.* **232**:76–86.

Berry, M. N., R. B. Gregory, A. R. Grivell, and P. G. Wallace (1983). Compartmentation of fatty acid oxidation in liver cells. *Eur. J. Biochem.* **131**:215–222.

Berthold, P. (1975). Migration: Control and metabolic physiology. In: *Avian Biology* (Farner, D. S. and J. R. King, Eds.), pp. 77–128. Academic Press, New York.

Blem, C. R. (1976). Patterns of lipid storage and utilization in birds. *Am. Zool.* **16**:671–684.

Brett, J. R. and T. D. Groves (1979). Physiological energetics. In: *Fish Physiology,* Vol. VIII (Hoar, W. S., D. J. Randall, and J. R. Brett, Eds.), pp. 279–352. Academic Press, New York.

Bursell, E. (1981). The role of proline in energy metabolism. In: *Energy Metabolism in Insects* (Downer, R. G. H., Ed.), pp. 135–154. Plenum, New York.

Caldwell, L. D. (1973). Fatty acids of migrating birds. *Comp. Biochem. Physiol.* **44B**:493–497.

Candy, D. J. (1980). Biological functions of carbohydrates. Wiley, New York.

Cherry, J. D. (1982). Fat deposition and length of stopover of migrant white-crowned sparrows. *Auk* **99**:725–732.

Christie, W. W. (1979). The composition, structure, and function of lipids in the tissues of ruminant animals. *Progr. Lipid Res.* **17**:111–205.

Cioni, M., G. Pinzauti, and P. Vanni (1981). Comparative biochemistry of the glyoxylate cycle. *Comp. Biochem. Physiol.* **70B**:1–26.

Congdon, J. D., L. J. Vitt, and W. W. King (1974). Geckos: Adaptive significance and energetics of tail autotomy. *Science* **184**:1379–1380.

Cooper, T. G. and H. Beevers (1969). β-Oxidation in glyoxysomes from castor bean endosperm. *J. Biol. Chem.* **244**:3514–3520.

Cowey, C. B. and J. R. Sargent (1977). Lipid nutrition in fish. *Comp. Biochem. Physiol.* **57B**:269–273.

Cowey, C. B. and J. R. Sargent (1979). Nutrition. In: *Fish Physiology,* Vol. VIII (Hoar, W. S., D. J. Randall, and J. R. Brett, Eds.), pp. 1–69. Academic Press, New York.

Derickson, W. K. (1976). Lipid storage and utilization in reptiles. *Am. Zool.* **16**:711–723.

Dial, B. E. and L. C. Fitzpatrick (1981). The energetic costs of tail autotomy to reproduction in the lizard *Coleonyx brevis* (Sauria: Gekkonidae). *Oecologia (Berl)* **51**:310–317.

Dils, R. R. (1983). Milk fat synthesis. In: *Biochemistry of Lactation* (Mepham, T. B., Ed.), pp. 141–157. Elsevier, Amsterdam.

Dole, V. P. (1965). Energy storage. In: *Handbook of Physiology,* Sec. 5: *Adipose Tissue* (Renold, A. E. and G. F. Cahill, Jr., Eds.), pp. 13–18. Am. Physiol. Soc., Washington, D.C.

Douce, R. and J. Joyard (1980). Plant galactolipids. In: *The Biochemistry of Plants,* Vol. 4 (Stumpf, P. K., Ed.), pp. 321–362. Academic Press, New York.

Downer, R. G. H. and J. R. Matthews (1976). Patterns of lipid distribution and utilisation in insects. *Am. Zool.* **16**:733–745.

Driedzic, W. R. and P. W. Hochachka (1978). Metabolism in fish during exercise. In: *Fish Physiology,* Vol. VII (Hoar, W. S. and D. J. Randall, Eds.), pp. 503–543. Academic Press, New York.

Drobney, R. D. (1980). Reproductive bioenergetics of wood ducks. *Auk* **97**:480–490.

Edney, E. B. (1977). *Water Balance in Land Arthropods.* Springer-Verlag, Berlin.

Falk-Petersen, S., R. R. Gatten, J. R. Sargent, and C. C. E. Hopkins (1981). Ecological investigations on the zooplankton community in Balsfjorden, Northern Norway: Seasonal changes in the lipid class composition of *Meganyctiphanes norvegica* (M. Sars), *Thysanoessa raschii* (M. Sars), and *T. inermis* (Krøyer). *J. Exp. Mar. Biol. Ecol.* **54**:209–224.

Fisher, K. C. and J. F. Manery (1967). Water and electrolyte metabolism in heterotherms. In: *Mammalian Hibernation III* (Fisher, K. C., A. R. Dawe, C. P. Lyman, E. Schönbaum, and F. E. Souths, Eds.), pp. 235–279. Oliver and Boyd, London.

Fitzpatrick, L. C. (1973). Energy allocation in the Allegheny Mountain salamander *Desmognathus ochrophaeus. Ecol. Monogr.* **43**:43–58.

Fitzpatrick, L. C. (1976). Life history patterns of storage and utilization of lipids for energy in amphibians. *Am. Zool.* **16**:725–732.

Frisch, R. E. and J. W. McArthur (1974). Menstrual cycles: Fatness as a determinant of minimum weight and height necessary for their maintenance or onset. *Science* **185**:949–951.

Galliard, T. (1980). Degradation of acyl lipids: Hydrolytic and oxidative enzymes. In: *The Biochemistry of Plants,* Vol. 4 (Stumpf, P. K., Ed.), pp. 85–116. Academic Press, New York.

Galster, W. A. and P. Morrison (1975). Gluconeogenesis in arctic ground squirrels between periods of hibernation. *Am. J. Physiol.* **228**:325–330.

Gilbert, L. I. (1967). Lipid metabolism and function in insects. *Adv. Insect Physiol.* **4**:69–211.

Goyal, S. P., P. K. Ghosh, and I. Prakash (1981). Significance of body fat in relation to basal metabolic rate in some Indian desert rodents. *J. Arid Environ.* **4**:59–62.

Groscolas, R. and C. Clément (1976). Utilisation des reserves énergétiques au cours du jeûne de la reproduction chez le Manchot empereur, *Aptěnodytes forsteri.* Comp. Rend. Séances Acad. Sci. Paris **282**:297–300.

Gurr, M. I. and A. T. James (1975). *Lipid Biochemistry: An Introduction,* 2nd ed. Chapman and Hall, London. 244 pp.

Hadley, N. F. and W. W. Christie (1974). The lipid composition and triglyceride structure of eggs and fat bodies of the lizard *Sceloporus jarrovi. Comp. Biochem. Physiol.* **48B**:275–284.

Harlow, H. J. (1981). Torpor and other physiological adaptations of the badger (*Taxidea taxus*) to cold environments. *Physiol. Zool.* **54**:267–275.

Harwood, J. L. (1980). Sulfolipids. In: *The Biochemistry of Plants,* Vol. 4 (Stumpf, P. K., Ed.), pp. 301–320. Academic Press, New York.

Hayward, J. S. 1965. Metabolic rate and its temperature-adaptive significance in six geographic races of *Peromyscus. Can J. Zool.* **43**:309–323.

Hitchcock, C. and B. W. Nichols (1971). *Plant Lipid Biochemistry.* Academic Press, New York.

Hochachka, P. W. and G. N. Somero (1973). *Strategies of Biochemical Adaptation.* W. B. Saunders, Philadelphia.

Huang, A. H. C., R. N. Trelease, and T. S. Moore, Jr. (1983). *Plant Peroxisomes.* Academic Press, New York.

Hulbert, W. C., M. Guppy, B. Murphy, and P. W. Hochachka (1979). Metabolic sources of heat and power in tuna muscles I. Muscle fine structure. *J. Exp. Biol.* **82**:289–301.

Idler, D. R. and W. A. Clemens (1959). The energy expenditure of Fraser River sockeye salmon during the spawning migration. *Int. Pac. Salmon. Fish. Comm., Prog. Rep.* 80 pp.

Johnson, D. W. (1973). Cytological and chemical adaptations of fat deposition in migratory birds. *Condor* **75**:108–113.

Jones, R. M. (1980). Metabolic consequences of accelerated urea synthesis during seasonal dormancy of spadefoot toads, *Scaphiopus couchi* and *Scaphiopus multiplicatus. J. Exp. Zool.* **212**:255–267.

Kammer, A. E. and B. Heinrich (1978). Insect flight metabolism. *Adv. Insect Physiol.* **13**:133–228.

Kilby, B. A. (1963). The biochemistry of the insect fat body. *Adv. Insect Physiol.* **1**:111–174.

King, J. R. (1972). Adaptive periodic fat storage by birds. *Proc. XVth Int. Ornithol. Congr.* pp. 200–217.

Krapu, G. L. (1981). The role of nutrient reserves in mallard reproduction. *Auk* **98**:29–38.

Lawrence, J. M. (1976). Patterns of lipid storage in post-metamorphic marine invertebrates. *Am. Zool.* **16**:747–762.

Lazarow, P. and C. de Duve (1976). A fatty acyl-CoA oxidizing system in rat liver peroxisomes; Enhancement by clofibrate, a hypolipidemic drug. *Proc. Nat. Acad. Sci. USA* **73**:2043–2046.

Lee, R. F. (1974). Lipid composition of the copepod *Calanus hyperboreus* from the Arctic Ocean. Changes with depth and season. *Mar. Biol.* **26**:313–318.

Lehninger, A. L. (1975). *Biochemistry,* 2nd ed. Worth Publishers, New York.

Le Maho, Y. (1977). The emperor penguin: A strategy to live and breed in the cold. *Am. Sci.* **65**:680–693.

Loveridge, J. P. and E. Bursell (1975). Studies on the water relations of adult locusts (Orthoptera, Acrididae). I. Respiration and the production of metabolic water. *Bull. Ent. Res.* **65**:13–20.

Macfarlene, W. V. (1964). Terrestrial animals in dry heat: Ungulates. In: *Handbook of Physiology,* Sec. 4: *Adaptation to the Environment* (Dill, D. B., E. F. Adolph, and C. G. Wiber, Eds.), pp. 509–539. Am Physiol. Soc., Washington, D.C.

Maiorana, V. C. (1975). Studies in the behavioral ecology of the plethodontid salamander *Batrachoseps attenuatus.* Ph.D. Dissertation, Univ. of Calif., Berkeley, California.

Martin, M. M. and T. J. Lieb (1979). Patterns of fuel utilization by the thoracic muscles of adult worker ants. The use of lipid by a Hymenopteran. *Comp. Biochem. Physiol.* **64B**:387–390.

Masoro, E. J. (1968). *Physiological Chemistry of Lipids in Mammals.* W. B. Saunders, Philadelphia.

McNab, B. K. (1968). The influence of fat deposits on the basal rate of metabolism in desert homoiotherms. *Comp. Biochem. Physiol.* **26**:337–343.

McNab, B. K. (1976). Seasonal fat reserves of bats in two tropical environments. *Ecology* **57**:332–338.

McPherson, R. J. and K. R. Marion (1982). Seasonal changes of total lipids in the turtle *Sternotherus odoratus. Comp. Biochem. Physiol.* **71A**:93–98.

Meier, A. H. and J. T. Burns (1976). Circadean hormone rhythms in lipid regulation. *Am. Zool.* **16**:649–659.

Montgomery, R., R. L. Dryer, T. W. Conway, and A. A. Spector (1974). *Biochemistry.* C. V. Mosby, St. Louis.

Moore, J. H. and W. W. Christie (1979). Lipid metabolism in the mammary gland of ruminant animals. *Progr. Lipid Res.* **17**:347–395.

Moreau, R. A. and A. H. C. Huang (1977). Gluconeogenesis from storage wax in the cotyledons of jojoba seedlings. *Plant Physiol.* **60**:329–333.

Mrosovsky, N. (1976). Lipid programmes and life strategies in hibernators. *Am. Zool.* **16**:685–697.

Mrosovsky, N. and D. S. Barnes (1974). Anorexia, food deprivation and hibernation. *Physiol. Behav.* **12**:265–270.

Neat, C. E., M. S. Thomassen, and H. Osmundsen (1981). Effects of high-fat diets on hepatic fatty acid oxidation in the rat. *Biochem. J.* **196**:149–159.

Needham, A. E. (1965). *The Uniqueness of Biological Materials.* Pergamon Press, London.

Nelson, R. A., J. D. Jones, H. W. Wahner, D. B. McGill, and C. R. Code (1975). Nitrogen metabolism in bears: Urea metabolism in summer starvation and in winter sleep and role of urinary bladder in water and nitrogen conservation. *Mayo Clin. Proc.* **50**:141–146.

Noble, R. C. (1979). Lipid metabolism in the neonate ruminant. *Progr. Lipid Res.* **18**:179–216.

Odum, E. P. (1965). Adipose tissue in migratory birds. In: *Handbook of Physiology,* Sec. 5: *Adipose Tissue* (Renold, A. E. and G. F. Cahill, Eds.), pp. 37–43. Am. Physiol. Soc., Washington, D.C.

Patton, R. L. (1963). *Introductory Insect Physiology.* W. B. Saunders, Philadelphia.

Pennycuick, C. J. (1969). The mechanics of bird migration. *Ibis* **111**:525–570.

Pinshow, B., M. A. Fedak, D. R. Battles, and K. Schmidt-Nielsen (1976). Energy expenditure for thermoregulation and locomotion in emperor penguins. *Am. J. Physiol.* **231**:903–912.

Pitts, C. W. and T. L. Hopkins (1965). Lipid composition of hibernating face flies. *Proc. North Cent. Branch Entomol. Soc. Am.* **20**:72–73.

Rauch, J. C. and H. W. Behrisch (1981). Ketone bodies: A source of energy during hibernation. *Can. J. Zool.* **59**:754–760.

Rice, D. W. and A. A. Wolman (1971). The life history and ecology of the gray whale (*Eschrichtius robustus*). Spec. Publ. #3, Am. Soc. Mammal.

Ricklefs, R. E. (1974). Energetics of reproduction in birds. In: *Avian Energetics* (Paynter, R. A., Ed.), pp. 152–297. *Pub. Nuttall Ornithol. Club,* No. 15.

Rogers, L. L. (1974). Shedding of foot pads by black bears during denning. *J. Mammal.* **55**:672–674.

Rose, F. L. (1967). Seasonal changes in lipid levels of the salamander *Amphiuma means. Copeia* **1967**:662–666.

Sacktor, B. (1975). Utilization of fuels by muscle. In: *Insect Biochemistry and Function* (Candy, D. J. and B. A. Kilby, Eds.), pp. 1–81. Chapman and Hall, London.

Sargent, J. R., R. F. Lee, and J. C. Nevenzel (1976). Marine waxes. In: *Chemistry and Biochemistry of Natural Waxes* (Kolattukudy, P. E., Ed.), pp. 49–91. Elsevier, Amsterdam.

Sawant, V. A. and A. T. Varute (1973). Lipid changes in the tadpoles of *Rana tigrina* during growth and metamorphosis. *Comp. Biochem. Physiol.* **44B**:729–750.

Schmidt-Nielsen, K. (1964). *Desert Animals.* Oxford Univ. Press, London.

Schmidt-Nielsen, K. (1979). *Animal Physiology: Adaptation and Environment,* 2nd ed. Cambridge Univ. Press, New York.

Seymour, R. S. (1973). Energy metabolism of dormant spadefoot toads (*Scaphiopus*). *Copeia* **1973**:435–445.

Smith, S. and S. Abraham (1975). The composition and biosynthesis of milk fat. *Adv. Lipid Res.* **13**:195–239.

Stryer, L. (1975). *Biochemistry.* W. H. Freeman, San Francisco.

Tashima, L. and G. F. Cahill, Jr. (1965). Fat metabolism in fish. In: *Handbook of Physiology,* Sec. 5: *Adipose Tissue* (Renold. A. E. and G. F. Cahill, Jr., Eds.), pp. 55–58. Am. Physiol. Soc., Washington, D.C.

Tinkle, D. W. and N. F. Hadley (1975). Lizard reproductive effort: Caloric estimates and comments on its evolution. *Ecology* **56**:427–434.

Vague, J. and R. Fenasse (1965). Comparative anatomy of adipose tissue. In: *Handbook of Physiology,* Sec. 5: *Adipose Tissue* (Renold, A. E. and G. F. Cahill, Jr., Eds.), pp. 25–36. Am. Physiol. Soc., Washington, D.C.

Valder, S. M., T. L. Hopkins, and S. A. Valder (1969). Diapause induction and changes in lipid composition in diapausing and reproducing faceflies. *J. Insect Physiol.* **15**:1199–1214.

Vitt, L. J., J. D. Congdon, and N. A. Dickson (1977). Adaptive strategies and energetics of tail autotomy in lizards. *Ecology* **58**:326–337.

Walsberg, G. E. (1983). Avian ecological energetics. In: *Avian Biology,* Vol. 7 (Farner, D. S. and J. R. King, Eds.), pp. 161–220. Academic Press, New York.

Ward, J. P., D. J. Candy, and S. N. Smith (1982). Lipid storage and changes during flight by triatomine bugs (*Rhodnius prolixus* and *Triatoma infestans*). *J. Insect Physiol.* **28**:527–534.

Wigglesworth, V. B. (1965). *The Principles of Insect Physiology,* 6th ed. Methuen and Co., London.

Willis, J. S. (1982). Intermediary metabolism in hibernation. In: *Hibernation and Torpor in Mammals and Birds* (Lyman, C. P., J. S. Willis, A. Malan, and L. C. H. Wang), pp. 124–139. Academic Press, New York.

Yom-Tov, Y. and A. Tietz. (1978). The effect of diet, ambient temperature and day length on the fatty acid composition in the depot fat of the European starling (*Sturnus vulgaris*) and the rock partridge (*Alectoris chucar*). *Comp. Biochem. Physiol.* **60A**:161–164.

Young, R. A. (1976). Fat, energy and mammalian survival. *Am. Zool.* **16**:699–710.

Zimmerman, D. C. and H. J. Klosterman (1965). Lipid metabolism in germinating flaxseed. *J. Am. Oil Chem. Soc.* **42**:58–62.

7

HEAT PRODUCTION
AND THERMAL INSULATION

In the preceding chapter we saw how energy derived from the oxidation of lipids is used by organisms to perform a variety of biological activities, many of which are essential for the survival of the species. Using palmitic acid (16:0) as an example, it was shown that 129 ATP, or approximately 942 kcal per mole, were formed in its complete oxidation to carbon dioxide and water. The conservation of energy as ATP, however, is not 100% efficient. In fact, only about 40% of the standard free energy of oxidation of palmitic acid is recovered as high energy phosphate—the remaining 60% is lost as heat, a percentage similar to that observed in glycolysis, the citric acid cycle, and oxidative phosphorylation. For most organisms, heat generated during the oxidative metabolism of foodstuffs is considered to be wasted energy, as actual amounts are quite small and most heat that is produced is rapidly lost to the surrounding environment. The situation for most birds and mammals is quite different. Here significantly higher metabolic rates, plus mechanisms for retaining heat, enable these two vertebrate groups to maintain a warm and relatively stable body temperature that is independent of ambient conditions. Certainly heat production from lipid metabolism is an important factor in the development of homeothermy and, as will be shown, plays an especially important role when these species are subjected to cold temperatures.

As indicated above, increased heat production via the oxidative degradation of lipids or any other foodstuff is effective in helping maintain high body temperatures in cold environments only if the thermogenic capacity is complemented with mechanisms for conserving heat. In birds and mammals such mechanisms may include behavior (huddling, seeking milder

194

microclimates), morphological adjustments to reduce the amount of exposed body surface, vasomotor changes that reduce blood flow to the extremities or peripheral parts of the body, and increased insulation. Of these mechanisms, only insulation relates to lipids, but their role here can be extremely important. Much of the stored lipid in adipose tissue is subcutaneous, and thus is anatomically interposed between the body core and the skin. Because the heat conductivity of fat is very low (only about one-third that of water), adipose tissue makes an ideal insulating material (Masoro, 1968). Moreover, any changes in the amount of subcutaneous adipose tissue can significantly influence the rate of heat loss.

Obviously the calorigenic and thermogenic (heat + insulation) properties of lipids are closely coupled. Any time lipids are oxidized to provide chemical energy in the form of ATP, heat is generated as a by-product. The decision to discuss these functions separately was made in order to essentially restrict coverage in this chapter to birds and mammals. Endogenous heat production resulting from lipid metabolism has been maximally developed in these two groups and, in the case of mammals, insulation provided by thick layers of adipose tissue enjoys a similar reputation. A separate section on heat production also provides an opportunity for an in-depth examination of brown fat, a special type of adipose tissue in mammals whose principal function is heat production. Before discussing heat production and insulation as they pertain to lipids, it is first necessary to define some of the temperature terminology to be used and to review some basic concepts involved in thermogenesis and heat exchange.

ENDOTHERMY AND ENERGY METABOLISM

Birds and mammals are usually categorized as being homeotherms or endotherms. Although both designations are correct, the two terms have separate meanings and should not be used interchangeably. Homeothermy implies having a body temperature that is relatively constant and independent of ambient temperature, whereas endothermy indicates that the heat required for the maintenance of this body temperature is produced by the animal's own metabolism, either as an obligatory by-product of ongoing biochemical processes or as heat produced specifically to prevent or retard cooling (Heinrich, 1981). Among terrestrial animals, only birds and mammals are capable of being continuously endothermic. Some representatives of lower groups, however, exhibit periodic endothermy during which time they may regulate their body temperature between a maximum and minimum set point. Certain insects (e.g., moths, bees, beetles, dragonflies) sustain body temperatures at levels that are significantly higher than air temperatures as a result of heat produced by their thoracic muscles during flight or during other activities in which these muscles are employed (Bartholomew, 1981). In many of these insect species the endothermically elevated temperatures are regulated by means of controlled heat transfer within the body or by varying rates of metabolic heat production. Since many of these insects utilize triacylglycerols as their principal fuel during flight (Chapter 6), the extra heat production is very much a functional attribute of lipids even in these intermittent endotherms. Some fast-swimming species of tuna and mackerel shark are also partially endothermic from time to time as a result of muscle metabolism (Carey, 1973; Guppy et al., 1979). In order to maintain a temperature gradient between the body core and the surrounding water, these fish have evolved a special countercurrent arrangement in the vascular supply to the powerful muscles used in swimming. This prevents complete thermal equilibration of the warm blood as it passes through the gills, which are bathed by the cooler water.

Even some plants can elevate their temperature through endogenous heat production. A classic example is *Philodendron selloum*, a member of the arum lily family. The inflorescence of this species consists of a spadix surrounded by a green spathe. At the time of pollination, the spathe unfolds for two days, exposing the male and female florets. The heat generated by the inflorescence (primarily the sterile male florets) is sufficient to warm the spadix to 38 to 46°C, even when ambient temperatures are as low as 4°C (Nagy et al., 1972). Moreover, *P. selloum* is able to regulate the temperature of the spadix within a narrow range by controlled but reversible reductions in heat production (Seymour et al., 1983). Supposedly this response protects the plant from thermal damage, and possibly facilitates the volatilization of chemicals that serve as insect attractants (see Chapter 9). The metabolic source of this heat is lipid contained in cells in the sterile florets. During maximum heating, the lipid reserves are progressively depleted. The absence of glyoxysomes or peroxisomes in *P. selloum* sterile florets suggests that the lipid is respired directly by mitochondrial β-oxidation rather than by glyoxysomal β-oxidation (Walker et al., 1983).

Endothermy in birds and mammals requires high internal heat production which, in turn, is reflected in inherently high rates of metabolism. Indeed, the standard metabolic rate of lower vertebrates (e.g., amphibians, reptiles) is typically 5 to 10 times lower than the basal metabolic rate of birds and mammals of the same size even when measurements are made at the same body temperature (37°C). The relationship between oxygen consumption and environmental temperature is summarized diagrammatically in the familiar curve of thermogenesis (Figure 7.1), which also provides a theoretical base from which additional concepts pertaining to thermoregulation in birds and mammals can be examined.

It is clear that body core temperature remains constant over a wide range of environmental temperatures. Within the zone of thermoneutrality, which is bounded by a lower and upper critical temperature, body temperature remains stable without the animal having to alter its oxygen consumption (i.e., heat production). The metabolic rate at this level is considered to be basal. As temperatures decrease below the lower critical temperature, ox-

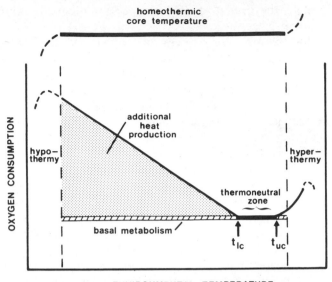

FIGURE 7.1. The relation of oxygen consumption to environmental temperature in a hypothetical bird or mammal. t_{lc} = lower critical temperature; t_{uc} = upper critical temperature. Based on Hoar (1975) and Gordon et al. (1977).

ygen consumption rises linearly, resulting in additional heat production and the maintenance of homeothermy. At some point the body temperature begins to decline even though maximum metabolic rates may not have yet been reached. If such temperatures are maintained, the animal will eventually die from cold exposure. Oxygen consumption also rises (exponentially) as environmental temperatures increase beyond the upper critical temperature. Body temperatures will begin to rise if heat loss mechanisms cannot cope with the heat gain, and eventually a level will be reached where death ensues.

Increased heat production at temperatures below the thermal neutral zone arises from muscular activity and nonshivering thermogenesis. The former includes exercise and shivering. Heat production from increased muscular activity during locomotion or work, even when these are performed under extreme cold, can be so great that the animal is confronted with the problem of heat dissipation. Both birds and mammals also obtain heat from shivering; its importance seems to vary from animal to animal, but it constitutes the predominant means of thermogenesis in birds (Gordon et al., 1977). Virtually all muscles are capable of shivering, and it is estimated that this thermogenic mechanism can elevate heat production up to five times the resting level (Jansky, 1965). As often noted, however, exercise eliminates shivering so that heat production from these two mechanisms is not additive. Heat production from vigorous exercise appears to be maximally developed in larger mammals, including humans, whereas shivering is a more effective heat source in smaller animals. One reason for the latter is that locomotor movements upset the insulative properties of fur in smaller animals, thus permitting a greater heat loss.

Nonshivering thermogenesis refers to the heat produced by tissues and organs as a by-product of metabolism, and not to processes that involve contraction of voluntary muscles (i.e., increased tone, microvibrations, clonic contractions, etc.). Because all living tissues produce some heat, an "obligatory" nonshivering thermogenesis is characteristic of both birds and mammals, and corresponds to the basal or standard metabolic rate (Jansky, 1973). In mammals, however, there is a second component called "regulatory" nonshivering thermogenesis, which refers to augmented heat production by certain metabolically active organs such as the liver, muscle, or brown fat. The term "nonshivering thermogenesis" usually refers to this component. It occurs in newborn mammals and in those that hibernate or become torpid; nonshivering thermogenesis can also be induced in some mammals when they are acutely exposed to cold. The status of nonshivering thermogenesis in birds is uncertain. It has never been clearly demonstrated, and if it does occur, does not assume the thermoregulatory importance of shivering.

More is said about metabolic heat production later in this chapter when the role of lipids in the response of birds and mammals to cold is considered. Before leaving this subject, however, a few additional comments warrant mention. First, whereas exercise and shivering are not additive, exercise and nonshivering thermogenesis are. Thus, both mechanisms can operate simultaneously, at least in mammals, to help sustain homeothermy in response to lowered ambient temperatures. Second, like muscular exercise, nonshivering thermogenesis also enables mammals to increase their energy metabolism above the resting level. The extent of this increase, or the "metabolic scope," is limited for birds and mammals, however, by virtue of the fact that they already have an inherently high basal metabolic rate below which their energy metabolism does not fall except during times of hibernation or torpor. As a result, total dependence on metabolically generated heat to maintain homeothermy in response to cold is not possible without the support of mechanisms for retaining the heat produced. This leads us to the next components of thermoregulation—insulation and conductance.

THERMAL CONDUCTANCE AND INSULATION

Thermal conductance is a measure of the ease with which heat is exchanged between an animal's body and its surroundings. Although this definition is relatively simple, the actual measurement of conductance is much more difficult because of the complexity of the physical factors involved in heat exchange between a terrestrial animal and its environment (McNab, 1980), and because conductance is dependent upon body size and the bird or mammal's circadian phase (Aschoff, 1981). Biologists often use the following equation to estimate conductance:

$$C = \frac{\dot{Q}}{T_b - T_a}$$

where C = conductance, $\dot{Q}$ = metabolic rate or heat production, and $T_b - T_a$ represents the temperature gradient between the body core and the environment. Depending upon how the units of metabolism are expressed, values for conductance can be in ml O_2 g^{-1} hr^{-1} $°C^{-1}$, cal hr^{-1} $°C^{-1}$, or watts $°C^{-1}$. Insulation is the reciprocal of conductance; when conductance is low, insulation is high and vice versa.

A clearer understanding of conductance (insulation) and its relation to thermoregulation can be obtained by referring once again to Figure 7.1. Within the thermal neutral zone, a bird or mammal can vary its thermal conductance by adjusting blood flow to the periphery, fluffing or depressing plumage or pelage, or by altering its posture. These changes enable the animal to maintain a constant body temperature without perceptible changes in its metabolic rate. At the upper critical temperature, conductance values are maximal; here heat dissipation is a primary concern. As environmental

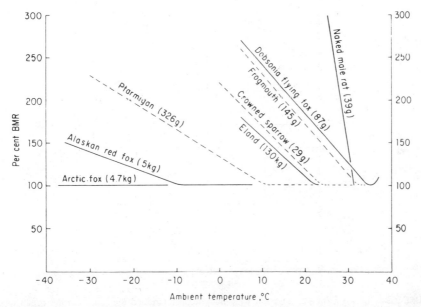

FIGURE 7.2. Response of metabolic rate to environmental temperature in birds (broken lines) and mammals (solid lines) adapted to different climates. Metabolic rates are plotted as percent of basal. Reprinted, with permission, from Gordon et al. (1977).

temperatures are lowered, conductance values decrease until at the lower critical temperature thermal conductance theoretically becomes minimal (insulation is now maximum). Below the lower critical temperature, conductance (insulation) is apparently constant (but see McNab, 1980), and the animal must increase its metabolic heat production in order to maintain homeothermy.

By comparing the metabolic rates of a variety of birds and mammals from different climates over a wide range of ambient temperatures, a further appreciation of the importance of insulation and its relation to increased metabolic heat production can be obtained. Such a plot is shown in Figure 7.2. To facilitate comparisons, the normal resting metabolic rate of each species has been assigned a value of 100%. Several patterns are clearly evident. The tropical species, regardless of size, have fairly narrow thermal neutral ranges and metabolic rates that rise rapidly below the lower critical temperature. With few exceptions, tropical mammals are poorly insulated (i.e., have high conductances) and must increase their metabolic rate dras-

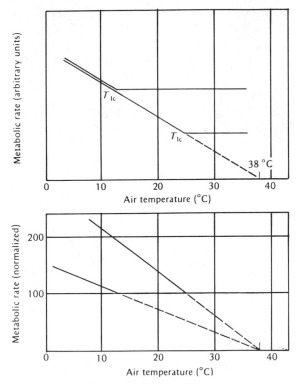

FIGURE 7.3. If two mammals have the same body temperature and the same thermal conductance (i.e., slopes of their metabolism curves are equal), but have different resting metabolic rates, they will also have different lower critical temperatures (top). When their metabolic rates are normalized to 100%, it incorrectly gives the impression that their conductance values are different (bottom). Reprinted, with permission, from Schmidt-Nielsen (1979).

tically with a rather meager drop in ambient temperature. In contrast, polar or cold-adapted birds and mammals (e.g., ptarmigan, Alaskan red fox) are well insulated and can maintain a constant body core temperature of about 37°C without having to increase their endogenous heat production even at extremely low environmental temperatures. Arctic species typically have a wider thermal neutral zone and a flatter metabolism/temperature slope below the lower critical temperature. The arctic fox is so effectively insulated that it shows no increase in metabolism until temperatures have dropped to nearly − 40°C; this mammal should be able to maintain homeothermy at − 70°C with less than a 50% increase in its metabolic rate (data from Scholander et al., 1950a). As noted by Schmidt-Nielsen (1979), however, plotting data in this manner can be misleading. If two mammals of similar size and with equal conductance values have different resting metabolic rates, they will have different lower critical temperatures. By assigning a value of 100% to the resting metabolic rate, it appears that their conductance values are different. The potential problem here is illustrated in Figure 7.3. The point is that both the resting level of metabolism and the conductance value are needed to correctly evaluate the relationship between energy metabolism and heat regulation.

FUR, FEATHERS, AND FAT

Although insulation in its broadest sense includes all methods by which bodily heat is conserved, the physical properties of fur, feathers, and fat certainly represent a major component of overall insulation in homeotherms. The thermal conductivity of these three biological insulating materials is given in Table 7.1. Thermal conductivity is an expression for how easily heat flows in a given material; high values indicate that the material is a good heat conductor, whereas low values indicate that the material is a poor conductor, or good insulator. Thermal conductivity values are also provided for the different types of media or substrates encountered by birds and mammals, and for commonly used insulating materials. The table also includes the temperature and density of the material at which thermal conductivity was measured. Conductivity usually increases at higher temperatures and with higher material densities; however, the extent of these changes depends strongly on the specific material (e.g., compare various states of water with wool).

Air has the lowest thermal conductivity value of all the materials in Table 7.1, indicating that its resistance to heat flow is high. This fact has a strong bearing on the insulative properties of many of the other materials listed. For example, because fur effectively traps large amounts of air between the hairs, its thermal conductivity is only slightly higher than that of air. Scholander et al. (1950b) were among the first scientists to quantify the insulative value of fur in arctic mammals. They found that insulation was correlated

TABLE 7.1
Thermal conductivity (λ) of some biological and physical materials.
Thermal conductivity units (SI) are watts per meter and degree Kelvin
(abbrev. W m^{-1} K^{-1}). 1 W m^{-1} K^{-1} = 2.388 × 10^{-3} cal sec^{-1} cm^{-1} K^{-1}.
The higher the thermal conductivity, the poorer the insulation

Material	Temperature (°C)	Density (kg m^{-3})	Thermal conductivity (W m^{-1} K^{-1})
Fur	20	176	0.037
Feathers	20	109	0.076
Fat	20	867	0.174
Air	20	1.29	0.025
Water	20	998	0.598
Snow	0	150	0.116
Snow	0	500	0.465
Ice	0	917	2.210
Wool	20	100	0.036
Wool	20	250	0.041
Wool	20	400	0.050
Cotton	20	81	0.058
Straw	20	140	0.050
Kapok	20	100	0.043
Paraffin oil	20	810	0.124
Paraffin wax	20	880	0.268

SOURCE: Values based on data in Raznjevic (1976) and Weast (1971).

with fur thickness, with polar mammals having a much better developed
fur coat than tropical mammals in most cases. Moreover, they noted an
increase in fur thickness during the winter. Materials such as wool, cotton,
and kapok likewise enclose a high proportion of air, and thus have been
used by humans to provide excellent insulation against the cold, as have
animal furs themselves (Schmidt-Nielsen, 1979). Smaller mammals are limited in the amount of fur they can bear because of hindrance to locomotion,
and must seek shelter in burrows and nests to supplement their insulation.
Again, the air-trapping features of grasses and straw (Table 7.1) provide
good insulation. Casey (1981) demonstrated a significant energy savings to
brown lemmings, which use a winter nest, in which the physical insulation
is combined with huddling to help reduce the rate of heat loss. Even snow
of the proper consistency (Table 7.1) is an effective insulating material.
Small mammals and birds, such as the ptarmigan, routinely burrow into the
snow to avoid the chilling wind and temperatures on the surface.

The thermal conductivity for feathers is about twice that of fur (Table
7.1), but the magnitude of this difference is somewhat misleading because
of the difficulty of producing the same ordered elevation of feathers during
the measurement that is exhibited by live birds. By fluffing the feathers, a

bird can create a substantial layer of stagnant air between its skin and the colder ambient air. It also gives the bird a more ''rounded'' shape, which improves its surface to volume ratio and hence its rate of heat loss. Thick plumage in species such as the willow ptarmigan and arctic gull apparently provides effective insulation against winter cold. In the grey jay, the decrease in the lower critical temperature from 36°C in summer to 7°C in winter is accompanied by a comparable increase in the insulative value of its feathers (Veghte, 1964). For many species, however, feather insulation by itself does not provide adequate protection from the cold and must be accompanied by changes in activity, microenvironments, and/or increased metabolism. This is especially true of smaller birds even though they may be winter accli- matized. Dawson and Carey (1976) found a difference in insulative capacity in goldfinches between winter and summer, but concluded that the difference was relatively minor in comparison with shifts in metabolic capacity. These authors noted, as did Irving (1960), that feathers serve first for flight and must meet aerodynamic requirements. These requirements impose restric- tions on the use of feathers for varying insulation much in the same way that extensive fur on small mammals would hinder locomotion.

With this background on fur and feather insulation, let us now consider the role played by adipose tissue as an insulating material. From a pure thermal conductivity standpoint, fat is not as effective as either fur or feathers because it lacks the capacity to trap air and thus benefit from air's low thermal conductivity. Nonetheless, fat's reputation as an excellent insulator is deserved when one considers under what conditions it is employed. The dense and thick fur that protects polar terrestrial mammals is of minimal value when the animal is submerged in water whose conductivity is about 24 times that of air (Table 7.1) and may increase to between 50 and 100 times greater as a result of convective heat transfer (Schmidt-Nielsen, 1979). A few species, such as the beaver (Scholander et al., 1950b) and the muskrat (Johansen, 1962), maintain a small layer of trapped air in the fur, which provides some protection against heat loss when swimming, but the fur of polar bears and most seals is completely wetted and essentially provides little or no insulation. To compensate, these species rely largely on sub- cutaneous fat (blubber). The fur-covered harp seal has a 2 to 6 cm thick layer of blubber, while in the polar bear the blubber, although unevenly distributed, can reach a thickness of 11 cm (Øritsland, 1970; Frisch et al., 1974). Extensive deposits of blubber are of course characteristic of hairless marine mammals like the elephant seal, walrus, and whales (Figure 7.4). Values for blubber thickness are greatest for whales, which reflects in part their overall massive body size. Average blubber thicknesses in right and Greenland whales exceed 50 cm, with the blubber accounting for approx- imately 36 to 45% of the animal's body weight (Slijper, 1979). Porpoises and some seals also have thick layers of blubber, with amounts ranging from 45 to 60% of their body weight in some individuals. A thick layer of subcutaneous fat is not totally restricted to marine mammals. The domestic

FIGURE 7.4. Eskimos removing the thick layer of blubber from the right whale. Photograph courtesy of Mary Nerini, National Marine Fisheries Service.

pig contains an envelope of fat that provides protection against cold during winter as well as a source of energy (Irving, 1969). Even humans, whose primary response to cold is one of avoidance, can benefit thermally from subcutaneous fat. Pugh and Edholm (1955) found that fat in swimmers of the English Channel was preferentially laid down in subcutaneous areas rather than in deep fat depots, apparently in response to these chilly waters.

The thickness of blubber in most marine mammals varies from season to season and over different regions of the body. In some species there are also changes associated with the development of thermoregulation. A newborn Weddell seal pup has a luxuriant coat of long fur (lanugo), which is comparable to that of arctic land mammals. This thick fur, which is replaced in four to six weeks by short fur like that of the adult, provides the necessary insulation for life on the ice and snow. While the lanugo is gradually being shed, the pup is accumulating a thick layer of insulating blubber, which eventually enables the pup to make the transition to water without special expenditures of metabolic heat (Kooyman, 1981). A similar correlation between the development of subcutaneous fat and tolerance of cold water was also shown for harp seal pups by Davydov and Makarova (1964). Infant fur-covered pups lost heat rapidly when immersed in cold water, but assumed the adult faculty of resisting ice water after molting and fattening.

The effectiveness of blubber as an insulator reflects its thermal conduc-

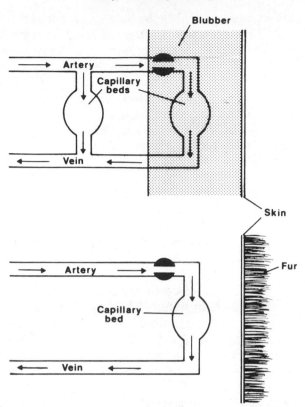

FIGURE 7.5. Comparison between insulation and circulatory changes in aquatic versus terrestrial mammals. Blubber is located internally relative to the surface of heat dissipation. Heat can be conserved by restricting blood flow to the capillary beds below the blubber or dissipated by allowing blood to flow to the periphery, bypassing the blubber layer. Fur, in contrast, is located outside the skin surface and its insulation value cannot be markedly changed by a circulatory bypass. Reprinted, with permission, from Schmidt-Nielsen (1979).

tivity plus the fact that it is an active tissue with a good, controllable blood supply. Irving and Hart (1957) observed that, when a harp seal was immersed in water (0°C), the temperature of the skin surface was very close to that of water, whereas at about 50 mm below the surface the temperature was 35°C. This temperature gradient is sustained largely by the blubber, which comprises most of the tissue over the 50 mm distance. During cold exposure, the seal and other mammals that depend on blubber insulation shunt warm arterial blood into the venous system under the blubber layer so that body heat does not reach the surface (Figure 7.5). With the skin surface and water temperatures nearly equal, there is little heat exchange, and the seal can remain in ice water without having to increase its metabolic rate. On the other hand, when the seal is actively swimming and must eliminate excess heat, vasodilation of peripheral arteries and opening of sphincters allow the warm blood to circulate through the blubber to the surface where the heat can be dissipated. Because terrestrial homeotherms have the insulator (fur or feathers) located externally to the skin surface, they cannot modulate heat loss from the skin surface to any great extent and must depend on other mechanisms for heat dissipation (Figure 7.5) (Schmidt-Nielsen, 1979).

Blubber's effectiveness as an insulator in marine mammals would be greatly reduced if during times of cold exposure body heat was easily lost through the poorly insulated flukes and flippers. These appendages lack the thick blubber layer found in other regions of the body, as they must remain flexible for propulsion and steering. To minimize heat loss and still provide a continuous supply of oxygen and nutrients, the blood vessels are arranged so that most of the heat in the warm arterial blood is transferred to "nearby" veins returning blood to the body core. This arrangement requires that cells in these appendages be able to function at near freezing temperatures. When there is a need to dissipate body heat, these countercurrent heat exchangers can be bypassed and the warm blood allowed to flow to the periphery and return via "far" veins.

It is well established that subcutaneous adipose tissue is responsive to changes in environmental temperature. The general pattern is for an animal to increase the synthesis of unsaturated fatty acids and thus fats with low melting points when exposed to cold (Hilditch and Williams, 1964). The type of fat present often varies continuously along the limbs of birds and mammals, with fat in the toes or footpads sometimes having a melting point 30°C lower than fats deposited proximally where insulating layers are thicker. The selective deposition of unsaturated fats in the distal parts of the extremities is considered to be useful and even essential if these appendages are to remain flexible while operating at near-freezing temperatures (Irving et al., 1957), and thus would appear to have special adaptive significance for polar birds and mammals. The deposition pattern, however, appears to be of more general occurrence, as a higher proportion of unsaturated fats in the extremities has also been observed in temperate and tropical species including humans. In addition to helping maintain flexibility, increased fluidity of depot fat in response to cold is thought to improve the animal's capacity to mobilize fatty acids from the depot when they are needed for energy. The selective deposition of unsaturated fats may also improve the insulative properties of the fat in cold, as the thermal conductivity of lipid is lower when it is fluid rather than solid (see paraffin, Table 7.1).

The existence of temperature gradients across thick layers of insulating subcutaneous adipose tissue have also prompted investigators to look for corresponding changes in the fatty acid composition of the triacylglycerols in outer versus inner regions of these fat depots. The adipose tissue on the back of the pig contains a higher proportion of unsaturated fatty acids (especially 18:1 and 18:2) than that deposited further into the tissue (Figure 7.6) (Christie et al., 1972). As a result, the physical state of the fat in the

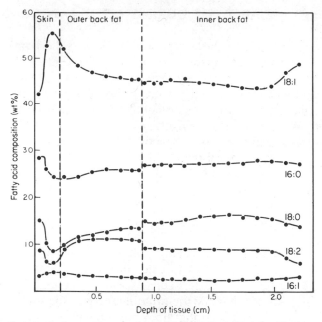

FIGURE 7.6. Variation in fatty acid composition with depth in pig back fat. The latter consists of two layers of adipose tissue separated by a thin layer of connective tissue. Myristic acid (14:0) (not shown) was a minor component and did not vary significantly through the tissue. Reprinted, with permission, from Christie et al. (1972).

cooler outer layers remains constant and similar to that in the warmer regions of the animal. However, in the walrus and several species of whales, the fatty acids of the outer blubber are more saturated than those of the inner blubber (Ackman et al., 1965; Ackman et al., 1975; West et al., 1979). Still, the overall percentage of unsaturated (one to six double bonds) fatty acids is so high that the blubber as a whole remains "soft" at temperatures down to 0°C. Ackman et al. (1975) suggest that the observed gradients in fatty acid saturation in the whales are more likely related to diet and rates of incorporation and turnover of the fatty acids in the blubber. That is, surplus polyunsaturated fatty acids from the diet are incorporated initially in the inner layers and lastly in the outer blubber of the dorsal sites. From a thermoregulatory standpoint, any increase of fat (saturated or unsaturated) would of course lead to increased blubber thickness and thus improved insulation. Interestingly, in the pig the softer fat found in the layer immediately below the skin surface is believed to be formed principally by increased desaturation of saturated fatty acids by specific enzyme systems located in this region rather than by selective deposition of unsaturated fatty acids formed elsewhere or obtained from the diet (Christie et al., 1972).

LIPID METABOLISM DURING COLD EXPOSURE

Basic mechanisms used by homeotherms to increase heat production on exposure to cold have been discussed earlier in the chapter. In birds, the extra heat is generated primarily by increased muscular activity, especially shivering thermogenesis. Presently there is no strong evidence that non-shivering thermogenesis makes a substantial contribution to heat production in birds (Dawson, 1975). Both shivering and nonshivering thermogenesis occur in mammals. Shivering is the major means of heat production during initial exposure to cold; nonshivering thermogenesis functions maximally after the mammal has become cold-adapted (Jansky, 1973). At very low temperatures (i.e., $-40°C$), both nonshivering thermogenesis and heat production from shivering are used for thermoregulation. Regardless of the mechanism used, however, lipids are an important fuel source for the cold-induced thermogenesis.

Although fat metabolism is usually linked with enhanced thermogenic capacities of birds when exposed to cold, details of its utilization are not well known. Carey et al. (1978) studied seasonal acclimatization to temperature in cardueline finches and their findings do provide some quantitative and rather convincing evidence that lipid metabolism is a key factor in the ability of these birds to withstand acute cold stress. During spring, summer, and fall, goldfinches quickly become hypothermic when exposed to extreme cold ($-70°C$), whereas the species when acclimatized to winter conditions can tolerate this temperature for as much as eight hours. The increased thermoregulatory capacity is primarily a result of greatly increased heat production, which in turn is largely the by-product of elevated lipid catabolism. Body lipid mass is about one-third higher in winter than in summer birds. Furthermore, winter birds are capable of much more rapid mobilization and utilization of stored fat than are their summer counterparts. Overnight (17 hours) fasting at $-10°C$ resulted in significant declines in the quantity of stored fat in winter birds, but only small changes in the amount of total body protein and in the amount of liver and pectoralis muscle carbohydrate. Glycogen stores in the pectoralis muscles, however, are significantly higher in winter than in summer birds; its catabolism likely supplements the heat produced from lipid oxidation when the birds are subjected to short-term, extreme cold ($-70°C$). The composition of stored fat also changes seasonally, with the proportion of unsaturated fatty acids in liver and pectoralis muscle increasing in winter birds. How this increase in unsaturation influences the thermogenic capacity of these or other avian species is uncertain, but it may increase rates of fatty acid uptake or catabolism, alter the configuration of the triacylglycerol molecule, or increase the rate at which triacylglycerols can be metabolized (Carey et al., 1978).

Numerous investigators have reported a marked increase in the rate of fat catabolism in mammals exposed to acute cold. In fact, for many years it was widely accepted that fat was the sole, or at least the preferred, fuel for cold-induced thermogenesis. Subsequent experimentation demonstrated, however, that all foodstuffs are utilized during shivering and nonshivering thermogenesis (Masoro, 1968). Nevertheless, fat's high caloric value, plus the fact that it can be mobilized rapidly in response to cold exposure, gives it special importance as a fuel substrate. The quickness by which fatty acids are released from fat stores suggests that this process is under nervous control. Indeed, it has been proven that fatty acids are mobilized under the influence of norepinephrine and regulated by the sympathetic nervous system (Jansky, 1973). Moreover, even though a mammal may not mobilize fat at an increased level after it has become cold-acclimated, it has the capacity to increase its fat mobilization and utilization if suddenly exposed to more intense cold or deprived of food. Responses to both of these emergencies involve an increased sensitivity to norepinephrine (Masoro, 1968).

Whether a diet rich in fats enhances a mammal's tolerance of cold remains pretty much an open question. In some studies, cold-stressed rodents exhibited better growth rates and greater resistance to hypothermia when fed high-fat diets, whereas in other studies high-fat diets resulted in higher mortality and lower rates of growth (see Bobek and Ginter, 1966; Kuroshima et al., 1974). Gordon and Ferguson (1980) found that the composition rather than quantity of dietary fat is more important for inducing cold resistance in white mice. When mice maintained at 0°C for six weeks were fed a diet supplemented with unsaturated fat, their survival time during acute cold exposure (−18°C) was 250% greater than that of mice fed a saturated fat diet. In humans it appears that lipid-rich diets are beneficial for life in cold environments. Men on polar expeditions and Eskimos selectively prefer higher fat in their diet than men in temperate climates (Folk, 1974). Despite a high-fat diet, Eskimos do not show ketosis, a fact which may reflect an increased capacity by members of this race to oxidize free fatty acids to carbon dioxide and water, an increased capacity to catabolize ketone bodies, or both.

BROWN FAT—A SPECIAL THERMOGENIC TISSUE

Thus far the role of lipids in heat production and insulation has been discussed in reference to white fat, which comprises the bulk of all adipose tissue. There is a second type of differentiated adipose tissue, called ''brown fat,'' that is found only in mammals (see, however, Oliphant, 1983) and that has unique heat-producing properties. Brown fat is especially prominent in species that hibernate. In fact, its ubiquitous occurrence in hibernators prompted some investigators early on to refer to brown fat as the ''hibernating gland.'' Subsequent studies have shown, however, that this tissue does not secrete any hormonelike substance that influences the hibernating process (Joel, 1965). Brown fat is also found in the young of many mammals including humans (Figure 7.7). In addition, many small adult mammals

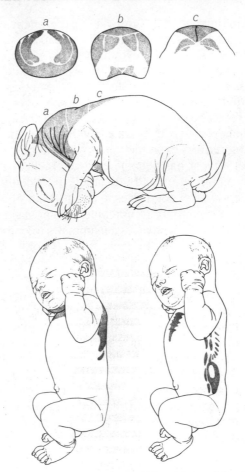

FIGURE 7.7.   Distribution of brown fat in newborn rabbit and human infant. Brown fat accounts for 5 to 6 percent of the body weight of the newborn rabbit, but in adult rabbits and humans it is virtually absent. From M. J. R. Dawkins and D. Hull, The production of heat by fat. Copyright © 1965 by Scientific American, Inc. All rights reserved.

increase their mass of brown fat in response to cold. The location of brown fat within the body varies somewhat according to species, but is typically restricted to the thoracic region. In newborn rabbits, guinea pigs, and humans, there are large deposits between the shoulder blades and around the neck (Figure 7.7); in hibernators such as the 13-lined ground squirrel, brown fat is intimately associated with the heart, major blood vessels, lungs, and spinal column (Joel, 1965). As will be shown, the anatomical position of the brown fat can be directly correlated with its heat-producing function.

Brown fat differs from white adipose tissue in both gross and microscopic structure. Major differences between the two tissue types have been briefly considered in Chapter 2. In contrast to the signet-ring cell with its single

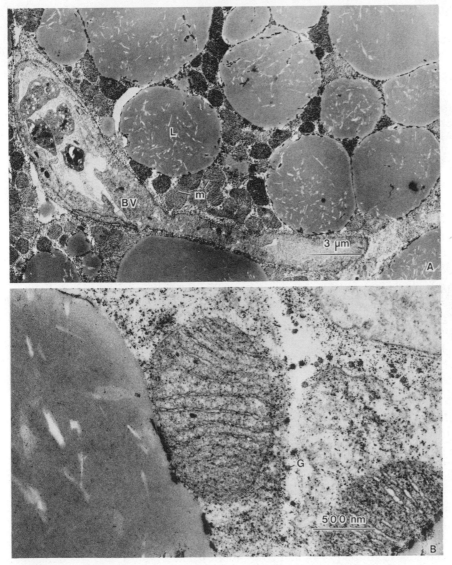

FIGURE 7.8. (Upper) Portions of brown fat cells from a four-day old rabbit. Large lipid bodies (L) surrounded by numerous mitochondria (m) are distributed randomly throughout the cytoplasm. A segment of a small blood vessel (BV) running between the cells is also present. (Lower) Higher magnification showing mitochondrion adjacent to lipid body and numerous glycogen granules (G) in the cytoplasm. Note that the mitochondrial cristae extend the entire width of this organelle in brown fat.

large spherical lipid inclusion that characterizes the white adipocyte, the brown fat cell is typically smaller, polygonal, and contains numerous small lipid inclusions and mitochondria, which impart a granular appearance (Figure 7.8). The lipid inclusions do not appear to be membrane-limited and, along with other cell organelles, are distributed randomly throughout the cytoplasm (Napolitano, 1965). Brown fat is highly vascularized; it is also richly innervated by sympathetic fibers. The abundance of blood vessels plus large amounts of cytochromes associated with the mitochondria are responsible for the brown color characteristic of this tissue. The intimate coupling of intracellular fat deposits with numerous mitochondria provides the potential for the rapid oxidation of fatty acids and the conversion of chemical energy into heat. Indeed, it has been shown that brown fat cells of rabbits can oxidize succinic acid, an intermediate product in the Kreb's cycle, at a rate 20 times greater than the oxidative capacity of white fat cells (Dawkins and Hull, 1965).

There are also differences in the chemical composition of white versus brown fat, some of which can be correlated with the respective structures of these two types of adipose tissue. Quantitatively, white fat possesses more triacylglycerols than brown fat because of its greater volume of included lipid, whereas the phospholipid content in brown fat is greater because of the abundance of mitochondrial membranes. Earlier reports (Fawcett, 1952) that the fatty acid complex of brown fat in the rat and mouse is more saturated than that of white fat appear to be accurate, but the differences are slight and likely have little functional significance. The fatty acid patterns in white and brown fat of the golden hamster, *Mesocricetus auratus,* are virtually identical. When this species was acclimated to cold, there was no change from controls for palmitic (16:0), palmitoleic (16:1), and stearic acids (18:0). Oleic (18:1) was elevated in both adipose tissue types, whereas linoleic acid (18:2) increased only in white fat (Minor et al., 1973). One significant compositional difference between the two types of adipose tissue, a difference corresponding to the more physiologically active nature of brown fat, is the abundance of glycogen granules in addition to lipid bodies in the cell cytoplasm (Figure 7.8). Glycogen normally is not observed in white adipose tissue unless the animal has been fasted.

Reports that brown fat plays an important role in cold thermogenesis began to appear in the early 1960s, when it was demonstrated that brown fat was responsible for generating much of the heat necessary to raise body temperatures back to euthermic levels during arousal from hibernation in mammals (Smith and Hock, 1963; Smalley and Dryer, 1963). Evidence for this thermogenic function was based on the disproportionate amount of blood flow to brown fat during the initial stages of arousal, the fact that warming took place even after curarization of skeletal muscle, and the increased temperature of the brown fat pads during the arousal process. The latter is dramatically illustrated in bats. These small mammals warm from about 8 to 36°C in 25 minutes, during which time the interscapular brown fat pads are the warmest portion of the body (Hayward et al., 1965) (Figure 7.9). A significant warming also occurs in the brown fat of arousing rodents; however, as noted by Lyman (1982), the temperature of fat pads does not

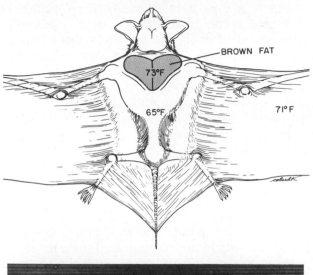

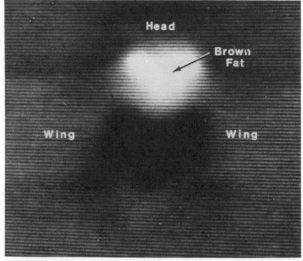

FIGURE 7.9. (Upper) Illustration of a bat as positioned for thermography showing location of the interscapular brown adipose tissue and the major temperature prevailing at the commencement of thermogenic scanning. (Lower) Thermogram of the dorsal surface of a bat during its arousal from hibernation. The higher the temperature and intensity of infrared radiation from the skin surface, the brighter the image. After Hayward and Lyman (1967), by permission.

usually exceed those recorded from the heart. The heat generated by the brown fat is transferred by vascular convection to the cervical portion of the spinal cord and the cardiorespiratory organs of the thorax, warming these so that their physiological and thermogenic activities can bring about complete arousal.

Changes in brown adipose tissue mass and the depletion of lipid from brown fat are offered as further evidence for the role of brown fat in heat production during arousal from hibernation. With few exceptions (see Feist and Quay, 1969), most species exhibit significant declines in both parameters. Spencer et al. (1966) reported losses of between 25 and 38% of brown fat glycerides at the end of arousal in golden-mantled ground squirrels, whereas in newly aroused 13-lined ground squirrels about 50% of the brown fat lipid was depleted (Joel, 1965). These data have also been used to quantify the amount of heat contributed by brown fat during the arousal process. Here the results are quite variable and inconclusive. For example, Joel (1965) estimated that if the lipid that disappeared was oxidized completely, it would provide enough calories to heat a 185-g ground squirrel back to its normal temperature. In contrast, Hayward and Ball (1966) calculated that the brown adipose tissue in big brown bats utilized only 5.7% of the total oxygen consumed during arousal. Recently, Cannon et al. (1981) found that about one-fourth of the interscapular brown fat disappears during arousal in the hamster (Figure 7.10). Assuming that all brown fat deposits in the body lose lipid to the same extent, they estimated that 220 mg would be lost from the animal during arousal. Since theoretically 275 mg of fat must be combusted to rewarm a hibernating hamster, the lipid lost accounts for nearly all the heat produced.

There are several reasons for the discrepancies in values for heat production by brown fat during arousal. All of these studies are beset with difficulties in accurately accounting for all of the brown fat mass, lack of reliable methods for measuring blood flow and oxygen consumption, and knowledge of other possible sources of heat for arousal from hibernation (Lyman, 1982). Until techniques are available that will enable investigators to make some of the physiological measurements *in situ* or to realistically simulate *in vitro* conditions to match those occurring in the intact animal, the precise amount of heat generated by brown fat during arousal is likely to remain a puzzle.

Mammals experience cold for the first time at birth. Many species are born immature (e.g., rats, rabbits, humans); they have little or no fur insulation, yet possess a high surface to volume ratio. Moreover, many cannot shiver at birth, and thus are unable to generate heat from muscular activity. For these species, heat production from the metabolic activity of brown fat is of critical importance. The extent of nonshivering thermogenesis in brown fat has been documented in newborn rabbits by Dawkins and Hull (1965). They found that newborn rabbits tripled their heat output when the environment was cooled to 25°C. With adequate fur insulation, this rate of heat production would protect the newborn rabbit at temperatures down to −30°C. At 25°C the temperature of the brown fat was 2.5°C higher. than muscle tissue in the back and 1.3°C higher than the body core. That brown fat was entirely or nearly entirely responsible for the threefold increase in heat

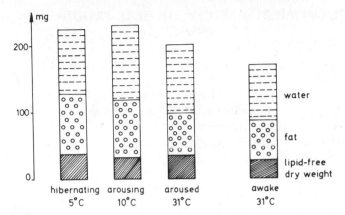

FIGURE 7.10.   The loss of lipid from the interscapular deposits of brown adipose tissue in hibernating, aroused, and awake hamsters. Reprinted, by permission of the publisher, from Physiological uncoupling in brown fat mitochondria, by B. Cannon, J. Nedergaard, and U. Sundin. In: *Survival in the Cold: Hibernation and Other Adaptations* (X. J. Musacchia and L. Jansky, Eds.), pp. 99–120. Copyright 1981 by Elsevier Science Publishing Co., Inc.

production was further demonstrated by removing 80% of this tissue and noting that the rabbits were no longer able to increase their heat production when exposed to cold (Dawkins and Hull, 1965).

In most hibernators, the amount of brown fat is greatest at the time of birth, but then progressively decreases so that little if any remains in the adult. Some adult nonhibernators (e.g., rats, mice, guinea pigs), however, which can increase their nonshivering thermogenesis in response to cold, show a concomitant increase in the amount of any brown fat that remains. This increase in brown fat mass is typically accompanied by a severalfold increase in mitochondrial content (Himms-Hagen, 1978). The quantitative contribution of brown fat to nonshivering thermogenesis in these species remains controversial, just as does the precise amount of heat generated by brown fat metabolism during arousal from hibernation. Based on data from several earlier studies, Jansky (1973) concluded that brown adipose tissue probably does not contribute more than 10% to the total nonshivering thermogenesis in cold-adapted rats, whereas Foster (1976), using tracer microspheres to measure blood flow, showed a thermogenic contribution of 60%. Although the quantity of heat produced remains unclear, there is little question that the response to cold is extremely rapid. Girardier (1980) cites studies that demonstrate that temperature gradients within the cardiovascular system resulting from the activation of brown fat deposits along the thoracic aorta appear less than two minutes after rats are transfered from 30 to 10°C, and that reduction of NAD^+ is observed within 2 to 3 seconds after stimulation of nerves innervating brown adipose tissue.

BIOCHEMICAL ASPECTS OF HEAT PRODUCTION IN BROWN FAT

The specific heat-generating mechanisms in brown fat that enable this unique tissue to steadily increase its heat production at birth, during acclimation to cold, and during arousal from hibernation are not yet firmly established. Nevertheless, in recent years sufficient progress has been made in this area so that there now exists at least a general consensus as to the principal biochemical processes operating during nonshivering thermogenesis in brown fat (see comprehensive and complementary reviews by Cannon et al., 1978; Himms-Hagen, 1978; and Horwitz, 1978). The following description of events responsible for the high oxidative capacity of brown fat and hence its remarkable heat-generating ability is based largely on information reported in these reviews, plus some more recent findings that generally support the basic sequence of biochemical events cited by the above authors.

Nonshivering thermogenesis in brown adipose tissue is due to the combustion of fatty acids. The initial step in this process involves the stimulation of the numerous sympathetic nerves within the brown fat and the subsequent release of norepinephrine from the nerve terminals (Figure 7.11). Although there is some evidence that norephinephrine binds with α-adrenergic receptors on the fat cell membrane (Mohell et al., 1980), the major pathway leading to the thermogenic response to brown adipose tissue is believed to be through activation of the β-adrenergic receptors, which may exceed 150,000 per adipocyte. Activation of the β-receptors causes an increase in the concentration of cyclic AMP, which serves as the second messenger. The cyclic AMP activates a specific protein kinase, which in turn is responsible for activating lipase enzyme. Both the formation of cyclic AMP and the activation of the protein kinase require ATP. The lipase hydrolyzes stored triacylglycerols within the brown fat cells, producing glycerol and free fatty acids (Figure 7.11).

The metabolic fate of these two products, especially the fatty acids, is central to the production of heat. Glycerol cannot be metabolized within the fat cell because of the absence of glycerol kinase, and thus is discharged into the blood and carried to the liver and muscle where it can be either metabolized or converted into storage compounds. Free fatty acids released from the triacylglycerol molecules can (1) be reesterified within the brown fat cell, (2) leave the cell along with glycerol and be oxidized at some other site in the body, or (3) enter the brown fat mitochondria for combustion. Many early reviews of the biochemistry of heat production by brown fat suggested the importance of reesterification as a means of converting the chemical bond energy of fatty acids into heat. However, the quantitative significance of such a purposeless cycle appears to be minimal, as it is possible to account for three fatty acids for every glycerol released (Cannon et al., 1978). Earlier reports also suggested that only a small percentage of the hydrolyzed fatty acids were released from the brown adipocytes. More

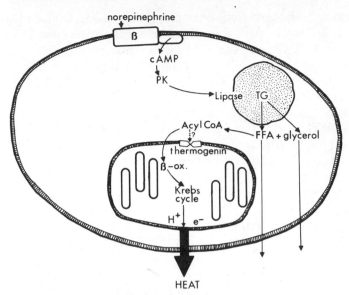

FIGURE 7.11. Schematic representation of the biochemical events of thermogenesis in brown adipose tissue. PK = protein kinase, TG = triacylglycerol, FFA = free fatty acid. Modified, by permission of the publisher, from Physiological uncoupling in brown fat mitochondria, by B. Cannon, J. Nedergaard, and U. Sundin. In: *Survival in the Cold: Hibernation and Other Adaptations* (X. J. Musacchia and L. Jansky, Eds.), pp. 99–120. Copyright 1981 by Elsevier Science Publishing Co., Inc.

recent evidence suggests that, on the contrary, the majority of the fatty acids formed are exported (Cannon et al., 1978). If so, then some of the heat attributed to brown fat is indeed a result of substrates delivered for combustion in other tissues. Nonetheless, the principal heat-producing mechanism occurs within the brown fat itself and involves the oxidative catabolism of fatty acids in the numerous mitochondria that characterize this tissue and perhaps to some extent combustion within brown fat peroxisomes (Alexson et al., 1980).

The sequence of steps by which fatty acids are oxidized within the brown fat cell is basically the same as described in Chapter 6. Before fatty acids can be combusted they must first be activated to fatty acyl-CoA (Figure 7.11). The acyl-CoAs enter the mitochondria, where they are initially converted to acetyl CoA by β-oxidation and subsequently to carbon dioxide and water via the Kreb's cycle. Because respiration is stimulated some 20 times over the resting state during thermogenesis, sufficient amounts of Kreb's cycle substrates (e.g., oxaloacetate) are required to react with the acetyl CoAs being formed from β-oxidation. This requires an ample supply of pyruvate, which is converted to oxaloacetate by an active pyruvate carboxylase enzyme. The pyruvate is believed to be derived from lactate in circulation or from glycolysis of stored glycogen granules in the brown fat.

The transfer of energy to heat occurs at the mitochondrial membrane.

The electrons from the various oxidation-reduction reactions in β-oxidation and the Kreb's cycle pass through the respiratory chain where they are eventually accepted by oxygen. At the same time, protons (H^+) are pumped out of the mitochondria, creating a high proton electrochemical gradient, which provides the driving force for ATP synthesis. In most tissues, mitochondrial oxygen consumption is closely coupled to ATP synthesis; that is, oxygen is consumed only when ATP is needed and respiration occurs as long as ADP is supplied to the ATP synthetase. Because of the efficiency of ATP synthesis, little heat is generated by this process until ATP is hydrolyzed or some product into which the energy has been incorporated is catabolized (Cannon et al., 1978). In order to generate the heat characteristically produced by brown fat, a partial uncoupling of respiration from ATP synthesis is necessary. This could be achieved by increasing the permeability of the mitochondrial membrane to protons, thus dissipating the proton electrochemical gradient and thereby permitting an increase in respiration that is independent of changes in the availability of ADP (see Nicholls, 1979). In brown fat mitochondria there is a specific protein, termed thermogenin (Cannon et al., 1981), which opens such channels in the mitochondrial membrane to allow the combustion of fatty acids without the concomitant harnessing of energy in the form of ATP (Figure 7.11). These channels of course would have to be regulated. During times when thermogenesis is unnecessary, the channel can be closed by certain purine nucleotides (e.g., GDP), which bind to the thermogenin and inhibit its activity. Similarly, there must exist antagonists to nucleotide binding that would open the channel in conjunction with the demands for heat production. Cannon et al. (1980) have postulated that acyl-CoA may serve as a physiological antagonist to purine nucleotides and, as such, function not only as a substrate but also as a stimulus for initiating mitochondrial respiration in brown fat (Figure 7.11).

Experimental data from *in vitro* measurements generally support the existence of a loose coupling mechanism as being primarily responsible for the large thermogenic response of brown fat mitochondria to norepinephrine. Still, there is little direct evidence that such biochemical processes operate as described *in vivo*. Furthermore, at least two other thermogenic mechanisms involving altered control of mitochondrial function have been proposed. These are an increased activity of plasma membrane Na^+K^+-ATPase by norepinephrine proposed by Horwitz and co-workers, and membrane depolarization caused by norepinephrine, which ultimately may regulate the cytosol level of Ca^{2+} and, in so doing, influence mitochondrial activity via the effect of this ion on glycerol-3-phosphate dehydrogenase (see Himms-Hagen, 1978; Nedergaard and Cannon, 1980). Although the independent contribution to heat production by brown fat mitochondria by either an ATPase type or ion-cycling type mechanism may be small in comparison with the loose coupling mechanism, the latter may require changes at the plasma membrane brought about by these first two mechanisms in order to function most effectively. It is quite possible that future studies will show a functional interrelationship between all three classes of thermogenic mechanisms.

CONCLUDING REMARKS

It is clear that lipids play a critical and dynamic role in the thermoregulation of birds and mammals, both as an insulating layer and as a source of heat. The properties of lipids that make them so effective in the above roles often provide the animal with auxiliary benefits that may be almost as important. The significance of adipose tissue as a metabolic source of energy has been considered in Chapter 6. Lipid deposits also form a good cushion and shock-absorber. The viscera of mammals are typically surrounded by an abundant blanket of fat. The high viscosity of lipids tends to reduce the rate at which they flow under shearing forces. As a result, there is less displacement of body organs during locomotion than if water were the packing fluid. To function in this manner, the lipids must be in a relatively fluid state themselves. This is the condition typically observed, regardless of environmental or body temperature. Body fat deposits also directly influence the locomotor ability of marine mammals, while at the same time protecting against excessive heat loss. In Cetacea, the thick blubber layer helps buoy the body mass and streamline the body, enabling these massive animals to move through the water with exceptional grace (Hertel, 1969). The contribution of lipids to buoyancy in plants and animals is considered in detail in the next chapter.

REFERENCES

Ackman, R. G., C. A. Eaton, and P. M. Jangaard (1965). Lipids of the fin whale *(Balaenoptera physalus)* from North Atlantic water. I. Fatty acid composition of whole blubber and blubber sections. *Can. J. Biochem.* **43**:1513–1520.

Ackman, R. G., J. H. Hingley, C. A. Eaton, J. C. Sipos, and E. D. Mitchell (1975). Blubber fat depositions in mysticeti whales. *Can. J. Zool.* **53**:1332–1339.

Alexson, S., J. Nedergaard, H. Osmundsen, and B. Cannon (1980). Peroxisomes in brown fat. In: *Advances in Physiological Sciences,* Vol. 32: *Contributions to Thermal Physiology* (Szelenyi, Z. and M. Szekely, Eds.), pp. 483–485. Pergamon Press, New York.

Aschoff, J. (1981). Thermal conductance in mammals and birds: Its dependence on body size and circadian phase. *Comp. Biochem. Physiol.* **69A**:611–619.

Bartholomew, G. A. (1981). A matter of size: An examination of endothermy in insects and terrestrial vertebrates. In: *Insect Thermoregulation* (Heinrich, B., Ed.), pp. 46–78. Wiley, New York.

Bobek, P. and E. Ginter (1966). The effect of lowered environmental temperature on lipid metabolism in rats fed on normal and high-fat, high-cholesterol diets. *Br. J. Nutr.* **20**:61–68.

Cannon, B., J. Nedergaard, L. Romert, U. Sundin, and J. Svartengren (1978). The biochemical mechanism of thermogenesis in brown adipose tissue. In: *Strategies in Cold: Natural Torpidity and Thermogenesis* (Wang, L. C. H. and J. W. Hudson, Eds.), pp. 567–594. Academic Press, New York.

Cannon, B., J. Nedergaard, and U. Sundin (1980). Physiological uncoupling in brown fat mitochondria. In: *Advances in Physiological Sciences,* Vol. 32: *Contributions to Thermal Physiology* (Szelenyi, Z. and M. Szekely, Eds.), pp. 479–481. Pergamon Press, New York.

Cannon, B., J. Nedergaard, and U. Sundin (1981). Thermogenesis, brown fat and thermogenin. In: *Survival in the Cold: Hibernation and Other Adaptations* (Musacchia, X. J. and L. Jansky, Eds.), pp. 99–120. Elsevier, Amsterdam.

Carey, C., W. R. Dawson, L. C. Maxwell, and J. A. Faulkner (1978). Seasonal acclimatization to temperature in cardueline finches II. Changes in body composition and mass in relation to season and acute cold stress. *J. Comp. Physiol.* **125**:101–113.

Carey, F. G. (1973). Fishes with warm bodies. *Sci. Am.* **228**:36–44.

Casey, T. M. (1981). Nest insulation: Energy savings to brown lemmings using a winter nest. *Oecologia* **50**:199–204.

Christie, W. W., D. M. Jenkinson, and J. H. Moore (1972). Variation in lipid composition through the skin and subcutaneous adipose tissue of pigs. *J. Sci. Fd. Agric.* **23**:1125–1129.

Davydov, A. F. and A. R. Makarova (1964). Changes in heat regulation and circulation in newborn seals on transition to aquatic form of life. *Trans. Suppl. No. 4, Part II, Fed. Proc.* **24**:T563–566.

Dawkins, M. J. R. and D. Hull (1965). The production of heat by fat. *Sci. Am.* **213**:62–67.

Dawson, W. R. (1975). Avian physiology. In: *Annual Review of Physiology* (Comroe, J. H., R. R. Sonnenschein, and I. S. Edelman, Eds.), pp. 441–465. Annual Reviews Inc., Palo Alto, California.

Dawson, W. R. and C. Carey (1976). Seasonal acclimatization to temperature in cardueline finches. I. Insulative and metabolic adjustments. *J. Comp. Physiol.* **112**:317–333.

Fawcett, D. (1952). A comparison of the histological organization and cytochemical reactions of brown and white adipose tissues. *J. Morphol.* **90**:363–405.

Feist, D. D. and W. B. Quay (1969). Effects of cold acclimation and arousal from hibernation on brown fat lipid and protein in the golden hamster (*Mesocricetus auratus*). *Comp. Biochem. Physiol.* **31**:111–119.

Folk, G. E., Jr. (1974). *Textbook of Environmental Physiology,* 2nd ed. Lea and Febiger, Philadelphia.

Foster, D. O. (1976). Calorigenic potential of rat brown adipose tissue and muscle reevaluated from blood flow measurements with tracer microspheres. *Physiologist* **19**:194.

Frisch, J., N. A. Øritsland, and J. Krog (1974). Insulation of furs in water. *Comp. Biochem. Physiol.* **47A**:403–410.

Girardier, L. (1980). Current topics in brown adipose tissue research. In: *Advances in Physiological Sciences,* Vol. 32: *Contributions to Thermal Physiology* (Szelenyi, Z. and M. Szekely, Eds.), pp. 469–474. Pergamon Press, New York.

Gordon, C. J. and J. H. Ferguson (1980). Role of dietary lipids in cold survival and changes in body composition in white mice. *J. Therm. Biol.* **5**:29–35.

Gordon, M. S., G. A. Bartholomew, A. D. Grinnell, C. B. Jorgensen, and F. N. White (1977). *Animal Physiology: Principles and Adaptations,* 3rd ed. Macmillen, New York.

Guppy, M., W. C. Hulbert, and P. W. Hochachka (1979). Metabolic sources of heat and power in tuna muscles. II. Enzyme and metabolite profiles. *J. Exp. Biol.* **82**:303–320.

Hayward, J. S. and E. G. Ball (1966). Quantitative aspects of brown adipose tissue thermogenesis during arousal from hibernation. *Biol. Bull.* **131**:94–103.

Hayward, J. S. and C. P. Lyman (1967). Non-shivering heat production during arousal from hibernation and evidence for the contribution of brown fat. In: *Mammalian Hibernation III* (Fisher, K. C., A. R. Dawe, C. P. Lyman, E. Schönbaum, and F. E. South, Eds.), pp. 346–355. Oliver and Boyd, Edinburgh.

Hayward, J. S., C. P. Lyman, and C. R. Taylor (1965). The possible role of brown fat as a source of heat during arousal from hibernation. *Ann. N.Y. Acad. Sci.* **131**:441–446.

Heinrich, B. (1981). Definitions and thermoregulatory taxonomy. In: *Insect Thermoregulation* (Heinrich, B., Ed.), pp. 4–6. Wiley, New York.

Hertel, H. (1969). Hydrodynamics of swimming and wave-riding dolphins. In: *The Biology of Marine Mammals* (Andersen, H. T., Ed.), pp. 31–63. Academic Press, New York.

Hilditch, T. P. and P. N. Williams (1964). *The Chemical Composition of Natural Fats* (4th ed.). Wiley, New York.

Himms-Hagen, J. (1978). Biochemical aspects of nonshivering thermogenesis. In: *Strategies in Cold: Natural Torpidity and Thermogenesis* (Wang, L. C. H. and J. W. Hudson, Eds.), pp. 595–617. Academic Press, New York.

Hoar, W. S. (1975). *General and Comparative Physiology,* 2nd ed. Prentice-Hall, New Jersey.

Horwitz, B. A. (1978). Neurohumoral regulation of nonshivering thermogenesis in mammals. In: *Strategies in Cold: Natural Torpidity and Thermogenesis* (Wang, L. C. H. and J. W. Hudson, Eds.), pp. 619–653. Academic Press, New York.

Irving, L. (1960). Birds of Anaktuvuk Pass, Kobuk and Old Crow. *U.S. Nat. Mus. Bull. No. 217.*

Irving, L. (1969). Temperature regulation in marine mammals. In: *The Biology of Marine Mammals* (Andersen, H. T., Ed.), pp. 147–174. Academic Press, New York.

Irving, L. and J. S. Hart (1957). The metabolism and insulation of seals as bare-skinned mammals in cold water. *Can. J. Zool.* **35**:497–511.

Irving, L., K. Schmidt-Nielsen, and N. S. B. Abrahamsen (1957). On the melting points of animal fats in cold climates. *Physiol. Zool.* **30**:93–105.

Jansky, L. (1965). Adaptability of heat production mechanisms in homeotherms. *Acta Univ. Carol.-Biol.* **1**:1–91.

Jansky, L. (1973). Non-shivering thermogenesis and its thermoregulatory significance. *Biol. Rev.* **48**:85–132.

Joel, C. D. (1965). The physiological role of brown adipose tissue. In: *Handbook of Physiology,* Sec. 5: *Adipose Tissue* (Renold, A. E. and G. F. Cahill, Jr. Eds.), pp. 59–85. Am. Physiol. Soc., Washington, D.C.

Johansen, K. (1962). Buoyancy and insulation in the muskrat. *J. Mammal.* **43**:64–68.

Kooyman, G. L. (1981). *Weddell Seal: Consummate Diver.* Cambridge Univ. Press, London.

Kuroshima, A., M. Kusabashi, K. Doi, T. Ohuo, and I. Fagiata (1974). Effect of cold adaptation and high fat diet on cold resistance and metabolic responses to acute cold exposure in rats. *Jap. J. Physiol.* **24**:277–293.

Lyman, C. P. (1982). Mechanisms of arousal. In: *Hibernation and Torpor in Mammals and Birds* (Lyman, C. P., J. S. Willis, A. Malan, and L. C. H. Wang, Eds.), pp. 104–123. Academic Press, New York.

Masoro, E. J. (1968). *Physiological Chemistry of Lipids in Mammals.* W. B. Saunders, Philadelphia.

McNab, B. K. (1980). On estimating thermal conductance in endotherms. *Physiol. Zool.* **53**:145–156.

Minor, J. G., G. E. Folk, Jr., and R. L. Dryer (1973). Changes in triglyceride composition of white and brown adipose tissues during developing cold acclimation of the golden hamster *Mesocricetus auratus. Comp. Biochem. Physiol.* **46B**:375–385.

Mohell, N., J. Nedergaard, and B. Cannon (1980). An attempt to differentiate between α- and β-adrenergic respiratory responses in hamster brown fat cells. In: *Advances in Physiological Sciences,* Vol. 32: *Contributions to Thermal Physiology* (Szelenyi, Z. and M. Szekely, Eds.), pp. 495–497. Pergamon Press, New York.

Nagy, K. A., D. K. Odell, and R. S. Seymour (1972). Temperature regulation by the inflorescence of *Philodendron. Science* **178**:1195–1197.

Napolitano, L. (1965). The fine structure of adipose tissues. In: *Handbook of Physiology,* Sec. 5: *Adipose Tissue* (Renold, A. E. and G. F. Cahill, Jr., Eds.), pp. 109–123. Am. Physiol. Soc., Washington, D.C.

Nedergaard, J. and B. Cannon (1980). A possible metabolic effect of membrane depolarization in brown adipose tissue. In: *Advances in Physiological Sciences,* Vol. 32: *Contributions to Thermal Physiology* (Szelenyi, Z. and M. Szekely, Eds.), pp. 475–477.

Nicholls, D. G. (1979). Brown adipose tissue mitochondria. *Biochim. Biophys. Acta* **549**:1–29.

Oliphant, L. W. (1983). First observations of brown fat in birds. *Condor* **85**:350–354.

Øritsland, N. A. (1970). Temperature regulation of the polar bear. *Comp. Biochem. Physiol.* **37**:225–233.

Pugh, L. G. C. and O. G. Edholm (1955). Physiology of channel swimmers. *Lancet* **2**:761–768.

Raznjevic, K. (1976). *Handbook of Thermodynamic Tables and Charts.* McGraw-Hill, New York.

Schmidt-Nielsen, K. (1979). *Animal Physiology: Adaptation and Environment.* Cambridge Univ. Press, London.

Scholander, P. F., R. Hock, V. Walters, F. Johnson, and L. Irving (1950a). Heat regulation in some arctic and tropical mammals and birds. *Biol. Bull* **99**:237–258.

Scholander, P. F., V. Walters, R. Hock, and L. Irving (1950b). Body insulation of some arctic and tropical mammals and birds. *Biol. Bull.* **99**:225–236.

Seymour, R. S., G. A. Bartholomew, and M. C. Barnhart (1983). Respiration and heat production by the inflorescence of *Philodendron selloum* Koch. *Planta* **157**:336–343.

Slijper, E. J. (1979). *Whales,* 2nd ed. Cornell Univ. Press, Ithaca, New York.

Smalley, R. L. and R. L. Dryer (1963). Brown fat: thermogenic effect during arousal from hibernation in the bat. *Science* **140**:1333–1334.

Smith, R. E. and R. J. Hock (1963). Brown fat: Thermogenic effector of arousal in hibernation. *Science* **140**:199–200.

Spencer, W. A., E. I. Grodums, and G. Dempster (1966). The glyceride fatty acid composition and lipid content of brown and white adipose tissue of the hibernator *Citellus lateralis. J. Cell Physiol.* **67**:431–444.

Veghte, J. H. (1964). Thermal and metabolic responses of the grey jay to cold stress. *Physiol. Zool.* **39**:171–184.

Walker, D. B., J. Gysi, L. Sternberg, and M. J. DeNiro (1983). Direct respiration of lipids during heat production in the inflorescence of *Philodendron selloum. Science* **220**:419–421.

Weast, R. C. (1971). *Handbook of Chemistry and Physics,* 51st ed. Chemical Rubber Co., Cleveland, Ohio.

West, G. C., J. J. Burns, and M. Modafferi (1979). Fatty acid composition of Pacific walrus skin and blubber fats. *Can. J. Zool.* **57**:1249–1255.

8

BUOYANCY

Although we marvel at the size of elephants, most whales are considerably larger. In fact, the weight of the blue whale can exceed the combined mass of 30 elephants (Slijper, 1979). Such massive body size is possible only in aquatic habitats where buoyancy provided by the much denser water (in comparison with air) helps counteract the gravitational pull on the body. Still, there are problems associated with buoyancy that must be overcome by aquatic organisms, especially large animals such as the whale. The specific gravity of protoplasm, exclusive of particularly dense materials like calcium carbonate shells or chitinous exoskeletons (Table 8.1), lies between 1.02 and 1.10, while seawater has a maximum value of about 1.026 (Hoar, 1966). Hence, unless the organism adjusts its density so that it equals water (i.e., achieves neutral buoyancy), the organism will continually sink or will have to swim to maintain a desired depth. To compensate, plants and animals have evolved a variety of mechanisms, some quite novel, for reducing their density or specific gravity. In this chapter we shall examine these mechanisms in a general sense, with emphasis on how organisms use low specific gravity lipids to regulate their density in fresh and seawater. Before considering these mechanisms, however, it is useful to briefly discuss the advantages and, in some cases, the necessity of maintaining neutral buoyancy.

ADVANTAGES OF NEUTRAL BUOYANCY

The principal advantages of maintaining neutral buoyancy, at least as they pertain to aquatic animals, are considered by Hochachka and Somero (1973). As they point out, most aquatic organisms are adapted and often restricted

TABLE 8.1
Densities of terrestrial versus aquatic habitats, common skeletal constituents, and principal organic compounds of organisms

Substance	Density
Habitats	
Air	0.00125
Fresh water	1.000
Seawater	1.026
Common skeletal constituents	
Chitin	1.2
Silica	2.32
Calcium carbonate (calcite)	2.71
Calcium phosphate (apatite)	3.2
Organic compounds	
Starch	1.53
Protein	1.33
Lipid[a]	0.91

Sources: Denton (1961), Hoar (1966), and Altman and Dittmer (1972).

[a]Densities of lipid classes presented in Table 8.2.

to a finite region of the water column in an ocean or a lake. The width of this zone or the depth experienced by an organism, however, varies greatly among different species. Furthermore, the vertical range or the preferred zone for an individual species can change significantly with life history stage, seasonally, and even diurnally.

For phytoplankton, the smallest of the aquatic organisms to be considered in this chapter, at least temporary residence in the euphotic zone is essential for their existence. Light intensity, temperature, and nutrient availability all vary considerably with depth in both lakes and in the sea. As a result, not only survival but also the potential growth of these photoautotrophs will be greatly influenced by their vertical position in the water column (Walsby and Reynolds, 1980). Despite their small size, most phytoplankton exhibit some means of regulating the density of their cell volume so that the mean daily residence time within the euphotic zone permits photosynthesis in excess of total daily respiratory and organic secretion losses incurred within and below this zone.

The distribution of zooplankton is largely determined by the drifting pasturage of phytoplankton. Since neither the depth nor the maximum concentration of phytoplankton is constant, zooplankton must regulate their body density so that the buoyancy provided is commensurate with the vertical position of the phytoplankton. In polar regions, where phytoplankton blooms occur only during two or three months of summer sunlight, large-scale vertical migrations are an integral part of the life history of some zooplankton species such as calanoid copepods. Nauplii larvae, which hatch from eggs laid by adults at depths of 300 to 400 meters, slowly swim to the surface, where they emerge as free-swimming copepodites just as the phytoplankton bloom begins (Benson and Lee, 1975). Although these copepods are excellent swimmers, by establishing neutral buoyancy they are able to maintain their vertical position with a minimum amount of energy expenditure. Thus more of the energy obtained from the abundant food supplies near the surface can be converted into fuel storage compounds (i.e., wax esters) for use during the long winter months when little if any feeding occurs. For those zooplankton species that exhibit diurnal vertical migration, descending to cooler water levels when not feeding may reduce metabolism and leave more energy available for growth and reproduction.

In larger and more motile aquatic animals, neutral buoyancy is essential for conserving energy and enhancing swimming and feeding efficiency. A fish swimming at normal cruising speeds (i.e., 3 to 4 body lengths per sec) must expend about 20% of the energy required simply to overcome sinking (Hochachka and Somero, 1973). By adjusting the body density for a given depth, the fish can significantly reduce this power expenditure and put more of the energy into horizontal movement or to meet other needs. Predatory sharks must maintain optimum maneuverability in order to capture their agile prey. This requires careful regulation of body density so that the loading of hydrodynamic lifting surfaces (e.g., pectoral fins) is maintained within the proper limits for the size range of a given species (Baldridge, 1972). Buoyancy regulation is perhaps even more critical for slow-moving species such as the cuttlefish, a marine cephalopod. It must regulate its density so that it can hover over the bottom of the sea with a minimal disturbance of surrounding water as it searches for prey. Unlike fast-swimming species, it generates little lift as a result of its forward motion. Finally, neutral buoyancy may enable species such as salmon and eels to orient in currents and/or preferred thermal zones and thus reduce the energy output during their long migrations.

MECHANISMS FOR ADJUSTING BUOYANCY

The ability to actively control buoyancy is obviously of great value to an aquatic organism. To float on the surface or remain at a specific depth without continuous energy expenditure, an organism must either reduce its density by excluding or reducing some of the heavier substances in the body or replace heavier substances with lighter materials. Although these options are, in theory, limited and quite simplistic, the mechanisms by which they are achieved are quite varied and often ingenious. In this next section, examples of such mechanisms are discussed for these two basic strategies, with emphasis on the advantages and disadvantages of each mechanism.

This is followed by a more in-depth examination of lipids as buoyancy agents in plants and animals, which includes a comparison of pertinent physical properties of different lipid classes and their effectiveness in density regulation compared with some of the nonlipid mechanisms described earlier. More complete information on the buoyancy question in general, especially the structure and function of nonlipid buoyancy structures or physiological processes in density adaptation, can be found in Denton (1961), Sculthorpe (1967), Smayda (1970), Hochachka and Somero (1973), and Schmidt-Nielsen (1979).

Reduction of Heavy Body Materials

One means of adjusting the specific gravity of an organism so that it more closely approximates that of its aquatic environment is by reducing the principal "sinking" components present in the body. These high density materials are typically associated with the skeleton or other supporting structures and include such compounds as calcium carbonate in the shell and exoskeleton of invertebrates, calcium phosphate in vertebrate bones, and silica in the cell wall of diatoms (Table 8.1). Protein, a major organic chemical constituent of the body, is also a relatively dense material (Table 8.1). Aquatic organisms which can make do with a lighter skeleton or less muscle can markedly improve their buoyancy. Certain mid-water pelagic oceanic fish have done just this. These species show reduced body ossification and have a protein content that amounts to about 5% of their dry weight compared with 16% for a typical coastal species (Denton and Marshall, 1958). For most teleosts removal of so much structural material would place them at a severe disadvantage, as the reduced muscle mass and associated supporting framework would not permit bursts of locomotor activity necessary to capture prey. These fish, however, live in the dark ocean waters where they attract prey with luminous lures rather than by vigorous swimming movements. Thus, the reduction of muscle and skeleton along the trunk and tail does not interfere with feeding efficiency. It should be noted that skeletal ossification and musculature remain strong and well developed in those structures for holding and swallowing prey (Denton and Marshall, 1958).

Altering the Ionic Composition of Body Fluids

There are appreciable differences in the densities of equimolar solutions of different inorganic ions. For example, solutions of NH_4^+ are lighter than those of the same concentration of Na^+; those of Na^+ are about 40% lighter than K^+; monovalent ions, in turn, are lighter than divalent ions such as Ca^{2+}, Mg^{2+}, and SO_4^{2-} (Walsby and Reynolds, 1980). Theoretically, aquatic organisms could realize some gain in buoyancy by reducing the ionic concentration of their body fluids in general, by excluding or reducing the amount of the heavier ions, or by replacing heavier ions with lighter ions and thus maintaining the same ionic concentration. Because of physiological constraints, not all of these options are available. Most marine invertebrates have body fluids that are isosmotic with the surrounding seawater. Removal of ions in amounts sufficient to affect buoyancy would create osmotic problems for these species as they would become hypotonic to sea water. Marine teleosts, in contrast, maintain fluids that are approximately 50% as concentrated as seawater and thus are hypotonic in their natural state. Although it is energetically expensive to maintain this ion gradient, some of the energy expended is recovered as a result of the reduced energy needed to maintain a desired vertical position. Nevertheless, any improvement in buoyancy in most adult teleosts due to hypotonicity is probably of minor importance because of their large skeleton and muscle mass. The latter is not a problem for the much lighter pelagic eggs of marine teleosts, which likely owe a portion of their buoyancy to tonicity lower than that found in adults.

Greater buoyancy gain can be achieved by eliminating heavier ions or by replacing these with lighter ions. The latter strategy is especially attractive in that the organism can maintain the necessary tonicity and thus avoid osmotic problems. The replacement of heavier ions with lighter ions is usually accomplished by selective accumulation of lighter ions from seawater or from products of metabolism. The floating ability of *Noctiluca,* a bioluminescent dinoflagellate, is due to the exclusion of most divalent ions, a high intracellular concentration of Na^+ relative to K^+, and relatively high concentrations of light NH_4^+ ions (Smayda, 1970). Similarly, many diverse invertebrate groups, especially gelatinous forms with a high water content (e.g., jellyfish), have reduced amounts of heavier ions such as Ca^{2+}, Mg^{2+}, and SO_4^{2-}. In many of these species the ion substitution occurs in a special region of the body that serves as a buoyancy tank. One of the largest of these tanks is the fluid-filled coelomic cavity of the deep sea squid *Cranchia scabra*. This chamber, which accounts for about two-thirds of the total volume of the squid, contains about 480 mM ammonium and only about 90 mM sodium, compared with 465 mM sodium in the body fluids (Denton, 1961). Despite its bulk, the fluid carried in the enlarged coelom is virtually incompressible and thus is not affected by pressure changes that occur during vertical movements made by the squid.

Elaboration of Gas Floats

At sea level, air has only 1/800th the density of an equal volume of water (Table 8.1). Thus, a chamber containing air or a gas mixture of comparable density is a most effective device for flotation. Such gas-filled spaces are used by a variety of aquatic plants and animals to achieve neutral buoyancy. Examples include the air-filled lacunae in the leaves of several aquatic vascular plants (Sculthorpe, 1967), the pneumatocysts of giant marine brown algae (Rigg and Swain, 1941), the gas float of the Portuguese man-of-war

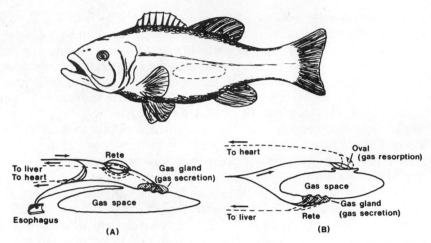

FIGURE 8.1. Location and two main types of swimbladders. (*A*) A physostome bladder from the eel *Anguilla vulgaris.* (*B*) A physoclist bladder from the perch *Perca fluviatilis.* Redrawn from Eckert and Randall (1983).

(Wittenberg, 1960), air-filled chambers of the shell of *Nautilus* and the cuttlebone of the cuttlefish (*Sepia,* a cephalopod) (Denton, 1961), and the swimbladders of fish (Denton, 1961; Alexander, 1966). In many of the above floats, the gas includes constituents that are normally toxic to higher animals (e.g., carbon monoxide) and requires specialized structures or unique metabolic processes for its secretion.

The teleost swimbladder is perhaps the most impressive of these floats in terms of developing and sustaining gas pressures. A true swimbladder is present in many shallow-water species (e.g., trout, salmon, perch) as well as in some marine forms that live at depths approaching 4000 m (Marshall, 1972). The swimbladder is located in the abdominal cavity, just below the spinal column (Figure 8.1). It develops from the esophageal region of the gut. In physostomatous fish, a connection (the pneumatic duct) is retained between the gut and the swimbladder, which permits the passage of air in either direction (Figure 8.1*A*). Although a gas-secreting gland is present, this arrangement also allows shallow-living species to fill their swimbladder by simply swallowing air at the surface. In physoclistous fish, the duct is lost during development (Figure 8.1*B*). These species secrete gas, usually oxygen, from a gas gland and resorb gas through a thin-walled, highly vascularized region of the swimbladder wall. Since the majority of deep sea fish are physoclists, they must be able to secrete gas into the bladder and retain this gas at tremendous pressures. Physiological mechanisms involved in gas secretion and absorption are discussed at length in Blaxter and Tytler (1978).

Whereas the swimbladder in fish provides an excellent solution to the problem of buoyancy, it also illustrates well some of the disadvantages of air-filled chambers, especially when they are enclosed by soft tissue. Air is highly compressible; its volume changes with pressure according to the gas laws. A fish moving from 0 to 10 m experiences a doubling of the hydrostatic pressure and a corresponding 50% reduction in bladder volume. Because the fish is now more dense, it will continue to sink unless it actively swims or secretes gas into the bladder to compensate for the pressure change. Upward migrations are potentially more dangerous. As the pressure is reduced, the air inside the bladder expands, causing the lift rate to increase. This, in turn, causes further expansion of the bladder. Unless the fish can reabsorb or eliminate the excess gas that accumulates, such rapid ascent can lead to the bursting of the swimbladder. The problem is especially critical for the physoclists, as they lack the pneumatic duct through which the excess gas can be rapidly vented to the exterior.

The high compressibility of air is also a problem for fish that live near the bottom in deep water. The greater the depth, the more difficult it is for fish to generate and maintain gas tensions within the swimbladder. To compensate, many abyssal species fill the bladder with lipids rather than air. The basis for this substitution lies in the much lower compressibility of lipid versus air and the higher solubility of oxygen in lipids than in water when under great pressure. These and other properties of lipids that make them effective buoyancy agents are now considered.

Deposition of Lipids

Lipids are used by a variety of plants and animals to attain or approach neutral buoyancy. Lipids have the lowest density of all of the major organic compounds in the body (Table 8.1). More significant, most lipid classes are less dense than seawater and indeed fresh water. The densities of some lipid classes present in the oils of marine organisms and the lift provided by the lipid are given in Table 8.2. Triacylglycerols and wax esters serve as buoyancy agents across a fairly broad phylogenetic spectrum of marine organisms, whereas the function of squalene and diacylglycerol ethers in density regulation is more restricted in its occurrence. Oleic acid accounts for up to 70% of the fatty acids in triacylglycerols and wax esters of marine fish and crustaceans (Hochachka and Somero, 1973). Despite its favorable specific gravity, a precise role for oleic acid in buoyancy control has not been specified. Generally, the lipid class appears to be a more important factor in determining density than either the length or the degree of saturation of the carbon chains that comprise the lipid. Pristane, probably the most effective of the naturally occurring buoyant agents as far as gravity is concerned (Table 8.2), is present in such limited amounts that its function in density regulation has to be questioned.

Although the differences in specific gravity among the lipid classes are

TABLE 8.2

Specific gravity and the upthrust generated by different lipids in sea water
of specific gravity 1.026[a]

Lipid	Specific gravity	Grams upthrust per ml lipid
Pristane	0.78	0.246
Squalene	0.856	0.170
Wax esters	0.858	0.168
Diacylglycerol ethers	0.891	0.135
Oleic acid	0.894	0.132
Triacylglycerols	0.916	0.110
Cholesterol	1.067	−0.041

SOURCES: Lewis (1970), Altman and Dittmer (1972), Weast (1981).
[a]Densities determined at 20 to 21°C (not specified for pristane).

not great, they can represent a substantial difference in the amount of lift
generated. For example, based on values in Table 8.2, wax esters would

$$CH_3CH(CH_2)_3\overset{\displaystyle \overset{CH_3}{|}}{C}H(CH_2)_3\overset{\displaystyle \overset{CH_3}{|}}{C}H(CH_2)_3\overset{\displaystyle \overset{CH_3}{|}}{C}H\,CH_3$$

Pristane (2,6,10,14-tetramethylpentadecane)

provide approximately 53% more lift than an equal volume of triacylglycerols. Sargent et al. (1976) have extended this kind of comparison to estimate the amount of wax esters or triacylglycerols required for a marine animal containing these lipids to be neutrally buoyant. For wax esters, 29% of
the total volume of the animal would have to be lipid compared to 37% for
triacylglycerols. For a 5.0-ml animal with a mass of approximately 5.2 g,
this means that an additional 430 mg of triacylglycerol would be needed.

In addition to their low specific gravity, lipids have other features that
make them ideal buoyancy agents. A case in point involves cholesterol. As
noted in the previous section, cholesterol is often a major constituent of the
lipids invested in the swimbladder of deep sea fish. Because its density is
greater than that of seawater (Table 8.2), cholesterol would appear to be
maladaptive in buoyancy control. Indeed, each ml of cholesterol would exert
a downward thrust of 0.041 g. However, cholesterol (and other lipids) are
virtually noncompressible, and thus make excellent substitutes for gases
when the latter are subjected to extreme hydrostatic pressures. Furthermore,
cholesterol may enhance the solubility of oxygen in the swimbladder or
facilitate the secretion of gas under such pressures (Phleger and Benson,
1971).

LIPIDS AND DENSITY REGULATION IN AQUATIC ORGANISMS

Let us now examine how aquatic organisms use the various lipids cited in Table 8.2, either independently or in association with other buoyancy mechanisms, to help maintain a preferred vertical position in the water column. Coverage continues to emphasize advantages and disadvantages of lipids over other mechanisms as well as one type of lipid over another. In addition, examples are selected that highlight special morphological structures with which these lipids are associated, metabolic processes required for the synthesis of often extensive amounts of unique lipid compounds, and the means by which a balance is achieved between the use of lipids for energy and reproductive activities versus buoyancy control.

Phytoplankton

The density of seawater is somewhat less than that of phytoplankton cytoplasm and significantly lower than that of cell walls, particularly those that contain silica. Thus, most phytoplankton must depend on some morphological or physiological mechanism(s) to remain suspended. Their small size (2 to 20 μm for most phytoplankton cells), cylindrical shape, and in some cases surface projections help retard the sinking rate. Even larger diatoms benefit from the fact that the volume of the cell wall relative to the total volume of the cell decreases with increasing cell size (Hutchinson, 1967). Some larger phytoplankton regulate their ionic composition to reduce the density of the cell fluids, and blue-green algae produce permanent gas-filled structures to provide positive buoyancy. Although gas bubbles in surface-dwelling phytoplankton would not be subjected to the extreme hydrostatic pressures experienced by swimbladders in fish living at great depths, bubbles small enough to be accommodated inside cells would have a large excess pressure generated by surface tension that would have to be sustained by high partial pressures of gases dissolved in the surrounding cytoplasm (Walsby, 1972).

The contribution of lipid substances to buoyancy in phytoplankton is at best uncertain. Many investigators have reported fat accumulations in these cells (especially diatoms), yet feel that their function in buoyancy regulation is unimportant or only partially effective. Even in those diatoms where lipids account for up to 40% of the dry weight, the cell's density is reduced by only 3.5% (Walsby and Reynolds, 1980). Other observations suggest that the concentration of lipids increases as the cells become older. If lipids are important in determining buoyancy, one might expect older cells to exhibit a decreased sinking rate or even ascend. With few exceptions, however, aging is accompanied by increased sinking rates. Nonetheless, enormous concentrations of the diatom *Coscinodiscus concinnus*, which reportedly cover over 8000 square miles of the North Sea, are attributed to the increased buoyancy produced by the species' high oil content. The slick created by these oil-laden diatoms and the resultant increase in the viscosity of the water create problems for birds similar to those caused by oil spills from ocean tankers (see Smayda, 1970).

Aquatic Vascular Plants

Buoyancy control is an adaptive requirement for two types of vascular hydrophytes: those plants that are free-floating and those that are attached to the substrate but possess floating leaves. Both groups of plants are essentially restricted to fresh water habitats and in neither group do lipids apparently play a significant role in density regulation. In all free-floating rosettes, buoyancy is endowed by the high proportion of air in the spaces contributed by the lacunate mesophyll. In some species, the lacunate tissue develops so excessively that bladderlike swellings similar to the floats produced by macroscopic brown algae are formed (Sculthorpe, 1967). Leaves in both types of plants often contain numerous hairlike projections that trap air in a manner similar to the plastron of certain aquatic insects. The trapped air not only helps provide flotation, but also protects the epidermal surface against excessive wetting. In addition, a wax bloom (see Chapter 5) is often present on the surface of floating leaves. Although the primary functions of this wax coating are probably to restrict water flux and protect against abrasion, the reduced density of the leaf relative to water because of the waxy cuticle may at least contribute to the positive buoyancy of these leaves.

Lipids probably contribute more significantly to the buoyancy of fruits and seeds produced by aquatic plants and in some cases terrestrial plants. As noted in Chapter 2, the cotyledons and endosperm of the seed and the surrounding fleshy fruit are often rich in triacylglycerols. In addition, the testa surrounding the endosperm and the outer layer of the fruit (pericarp) are often impregnated with waxy substances. The buoyancy provided by these lipid inclusions is enhanced by the presence of air-filled cavities in the seed and/or fruit and sometimes an accumulation of suberized corky tissue in the mesocarp. Despite these buoyancy mechanisms, most tropical seeds and fruits do not float in either fresh or seawater or do so for only a few hours to several days (Gunn and Dennis, 1976). Even this short period of time, however, may be adequate to carry the propagules away from the more competitive parental habitat.

PLANKTONIC CRUSTACEANS

In Chapter 6 it has been noted that stored lipid in pelagic marine invertebrates is essential for both maintenance energy and gametogenic activity. It appears that these lipid stores also have a direct bearing on the buoyancy of many

FIGURE 8.2. Photograph of *Euchaeta japonica,* a deep water, omnivorous copepod, showing a large sac filled with wax esters and some triacylglycerols. This female also bears a wax ester-rich cluster of eggs. Courtesy of A. A. Benson.

of these organisms. Of particular interest here are the copepods, which store up to 90% of their lipids in the form of wax esters. From Table 8.2 it is clear that wax esters are very efficient in generating buoyancy. Data have already been presented that indicate that wax esters can provide substantially more lift than an equal volume of triacylglycerols. Wax esters are also excellent buoyancy agents because of their resistance to compression and thermal expansion. This is critical to small-bodied organisms like copepods, which make extensive vertical migrations from depths where temperatures are low and pressure great to surface waters where temperatures are higher and pressures reduced. Sargent et al. (1976) estimated that a copepod rich in wax esters that migrates from 500 m (3°C) up to 200 m (10°C) would experience a volume change of about 2%, an increase that seems small enough to be physiologically safe.

The two major depot lipids involved in the buoyancy regulation of copepods vary in their location within the body. Triacylglycerols are generally distributed throughout the body, whereas wax esters are located in a specific region (Benson and Lee, 1975). In calanoid copepods, which are abundant in temperate and polar waters, the wax esters are contained in a large, centrally located sac (Figure 8.2). Because wax esters of marine organisms tend to be highly unsaturated molecules, they are usually in a liquid state even at the near freezing temperatures encountered in surface waters. During winter months, females divert much of their stored wax into the production of eggs. The lipids in eggs, which often are predominantly triacylglycerols rather than wax esters, give the eggs a positive buoyancy and cause them to float toward the surface when released from the female. In about two and one-half days the eggs hatch and the nauplii swim to the surface, where they feed on phytoplankton as they go through six molting stages on the way to becoming adult copepods (Benson and Lee, 1975).

The use of wax esters to achieve neutral buoyancy by copepods is potentially complicated by the fact that these lipids are also used for metabolic fuel, and therefore are subject to fluctuations in abundance. A copepod would become less buoyant as wax esters are used unless it employed other compensatory mechanisms to maintain a constant body density. On a short-term (day-to-day) basis, this problem may be avoided, or at least minimized, by an enzymatic control of lipid metabolism, which favors the breakdown of triacylglycerols rather than wax esters (Chapter 6). When wax ester stores become depleted, the copepod must migrate to the surface to feed and replenish its lipid levels. Once neutrally buoyant again, the copepod can descend to any given depth in the water column and remain stationary with minimal energy output (Sargent et al., 1976). The finding that oxygen consumption is lowest in *Calanus* copepods having the highest lipid levels is consonant with this concept (Conover and Corner, 1968). For other mid-water plankton crustaceans, however, neutral buoyancy *per se* may not be as important a mechanism for conserving energy. This is largely based on the findings that between 400 and 700 m relative buoyancy and respiratory rates are independent and that lipid content decreases in the deepest living species to the extent that few of them achieve neutral buoyancy (Childress and Nygaard, 1974).

Fish

Marine fish use a variety of mechanisms to achieve neutral buoyancy. These include (1) a swimbladder filled with gas or lipid, (2) storage of lipid in soft body tissues such as flesh or liver, (3) poorly ossified bone and reduced musculature, (4) constant swimming, and (5) oil-filled bones (Lee et al., 1975). The mechanism(s) of choice is largely determined by the taxonomic group to which the fish belongs. Elasmobranchs (e.g., sharks, skates, rays) are cartilagenous fish that exhibit almost no bone. Even more critical to the buoyancy question is the absence of a swimbladder in these fish. Teleosts, the true bony fish, have a skeleton composed partly or chiefly of bone. Most teleosts species also have a well-developed swimbladder. In both groups, however, neutral buoyancy is at least aided by the presence of lipids in various body organs.

(a) Elasmobranchs. The density of sharks varies widely among different species. Many sharks are quite dense and must swim constantly to

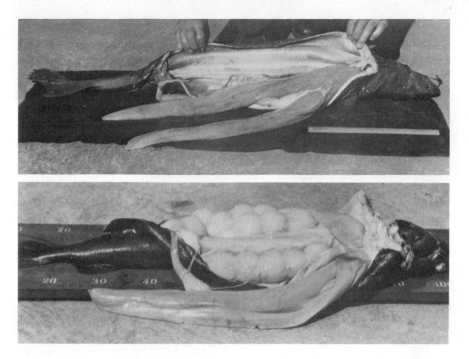

FIGURE 8.3. Two shark specimens (*Centrophorus squamous*, upper; *Centroscymnus coelolepis*, lower) opened to display the enormous liver. A large weight of neutrally buoyant eggs is also present in *C. coelolepis*. Reprinted, by permission, from Corner et al. (1969).

maintain a certain depth. Other species, especially deep sea members of the family Squalidae, are very close to neutral buoyancy. It has been shown that these species obtain a great deal of lift from oils deposited in their enormous liver (Figure 8.3). Corner et al. (1969) found that the liver of deep sea sharks accounted for between 20 and 30% of the total weight of the fish (compared with 1 to 2% in teleosts), with oil comprising over 80% of the liver weight. Comparable data were reported for Florida sharks by Baldridge (1972). Quantitative relationships between liver size, oil content, and effective buoyancy were examined by Bone and Roberts (1969). They found that the larger the relative size of the liver, the less dense the fish, and that fish with relatively large livers obtain more lift per gram of liver than do fish with relatively small livers (Figure 8.4). The latter relationship reflects the fact that an increase in liver size results only from an increase in oil content. Bottom-living elasmobranchs such as rays and skates have smaller livers with a lower oil content. In the ray *Raja*, the size of the liver averaged less than 8% of the total body weight and contained less than 50% oil (Schmidt-Nielsen, 1979).

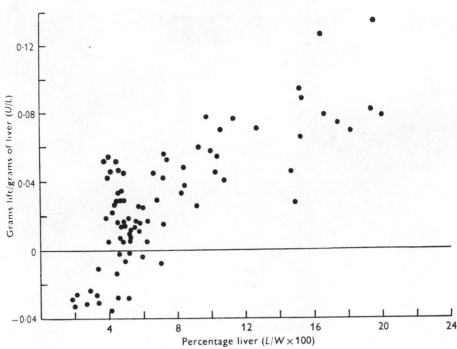

FIGURE 8.4. Variation in the static lift provided by livers of different relative size. Note that livers weighing less than 4% of a shark's body weight do not normally provide any static life. Reprinted, by permission, from Bone and Roberts (1969).

The upthrust generated by liver oils depends not only on their quantity but also on the types of oil present. The chemical composition of the liver oils of sharks varies considerably with species; however, major components generally include triacylglycerols, diacylglycerol ethers, and the hydrocarbon squalene (see Chapter 1), the latter being especially common in deep sea sharks. The abundance of squalene is noteworthy for several reasons. Squalene is a very low density lipid (specific gravity = 0.86, Table 8.2). Only pristane, another hydrocarbon that sometimes co-occurs with squalene but in much smaller amounts, has a lower specific gravity (0.78). It has been calculated that 1 g of squalene will give approximately 80% more lift than 1 g of triacylglycerols (Corner et al., 1969). Thus, even though the livers of these sharks are enormous, they would have to be very much larger to accommodate the quantity of triacylglycerols required to produce an equivalent buoyancy. Squalene is also unique from a metabolic standpoint. Unlike triacylglycerols, squalene cannot be converted into acetate units. The only known route of squalene metabolism is to cholesterol and bile salts (Sargent et al., 1973). Its deposition in the liver provides the shark with a pool of lipid that can function as a source of lift without continually being drawn upon for metabolic energy. Finally, the synthesis of 1 g of squalene requires 2.5 g of acetic acid, whereas to obtain the same lift the shark would have to use 3.5 g of acetic acid in making cod-liver oil (primary triacylglycerols) (Corner et al., 1969). Thus, not only is squalene superior in terms of lift generated, it is also economical to manufacture.

Some sharks regulate the composition of their liver oils to precisely control their body density. For example, the two primary low density lipids in the liver of the dogfish shark are triacylglycerols and diacylglycerol ethers. The latter consist of an aliphatic alcohol linked to the glycerol molecule in the alpha position, with the two remaining hydroxyl groups of glycerol esterified to fatty acids (see Chapter 1). Although the specific gravities of these two lipid classes are similar (Table 8.2), 1 g of diacylglycerol ethers gives approximately 14% more lift in seawater than 1 g of triacylglycerol. When the body weight of the sharks was artificially increased by suspending lead weights from their pectoral fins, the weighted sharks exhibited a significant increase in the ratio of diacylglycerol ethers to triacylglycerols (Malins and Barone, 1970). By selectively synthesizing more of the less dense lipid, the weighted fish are able to offset the additional weight and maintain neutral buoyancy.

(b) Teleosts. Most true bony fishes use lipids to help them control their body density. For species with an air-filled swimbladder, these lipids may play only a supplemental role, whereas in species in which the swimbladder is absent or rudimentary, lipids deposited in the soft parts of the body and/or incorporated into skeletal structures are primarily responsible for maintaining neutral buoyancy.

Unlike sharks, which tend to concentrate their lipid stores in the liver, the lipids used in buoyancy control by teleosts tend to be distributed throughout various body tissues and organs. Some mid-water pelagic oceanic fish deposit fat in the mesentaries and exhibit a well-developed system of fat sinuses under the skin. In addition, many of these same species possess a fat-filled swimbladder. Massive amounts of lipid are found in the muscle of some lantern fishes and gempylids. The predominant lipids in these species are wax esters, which probably serve as buoyancy agents and are used only secondarily, if at all, as an energy source by the muscle fibers (Nevenzel, 1970). When all fat reserves are taken together, they occupy approximately 15% of the body volume. It is estimated that marine fishes would be neutrally buoyant if 21 to 24% of their fresh weight were lipid. Therefore, even without a swimbladder these fish are close to neutral buoyancy.

One marine fish that is especially rich in lipids is the castor oil or scour fish, *Ruvettus pretiosus*. It achieves neutral buoyancy by storing large amounts of low-density lipid in the dermis and bones of the skull. The surface of *Ruvettus* (except for the anterior) is covered with a regular array of large ctenoid scales and contains a system of large subdermal spaces that communicate to the exterior via pores (Figure 8.5). The ctenoid scales contain

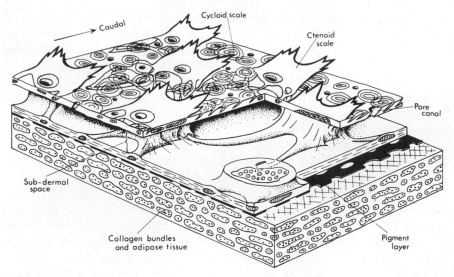

FIGURE 8.5. Diagram showing the complex integument of *Ruvettus pretiosus*. The ctenoid scales and the outer portion of the dermis overlying the subdermal spaces are rich in low-density lipids. Reprinted, by permission, from Bone (1972). Copyright by the American Society of Ichthyologists and Herpetologists.

large numbers of oil-filled cells, as does the outer portion of the dermis overlying the subdermal spaces (Bone, 1972). These lipid deposits account for the low density of the outer portion of the integument. In addition, the bones of the skull contain numerous oil sacs, causing the center of buoyancy to be located anterior to the center of gravity. As a result, the fish rests in the water with its head tilted at 45° to the horizontal, a position that may enhance its ability to capture prey (Bone, 1972). Oil-filled bones were also found by Lee et al. (1975) in three species of marine fish (two stromateoids and the sablefish) that lack a swimbladder. In one of the stromateoids, *Peprilus simillimus,* the dry weight of the skull contained 68% lipid with triacylglycerol accounting for 97% of the bone oil. The skull bones in two other marine teleosts with fat-filled swimbladders contained only 0.2% lipid (Lee et al., 1975). Although the evidence strongly suggests that bone lipid functions as a buoyancy agent, the number of fish examined thus far is too low to permit definitive correlations between the presence or absence of oil-filled bones and the structure, habitat, or behavior of the fish.

The advantages and disadvantages of the principal mechanisms used by teleosts to control buoyancy can be further compared by considering their merits relative to energetic costs. In other words, which mechanism—hydrodynamic compensation (i.e. swimming or movement of fins), buoyancy provided by lipids, or buoyancy provided by a swimbladder—requires the least amount of energy expenditure by the fish? Alexander (1972) addressed this very question for species that maintain a constant depth, and also for species that make extensive vertical migrations. For fish in the first category, he estimated energy output in terms of oxygen consumption of 25 ml kg^{-1} h^{-1} for hydrodynamic compensation, 10 to 17 ml kg^{-1} h^{-1} for buoyancy derived from lipids, and 2 ml kg^{-1} h^{-1} for buoyancy due to a swimbladder. Although the data on which these calculations are based are not firm, the values clearly suggest that it is metabolically less expensive to maintain buoyancy at a specific depth (particularly one near the surface) by use of a swimbladder, and that lipids are more economical than hydrodynamic compensation. The range of metabolic values stated for lipid-derived buoyancy reflect in part the density and thus the amount of lipid used for achieving neutral buoyancy. Alexander estimated that a typical marine fish whose volume to body mass ratio approximates 50 cm^3 per kg would need either about 500 ml of triacylglycerols or 300 ml of wax esters per kg of lipid-free body weight. These amounts would increase the volume of the body by about 54% (triacylglycerols) and 32% (wax esters). Because the use of wax esters results in a smaller increase in the bulk of the fish, the drag that it has to overcome when it swims is also comparably reduced. Ultimately this difference in work load should manifest itself in a lower rate of oxygen consumption.

The above calculations refer to life at a constant depth. Maintenance of buoyancy is potentially a greater problem for deep sea fish that make substantial vertical migrations, especially for those with a swimbladder. As a fish descends, it must either secrete gas into the swimbladder to compensate for compression or compensate hydrodynamically for increased density. Both of these processes require energy. The secretion of oxygen into the swimbladder under great pressure is likely quite expensive metabolically. Moreover, energy is also needed to replace gas lost through diffusion (Alexander, 1972). Some savings of energy can probably be realized by reducing or eliminating the swimbladder and replacing it with low-density lipids, or by filling the swimbladder with lipids. Fish that achieve neutral buoyancy as a result of extensive body lipid stores experience little change in body density during vertical excursions. A fish can swim upward with as little effort as it swims horizontally, since its potential energy is not changed (Alexander, 1972). If a functional swimbladder is retained, its investment with lipids will provide greater resistance to compression and allow the swimbladder to make a greater contribution to buoyancy at the high pressures associated with great depths. The presence of lipids may also reduce the metabolic cost of gas secretion by enhancing the diffusion of oxygen into the swimbladder (Hochachka and Somero, 1973). Unfortunately, the energetic demands of these physical and metabolic processes are not well enough known to permit quantitative comparisons.

Marine Mammals

For most marine mammals, buoyancy control is probably secondary to the physiological problems associated with deep diving. Although they may have to swim actively to remain on the surface or to reach various depths at which they feed, many species shuttle between land (ice) and the water, and thus must expend this energy only periodically. The presence of a thick blubber layer and their mode of respiration are two positive features for reducing body density. Air-filled lungs provide considerable buoyancy. When fully expanded they enable these animals to significantly reduce the energy needed to remain afloat. Blubber oil, whose specific gravity is substantially lower than that of sea water, must also provide some degree of uplift. Even massive whales, which are continuously pelagic, likely carry the correct amount of air and fat to be neutrally buoyant at a specific depth, although they will have to compensate hydrodynamically to move vertically in the water column.

There is one marine mammal that not only precisely controls its body density during deep dives, but apparently uses lipids to achieve neutral buoyancy at the various depths. This is the sperm whale, a toothed species whose range extends from warm equatorial waters to the polar seas of the Antarctic. Because this species floats after being killed, it has been a target of whalers for centuries, since it could be captured by harpoons and handled by men in small boats. These massive animals feed almost exclusively on cephalopods, which are caught at depths greater than 1000 m. Males may remain submerged for nearly one hour; dive times for females and calves are somewhat shorter. The depth and duration of the dives, plus the observation that individuals tend to emerge close to the point of submergence, suggests that when the whale reaches the preferred depth it must almost lie still in the water. To do this, it must be neutrally buoyant. How this is accomplished has been the subject of extensive investigation by Malcolm Clarke of the Plymouth Laboratory in England. The following description of the mechanisms used by the sperm whale to control its buoyancy is based on findings reported in a series of papers by Clarke (1978a,b,c, 1979).

Morphologically, sperm whales are distinguished by an enormous head, which can represent 25% of the animal's length and 33% of its total weight. The structure responsible for most of this bulk is the spermaceti organ located above the upper jaw (Figure 8.6). The spermaceti organ contains spermaceti oil, a complex mixture of wax esters and triacylglycerols (Morris, 1975). The organ can be divided into two compartments: (1) the spermaceti sac, which is located in the upper portion of the snout and is surrounded by the tough, fibrous "case," and (2) the "junk," a series of coffin-shaped blocks located in the lower region of the snout. The collection of up to four tons of oil from a spermaceti organ of an adult male attests to the magnitude of this structure.

The spermaceti organ—specifically spermaceti oil—must be able to vary its density to function in buoyancy control. Field observations plus laboratory tests indicate this is indeed the case. At 33°C, the temperature of the oil in whales resting at the surface, the oil is liquid. When temperatures drop below 31°C (i.e., during a deep dive), the oil begins to crystallize. In the solid state the oil is more dense and thus occupies less volume. Conse-

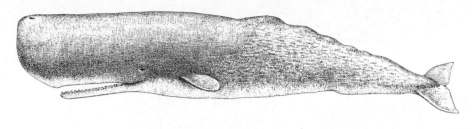

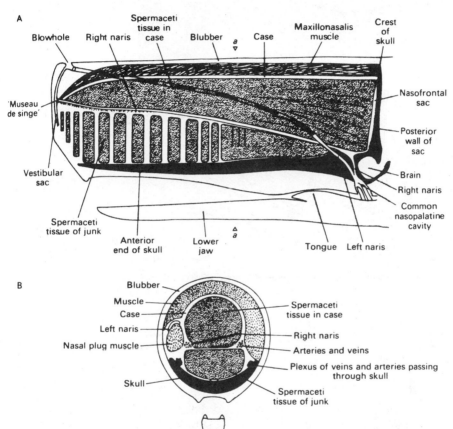

FIGURE 8.6. (Upper) Drawing of the sperm whale *Physeter catodon*. Adult males are typically 60 feet long and weigh 50 tons. (Lower) Diagrammatic (*A*) sagittal and (*B*) transverse sections of the head of a sperm whale adult. The transverse section is taken on line a-a. Reprinted, by permission, from Clarke (1978a).

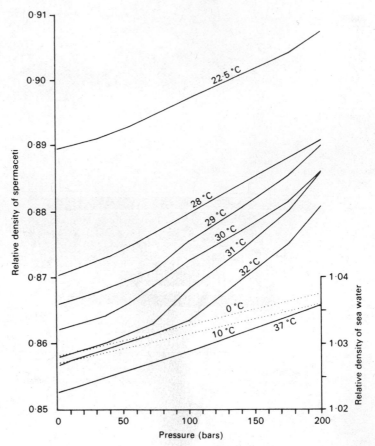

FIGURE 8.7. The relative density of spermaceti oil at various temperatures and pressures. 1 bar ≈ 1 atmosphere. The right-hand scale relates to the relative density of sea water at 0 and 10°C shown as dotted lines. Reprinted, by permission, from Clarke (1978b).

quently, less seawater is displaced, and the buoyancy generated is reduced. Increased pressure also causes oil density to increase and thus leads to reduced buoyancy. The combined effects of temperature and pressure on the density of spermaceti oil are depicted in Figure 8.7. The effect of temperature on the density of seawater is also an important factor here. The colder water found below the surface has a higher density and therefore generates more uplift than the warmer surface water, especially in equatorial regions. If a whale is to remain neutrally buoyant at depths of 200 m or more, the increased density of the spermaceti oil is necessary to counter the increased lift provided by colder water at these depths.

Is the sperm whale capable of dissipating enough heat from the spermaceti organ during the descent to maintain neutral buoyancy at depths greater than 200 m over its entire geographical range? Furthermore, is sufficient heat available to the whale in the diving cycle for reheating the oil? Anatomical studies and physiological estimates performed by Clarke suggest that the answer to both questions is yes. The tissues that house the spermaceti oil are highly vascularized and thus provide a means for rapid heat exchange. This exchange is enhanced by the countercurrent arrangement of the large arteries and veins that serve the spermaceti organ, and also by asymmetrical nasal passages (see Figure 8.6). The right nasal passage, which is often more than a meter wide, passes through the core of the spermaceti organ; it also forms two sacs, which cover the front and rear of this organ. Because spermaceti tissue lies in intimate contact with the right nasal passage, the spermaceti organ can be cooled when water enters the nasal passage. Clarke's calculations show that heat loss through the skin of the whale's snout combined with that dissipated through the nasal passage should be rapid enough to enable the whale to reach neutral buoyancy within three to five minutes of attaining the desired depth. At this point, vasoconstriction would prevent any further cooling of the oil. During the time that the whale is submerged, heat produced by the large muscle masses will increase the body temperature of the whale. By conducting the warm blood from the interior of the body forward to the spermaceti tissue just before beginning the ascent, a mechanism exists for heating the solidified spermaceti oil and converting it back to a fluid state. The oil, which is now less dense, increases in volume. This causes the whale's buoyancy to shift from neutral to positive, thereby facilitating its rise to the surface.

While the anatomical pecularities of the snout and the physical properties of spermaceti oil strongly suggest that sperm whales could use the spermaceti organ to control their buoyancy, measurements of temperature or density of the oil *in situ* in a diving sperm whale are needed to verify the hypothesis proposed by Clarke. There is also the possibility that whales use other structures (e.g., the lungs) or processes in regulating body density, and it appears highly probable that the spermaceti organ may also function in other capacities (e.g., sound production, see Chapter 9). However neutral buoyancy is achieved, its adaptive value to this species is evident. By being able to remain more-or-less motionless at a specific depth, the whale can conserve energy and thus remain submerged for longer periods. The quiet, stationary behavior may also facilitate the capture of squid at these great depths. If for some reason the whale should become exhausted while submerged, the lift provided by the rewarmed spermaceti organ would enable the whale to rise to the surface without having to expend the energy associated with swimming. Finally, the heat exchange capacities of the spermaceti organ, which operate during buoyancy control, may be very important in the thermoregulation of these large animals, especially while they are swimming in warm equatorial waters.

CONCLUDING REMARKS

The low density and incompressibility of lipids make them ideal compounds for use in buoyancy control. How and to what extent lipids are employed in density regulation, however, varies widely among plants and animals. The extensive deposits of wax esters in certain marine zooplankton and the enormous fatty liver of sharks are probably the primary means by which neutral buoyancy is achieved in these animals. On the other hand, lipid-based density control in teleosts is often secondary to that achieved through the presence of air-filled floats or reduction in heavy body substances. Although the effectiveness of lipids in providing buoyancy for plants and animals is less than that obtained by the use of gas-filled floats, lipids are superior in terms of their pressure independence and energy requirements. Except for the high initial cost of procurement and synthesis, lipids require little if any maintenance, whereas gas floats (e.g., swimbladders) need a constant supply of energy to assure proper gas volumes, especially in species that perform large-scale vertical migrations. Even rigid-walled gas floats, which are perhaps the most effective of all buoyancy devices, probably require some energy for their maintenance (Schmidt-Nielsen, 1979).

We have seen two basic patterns by which lipids are used to control buoyancy. One is to simply increase their overall concentration in the body. Although the resultant lower density improves an organism's flotation, it may be energetically inefficient to synthesize a mixture of lipids when all compounds are not equally effective buoyancy agents. In addition, it is likely more difficult for the organism to adjust lipid levels to meet changes in buoyancy requirements if these compounds are distributed throughout the body. The second option is to deposit large quantities of a specific low-density lipid class (e.g., wax esters) or molecule (e.g., squalene) in a specialized body compartment or organ. This strategy satisfies some of the criticisms levied against option one; it also produces maximum lift for an organism while minimizing the weight gain or the body volume required to house the lipid compound. Furthermore, many of the low-density lipid compounds, either by virtue of their chemical nature or location in the body, are "immune" from metabolic degradation. As a result, the competition for their use as a fuel substrate versus their function in buoyancy control is largely avoided. Clearly, organisms can use lipids as they do other mechanisms to fine tune their body density in response to buoyancy demands. The biochemical processes operative in this regulation, plus the interaction between lipids and other means for controlling buoyancy, are just beginning to be understood and represent an important area for future research.

REFERENCES

Alexander, R. McN. (1966). Physical aspects of swimbladder function. *Biol. Rev.* **41**:141–176.

Alexander, R. McN. (1972). The energetics of vertical migration by fishes. *Symp. Soc. Exp. Biol.* **26**:273–294.

Altman, P. L. and D. S. Dittmer (1972). *Biology Data Handbook,* Vol. 1, 2nd ed. Fed. Am. Soc. Exp. Biol, Bethesda, Maryland.

Baldridge, H. D., Jr. (1972). Accumulation and function of liver oil in Florida sharks. *Copeia* **1972**:306–325.

Benson, A. A. and R. F. Lee (1975). The role of wax in oceanic food chains. *Sci. Am.* **232**:76–86.

Blaxter, J. H. S. and P. Tytler (1978). Physiology and function of the swimbladder. *Adv. Comp. Physiol. Biochem.* **7**:311–367.

Bone, Q. (1972). Buoyancy and hydrodynamic functions of integument in the castor oil fish, *Ruvettus pretiosus* (Pisces: Gempylidae). *Copeia* **1972**:78–87.

Bone, Q. and B. L. Roberts (1969). The density of elasmobranchs. *J. Mar. Biol. Ass. U.K.* **49**:913–937.

Childress, J. J. and M. Nygaard (1974). Chemical composition and buoyancy of midwater crustaceans as function of depth of occurrence off southern California. *Mar. Biol.* **27**:225–238.

Clarke, M. R. (1978a). Structure and proportions of the spermaceti organ in the sperm whale. *J. Mar. Biol. Ass. U.K.* **58**:1–17.

Clarke, M. R. (1978b). Physical properties of spermaceti oil in the sperm whale. *J. Mar. Biol. Ass. U.K.* **58**:19–26.

Clarke, M. R. (1978c). Buoyancy control as a function of the spermaceti organ in the sperm whale. *J. Mar. Biol. Ass. U.K.* **58**:27–71.

Clarke, M. R. (1979). The head of the sperm whale. *Sci. Am.* **240**:128–141.

Conover, R. J. and E. D. S. Corner (1968). Respiration and nitrogen excretion by some marine zooplankton in relation to their life cycles. *J. Mar. Biol. Ass. U.K.* **48**:49–75.

Corner, E. D. S., E. J. Denton, and G. R. Forster (1969). On the buoyancy of some deep-sea sharks. *Proc. Roy. Soc. B.* **171**:415–429.

Denton, E. J. (1961). The buoyancy of fish and cephalopods. *Prog. Biophys.* **11**:178–234.

Denton, E. J. and N. B. Marshall (1958). The buoyancy of bathypelagic fishes without a gas-filled swimbladder. *J. Mar. Biol. Ass. U.K.* **37**:753–767.

Gunn, C. R. and J. V. Dennis (1976). *World Guide to Tropical Drift Seeds and Fruits.* Demeter Press, New York.

Hoar, W. S. (1966). *General and Comparative Physiology.* Prentice-Hall, New Jersey.

Hochachka, P. W. and G. N. Somero (1973). *Strategies of Biochemical Adaptation.* Saunders, Philadelphia.

Hutchinson, G. E. (1967). *A Treatise on Limnology,* Vol. 2. Wiley, New York.

Lee, R. F., C. F. Phleger, and M. H. Horn (1975). Composition of oil in fish bones: Possible function in neutral buoyancy. *Comp. Biochem. Physiol.* **50B**:13–16.

Lewis, R. W. (1970). The densities of three classes of marine lipids in relation to their possible role as hydrostatic agents. *Lipids* **5**:151–152.

Malins, D. C. and A. Barone (1970). Glyceryl ether metabolism: Regulation of buoyancy in dogfish *Squalus acanthias. Science* **167**:79–80.

Marshall, N. B. (1972). Swimbladder organization and depth ranges of deep-sea teleosts. *Soc. Exp. Biol. Symp.* **26**:261–272.

Morris, R. J. (1975). The lipid structure of the spermaceti organ of the sperm whale (*Physeter catodon*). *Deep-Sea Res.* **20**:911–916.

Nevenzel, J. D. (1970). Occurrence, function and biosynthesis of wax esters in marine organisms. *Lipids* **5**:308–319.

Phleger, C. F. and A. A. Benson (1971). Cholesterol and hyperbaric oxygen in swim-bladders of deep sea fishes. *Nature* **230**:122.

Rigg, G. B. .and L. A. Swain (1941). Pressure-composition relationships of the gas in the marine brown alga, *Nereocystis luetkeana*. *Plant Physiol.* **16**:361–371.

Sargent, J. R., R. R. Gatten, and R. McIntosh (1973). The distribution of neutral lipids in shark tissues. *J. Mar. Biol. Ass. U.K.* **53**:649–656.

Sargent, J. R., R. F. Lee, and J. C. Nevenzel (1976). Marine waxes. In: *Chemistry and Biochemistry of Natural Waxes* (Kolattukudy, P. E., Ed.), pp. 49–91. Elsevier, Amsterdam.

Schmidt-Nielsen, K. (1979). *Animal Physiology: Adaptation and Environment,* 2nd ed. Cambridge Univ. Press, London.

Sculthorpe, C. D. (1967). *The Biology of Aquatic Vascular Plants.* St. Martin's Press, New York.

Slijper, E. J. (1979). *Whales.* Cornell Univ. Press, Ithaca, New York.

Smayda, T. J. (1970). The suspension and sinking of phytoplankton in the sea. *Oceanogr. Mar. Biol. Ann. Rev.* **8**:353–414.

Walsby, A. E. (1972). Structure and function of gas vacuoles. *Bacteriol. Rev.* **36**:1–32.

Walsby, A. E. and C. S. Reynolds (1980). Sinking and floating. In: *The Physiological Ecology of Phytoplankton* (Morris, I., Ed.), pp. 371–412. Univ. Calif. Press, Berkeley.

Weast, R. C., Ed. (1981). *Handbook of Chemistry and Physics,* 62nd ed. The Chemical Rubber Co., Cleveland, Ohio.

Wittenberg, J. B. (1960). The source of carbon monoxide in the float of the Portuguese man-of-war, *Physalia physalis* L. *J. Exp. Biol.* **37**:698–705.

9

COMMUNICATION

Communication, in one form or another, is a very widespread and essential biological phenomenon. For social animals, such as bees and humans that live in complex societies, communication provides a mechanism through which individuals interact with one another and by which members are organized according to their status and functions (Shorey, 1976). Communication is also a necessity for most solitary individuals. It is often less obvious because the transmission of information may occur only at critical times in the life of these animals (e.g., premating communication) and because often it is more difficult to detect. The extent to which animals use communication in their daily lives is directly related to the degree of development of their sensory-neural system. Animals with simple nervous systems will have fewer channels of communication available to them than animals with well-developed senses and a highly evolved nervous system (Frings and Frings, 1977). Still, communication via a single channel (e.g., chemical communication) in a one-celled organism may be more critical to the well-being of that individual than information transmitted via a suite of communicative channels in members of more advanced phyla.

While the importance of communication cannot be denied, few scientists agree as to what exactly constitutes "biological communication." Numerous definitions have been proposed. These range from rather straightforward statements such as the "response of organisms to a stimulus" to more complex, restrictive definitions in which intraspecificity, reciprocity, and the presence of specialized sensory structures are required criteria (see Burghardt, 1970). It is not the purpose of this chapter to either defend or criticize these definitions, but rather to select one around which the discussion of lipids relative to communication can be built. Wilson (1970) has defined communication as "an action on the part of one organism (or cell) that

247

alters the probability pattern of behavior in another organism (or cell) in an adaptive fashion.'' This particular definition is appealing because of its breadth and because it does not eliminate plant-animal interactions, nor does it restrict the level of organization at which the communication can occur. Still, some of the interactions that are described in the following sections may not qualify as true communication based on this definition. These interactions, however, regardless of how defined, do exist in nature and strengthen the evidence for the unique and varied role of lipids in information transfer in biological systems.

The stimulus modalities and associated sense organs used by animals in communication include chemical (olfactory or gustatory), mechanical (tactile or sonic), and radiational (light perception or visual). Although a case could probably be made for the role of lipids in each of the above, the two modalities for which such a role has been firmly established are chemical and mechanical. In chemical communication, lipids or their derivatives often serve as the actual transmitted signal, whereas in mechanical communication lipid-containing structures are responsible for the transmission of sound waves. As in previous chapters, emphasis here is on those physical and chemical properties of lipids that make them well suited for these communicative functions. The enormous literature on chemical communication necessitates that only selected examples be used for illustrative purposes. Coverage in this section, however, has been purposely broadened to include the role of lipids in communication at the cellular level and between plants and animals.

CHEMICAL COMMUNICATION

Occurrence and Basic Terminology

Chemical communication is the most widespread mode of information transfer among organisms. With the exception of birds and higher primates, it also appears to be the major communication channel used by animals (Shorey, 1976). The response of primordial single-celled organisms to chemical substances released into the environment indicates that this form of communication appeared early in animal evolution and may indeed be responsible for the high level of sociality found in certain insect and mammalian groups. Wilson (1970) has considered the interesting possibility that these chemical releasers, or pheromones, are in a special sense the lineal ancestors of hormones.

Rapidly expanding research with chemical-releasing stimuli has resulted in many synonymous and confusing terms pertaining to chemical communication. Nordlund (1981) has reviewed the existing terminology and proposed a system based on function that is satisfactory for the intended coverage here. He uses the term ''semiochemicals,'' which was first proposed by Law and Regnier (1971), for chemicals that mediate interactions between organisms, and divides the semiochemicals into two major categories: pheromones and allelochemics. The former term has gained wide acceptance among scientists worldwide, although like communication no single definition has been universally agreed upon. Nordlund defines pheromone as a substance that is secreted by an organism to the outside and that causes a specific reaction in a receiving organism of the same species (e.g., a hydrocarbon produced by a female moth that attracts male moths from a distance). Allelochemic is defined as a chemical that is significant to organisms of a species different from its source for reasons other than food. The major distinction between these two terms is whether the interactions are intraspecific (pheromone) or interspecific (allelochemic). Nordlund further subdivides allelochemic responses into four groups: allomones, kairomones, synomones, and apneumones. The definition for each of the first three terms has a common base—a substance produced or acquired by an organism that, when it contacts an individual of another species in the natural context, evokes in the receiver a behavioral or physiological reaction. In the case of allomones, the reaction is adaptively favorable to the emitter but not to the receiver. Defensive secretions produced by plants and animals fall under this category. In the case of kairomones, the reaction is adaptively favorable to the receiver but not to the emitter. Secondary plant substances that are used for aggregation or feeding by herbivores on the emitting plants are examples of kairomonal stimulants. Synomones refer to chemicals that mediate mutualistic interactions; that is, the response is adaptively favorable to both the emitter and receiver. Substances such as floral scents that attract pollinators of various types would qualify as synomones. Apneumones differ from these other allelochemic groups in that the chemicals mediating the interactions between individuals of different species originate from a non-living source (Nordlund, 1981). Interactions of this nature are poorly known, and are not considered in the discussion that follows.

Another series of terms is often used in conjunction with the different categories of semiochemicals. These terms, proposed by Dethier et al. (1960), describe the manner in which the chemicals modify animal behavior. Response categories include arrestant, attractant, repellent, stimulant, and deterrent. Thus, the hydrocarbon released by the female moth that draws in male moths functions as a sex attractant pheromone, while the same or different chemical that prompts copulatory behavior serves as a mating stimulant pheromone. Similar combinations are also used to more precisely describe the action of allelochemics.

Semiochemicals: Physical and Chemical Features

Semiochemicals may be transmitted as liquids or gases that are either smelled or tasted by an animal. A chemical signal released as a gas diffuses from the point of emission, creating a zone within which the concentration of the molecules is sufficient to elicit a behavioral or physiological response on the part of the receiver. This zone is referred to as the "active space" (Wilson and Bossert, 1963); its shape can be spherical or ellipsoidal (Figure 9.1), depending on whether the chemical is released into still air or into wind. Among the factors that determine the size of the active space are the rate at which the signal is released by the emitter, the minimal concentration of molecules detectable by the responder, and properties of the medium through which the molecules must diffuse.

A highly variable and very important factor in estimating active space volumes is the response threshold to the semiochemical. Several investigators have noted the almost unbelievable signal power of some chemical attractants. For example, Wilson (1963) reports that male silkworm moths respond to a concentration of the sex attractant bombykol that is no more than a few hundred molecules per cubic centimeter. The acute sensitivity

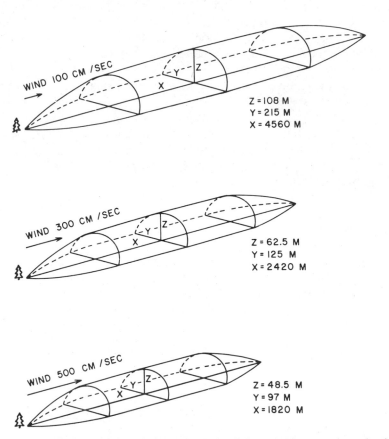

WIND 100 CM/SEC

Z = 108 M
Y = 215 M
X = 4560 M

WIND 300 CM/SEC

Z = 62.5 M
Y = 125 M
X = 2420 M

WIND 500 CM/SEC

Z = 48.5 M
Y = 97 M
X = 1820 M

FIGURE 9.1.   Predicted active spaces of a pheromone in the air downwind from a continuously releasing source. As wind shifts from moderate to strong, increased turbulence contracts the active space. Reprinted, by permission, from Wilson and Bossert (1963).

to the released chemicals enables the signal to be detected at a considerable distance from the point of release. Using a formula that takes into account the release rate of the semiochemical, the response threshold, and the wind velocity, Wilson and Bossert (1963) estimated that the mean maximum communication distance between a pheromone-releasing female and a responding gypsy moth male is 4.6 km when wind velocity is 1 m per sec. Field tests using mark-recapture techniques have produced potential communication distances of 1 km to sometimes over 10 km (Shorey, 1976).

Many semiochemicals that function over a long distance serve as sex attractants or alarm substances. Their perception relies upon smell. Other semiochemicals, especially defensive compounds, feeding deterrents, or phagostimulants, require actual physical contact or near contact between the communicating partners. Since these chemicals diffuse scarcely or not at all outside the organism that produces them, their active space is essentially stationary (Daloze et al., 1980). Communication proceeds through taste, and often requires actual gustation.

The mode of action and dispersal of airborne semiochemicals impose certain restraints on the chemical properties of these messengers. A set of "rules" pertaining to the molecular size of pheromone molecules was proposed by Wilson and Bossert (1963). They predicted that the majority of pheromones should contain between 5 and 20 carbons and have molecular weights between 80 and 300. There are several reasons for this predicted size range. First, the number of compounds containing five or less carbon atoms is relatively small, and even fewer of these compounds can be manufactured and stored by organisms. Above the lower limit the molecular diversity increases very rapidly. For example, the number of possible kinds of hydrocarbons increases in an approximately exponential manner as the number of carbon atoms is increased. With the addition of an oxygen atom to produce alcohols, aldehydes, ketones, and so on, the molecular diversity becomes substantially greater. Also contributing to the greater diversity is a concomitant increase in the number of possible structural and geometric isomers. The saturated acyclic monoterpene alcohol whose formula is $C_{10}H_{22}O$ has nine isomeric forms, many of which possess a distinct odor (Bedoukian, 1970).

In addition to molecular diversity, Wilson and Bossert also considered the volatility and diffusability of molecules, as well as their stimulative efficiency in predicting an optimum size range for airborne semiochemicals. In a homologous series, smaller molecules have higher vapor pressures and hence greater volatility than larger molecules. Smaller molecules also have higher diffusion coefficients, although this is not a significant factor when turbulent transfer is involved. One problem with very small molecules, however, is that they often appear to have a lower stimulative efficiency. In one series of esters tested on flies, doubling of molecular weight resulted in as much as a thousandfold increase in efficiency (Wilson, 1963). In the ant *Acanthomyops claviger,* long chain *n*-alkanes (i.e., C_{11}, C_{13}) are far more

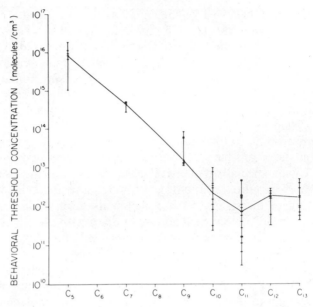

FIGURE 9.2. Estimates of behavioral threshold concentrations of members of the alkane series from pentane (C_5H_{12}) to tridecane ($C_{13}H_{28}$) in the ant *Acanthomyops claviger*. Reprinted with permission from F. E. Regnier and E. O. Wilson, The alarm-defence system of the ant *Acanthomyops claviger*. *J. Insect Physiol.* **14.** Copyright 1968, Pergamon Press, Ltd.

efficient in eliciting an alarm reaction than their lower homologs (Figure 9.2). As the upper limit is approached, any further benefits due to diversity and stimulative efficiency are outweighed by negative factors. Energetically, it becomes quite costly to synthesize, transport, and store large molecules. Moreover, the volatility decreases to the point where the gaseous diffusion of such molecules becomes impracticable (Wilson and Bossert, 1963).

The predictions pertaining to the expected size of pheromone molecules were made at a time when the number of these compounds and the knowledge of their chemistry was quite limited. The subsequent discovery of numerous semiochemicals in phylogenetically diverse organisms, plus the advent of modern analytical techniques, have resulted in some exceptions to these "rules." Still, the majority of semiochemicals known today fit the criteria proposed by Wilson and Bossert. There are hydrocarbon pheromones with molecular weights greater than 300 (see Howard and Blomquist, 1982), but in most cases these appear to function principally as contact stimulants rather than long-distance messengers.

Lipid-based Semiochemicals

The majority of semiochemicals identified thus far are lipids, lipid derivatives, or miscellaneous compounds that are more closely related to lipids by virtue of their insolubility in water than they are to the other major organic groups in biological systems. Why the preponderance of lipids? For one thing, lipids as a group most closely satisfy the "requirements" described above for airborne chemical messengers. There is a tremendous variety of lipid molecules that fall within the predicted optimum size range. This molecular diversity is greatly increased by the ability of lipids to form structural and geometric isomers, and by alterations in molecular structure resulting from the substitution of functional groups. Lipids are relatively stable compounds and are not likely to be enzymatically deactivated or detoured into other chemical reactions. Moreover, the chemical nature of lipids, in combination with an appropriate chain length, gives many of these molecules a high degree of volatility. Neither carbohydrates nor proteins are as well suited. Carbohydrates, especially the monosaccharides, offer considerable potential for molecular diversity within the stated molecular size range; however, because of their many hydroxyl groups they are essentially nonvolatile. Proteins also have an essentially zero vapor pressure (and hence zero volatility) and can only be transmitted in air if adsorbed onto dust particles or moisture bubbles (Wilson, 1970). The molecular weight of both carbohydrates (polysaccharides) and proteins greatly exceeds the upper limits predicted for semiochemicals, although as we shall see, proteins do function as signal compounds in aquatic environments.

The major lipid classes and chemical structures of some representative semiochemicals are given in Figure 9.3. Nonisoprenoid hydrocarbons include saturated and unsaturated straight-chain and branched compounds. These hydrocarbon molecules are produced by specific glands or are part of the cuticular wax complex of arthropods, where they may also function in waterproofing (Chapter 5). Fatty acids have been reported as sex attractant pheromones in honey bees (Gary, 1962) and carpet beetles (Silverstein et al., 1967). Overall they do not appear to play a very important role in chemical communication, perhaps because fatty acids are much less volatile than the corresponding paraffin. Short-chain (C_1 through C_{10}) carboxylic acids, however, are potent olfactants and function as predator deterrents (Blum, 1981). Fatty acid derivatives (e.g., alcohols, aldehydes, ketones, acetate esters), in contrast, are utilized widely as signaling agents. Alcohols serve as sex pheromones in Lepidoptera (butterflies and moths) and Coleoptera (beetles). The alcohol *trans*-10, *cis*-12-hexadecadienol-1-ol, which is secreted by the female silkworm moth, *Bombyx mori,* has the distinction of being the first sex attractant isolated and identified (Butenandt et al., 1959). Short-chain alcohols tend to be fairly toxic and volatile, and thus are often found as part of the defensive secretions of arthropods, especially members of the Hymenoptera (bees and ants) (Blum, 1981). Alkyl aldehydes and ketones correspond to the fatty acids and monohydroxy alcohols, respectively. Although there are a few examples of these compounds functioning as pheromones (e.g., the alkyl ketones civetone from the civet cat and muscone from musk deer), their principal role appears to be as defensive

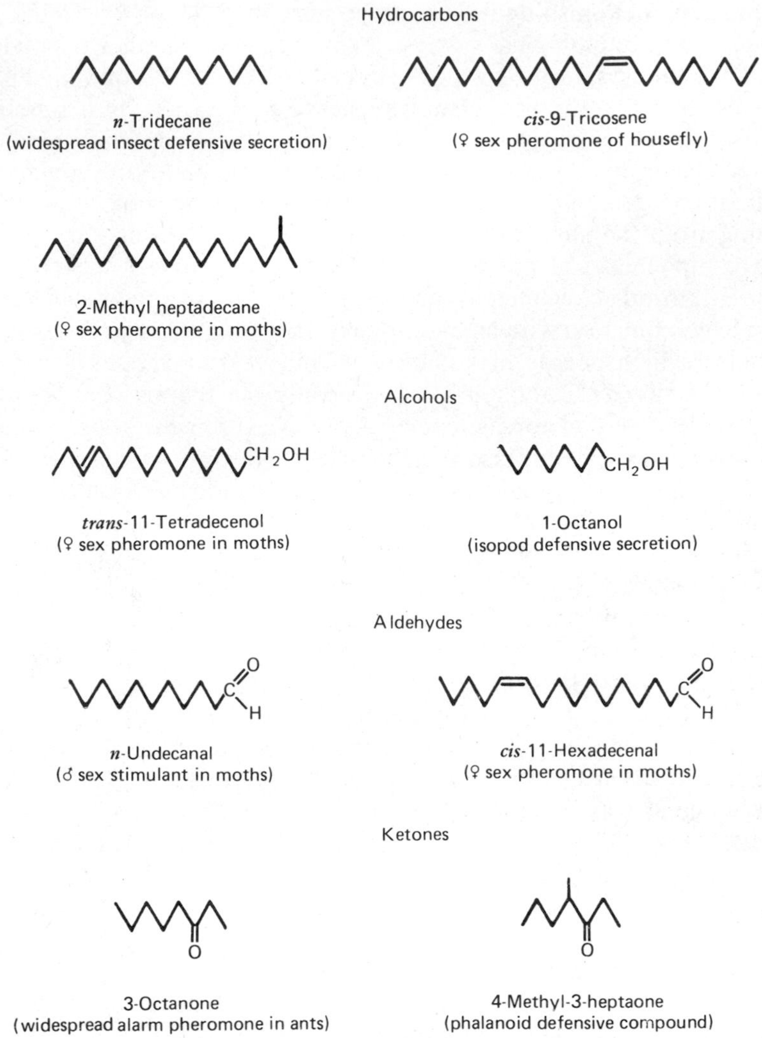

Hydrocarbons

n-Tridecane
(widespread insect defensive secretion)

cis-9-Tricosene
(♀ sex pheromone of housefly)

2-Methyl heptadecane
(♀ sex pheromone in moths)

Alcohols

trans-11-Tetradecenol
(♀ sex pheromone in moths)

1-Octanol
(isopod defensive secretion)

Aldehydes

n-Undecanal
(♂ sex stimulant in moths)

cis-11-Hexadecenal
(♀ sex pheromone in moths)

Ketones

3-Octanone
(widespread alarm pheromone in ants)

4-Methyl-3-heptaone
(phalanoid defensive compound)

FIGURE 9.3.   Major lipid classes and chemical structures of some representative semiochemicals in plants and animals.

allomones, especially among ant species. Many of the aldehyde and ketone compounds present in the defensive exudates are terpenoid compounds (see below).

Certainly the most abundant chemical form of sex attractants are the acetate esters (i.e., short- to medium-chain alcohols esterified to acetate). Of the 30 compounds identified as sex attractant pheromones in female Lepidoptera, 18 (60%) are acetates having one or two double bonds and most having 12, 14, or 16 carbons in their alcohol moiety (Tamaki, 1977). Esters often accompany classes of more reactive allomones (e.g., aldehydes, carboxylic acids) in the defensive exudates of arthropods, and are believed

Acetate esters

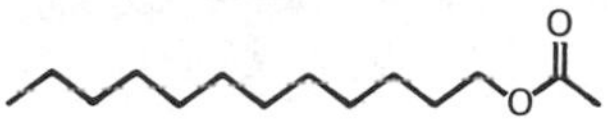

Dodecyl acetate
(♀ sex attractant pheromone
in moths)

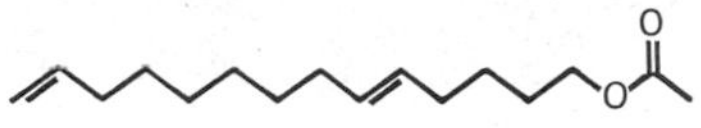

5-13-Tetradecadienyl acetate
(contact defensive secretion in
lepidopteran larvae)

Terpenoids (including steroids)

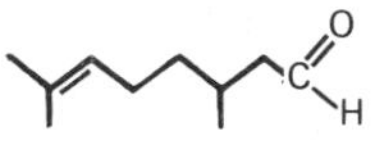

Citronellal
(terpene aldehyde;
ant defensive secretion)

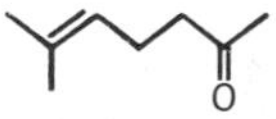

6-Methyl-5-hepten-2-one
(terpene ketone;
pollinator attractant produced by
orchid flowers)

α-Pinene
(bicyclic monoterpenoid; insect
attractant produced by pine trees;
termite defensive compound)

β-Pinene
(bicyclic monoterpenoid; insect
repellent produced by pine trees;
termite defensive compound)

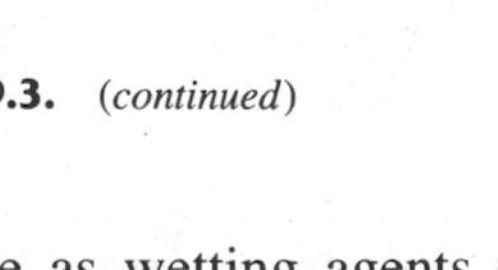

5α-androst-16-en-3-one
(steroid; musk odor produced
by female pig)

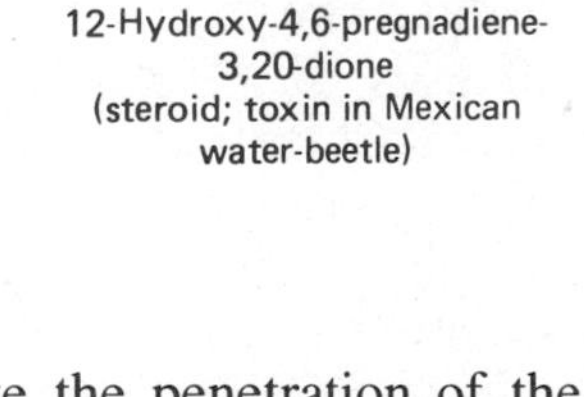

12-Hydroxy-4,6-pregnadiene-
3,20-dione
(steroid; toxin in Mexican
water-beetle)

FIGURE 9.3. (*continued*)

to serve as wetting agents or to facilitate the penetration of the primary irritants, thereby increasing the defensive "punch" of the secretion (Blum, 1981). There are a large group of esters in which the alcohol moiety is combined with a short-chained fatty acid (i.e., wax esters). They are quite volatile and almost without exception have strong and attractive odors despite the unpleasantness of some of the free acids (butyric, valeric, caproic) or free alcohols (amyl) of which they are composed (Needham, 1965). Such wax esters are primarily found in fruits, where they no doubt function as feeding attractants, which in turn leads to seed dispersal.

A large number of semiochemicals in plants and animals belong to a diverse collection of compounds known as isoprenoids or terpenoids (see Chapter 1). Biosynthetic studies have shown that these compounds are derived from mevalonic acid or a closely related precursor. The basic terpenoid skeleton is formed from isoprene or isopentane units:

$$\overset{\displaystyle CH_3}{\overset{|}{-CH_2-CH-CH_2-CH_2-}}$$

Although they are linked together in various ways and with various types of ring closures, degrees of unsaturation, and functional groups, the commonest arrangement appears to be "head-to-tail" (Robinson, 1980):

$$\overset{\displaystyle C}{\overset{|}{-C-C}}-C-\overset{}{C}\vdots \overset{\displaystyle C}{\overset{|}{C-C}}-C-C-$$

Major classes of isoprenoids, based on the number of carbons in the parent chain, include the monoterpenes (C_{10}), sesquiterpenes (C_{15}), diterpenes (C_{20}), sesterterpenes (C_{25}), triterpenes (C_{30}), and tetraterpenes (C_{40}). Steroids, an important lipid group that is usually given separate class status, are structurally triterpenes.

Of the various terpenoid classes, monoterpenes have the most important role in chemical communication. These compounds offer the potential for high information content as a result of their high volatility and considerable structural complexity. Monoterpenes occur in a variety of open-chain and cyclic structural types and derivatives (see Figure 9.3). The addition of an oxygen-containing functional group typically increases the scent as well as the compound's solubility in water. In insects, monoterpenes (e.g., terpenoid alcohols) function as defensive substances or as sex or alarm pheromones. These compounds are typically synthesized *de novo,* although some that appear in defensive secretions may be initially acquired in the animal's diet. The numerous monoterpenes identified in plants often function in attracting pollinating insects and seed-dispersing animals, in repelling insect pests and browsing animals, and in inhibiting the germination or the growth of potential competitors (Croteau, 1981). The principal function of monoterpenes, and for that matter most lower terpenoids, in both plants and animals appears to be ecological rather than physiological.

Multicomponent Semiochemical Systems

In the early studies on insect attractants, the pheromone in question was reported to be a single chemical. The concept of "one insect–one specific compound" became strongly implanted among researchers in the developing pheromone field. Field trials based on these single compounds, however, were almost invariably disappointing. Moreover, the purer the compound, the more disappointing were the results (Silverstein, 1977). The first demonstration of a multicomponent attractant pheromone was provided by Silverstein et al. (1966), who showed that the pheromone produced by the male bark beetle, *Ips paraconfusus,* which attracts both males and females, was a mixture of three terpene alcohols, none of which was particularly active in the field by itself. Subsequent studies have shown that, although there are pheromones that consist of a single active compound, multicomponent systems predominate. Most of the systems so far identified in insects are $n = 2$ systems; $n = 3$, $n = 4$, and even $n = 10$ systems have been reported, but in few cases has the biologic activity of each component been identified (Tamaki, 1977). With few exceptions, defensive secretions also appear to be remarkably complex mixtures. Brand et al. (1973) found that the defensive secretion of the beetle *Drusilla canaliculata* contains no less than 14 components, while Tschinkel (1975b) identified 13 compounds in the defensive secretion of the tenebrionid beetle *Eleodes beameri.*

The effectiveness of a pheromonal signal is determined by not only the number and types of compounds present, but also by the relative quantity of each component. An example of the importance of a specific ratio of pheromonal components has been demonstrated in the armyworm moth, *Spodoptera litura,* by Yushima et al. (1974) (Table 9.1). The *cis*-9, *trans*-11-tetradecadienyl acetate alone showed only weak activity as an attractant in the field. The activity was greatly enhanced by adding *cis*-9, *trans*-12-tetradecadienyl acetate, which is a naturally occurring synergist in the multicomponent sex-pheromone system of this species. In the 15 species of Lepidoptera for which multicomponent sex attractant pheromones have been found, the ratio of components is critical for biological activity in all but two species (Tamaki, 1977).

Although a blend of compounds appears to have more potential for information transfer than is possible with a single compound, the contribution

TABLE 9.1
Attraction of male *Spodoptera litura* to different mixtures of
pheromonal components

Ratio of components $(c9,t11\text{-}14\!:\!Ac/c9,t12\text{-}14\!:\!Ac)$[a]	Number of males caught by 3 traps for 6 nights
0/10	0
5/5	90
8/2	253
9/1	238
19/1	305
39/1	318
10/0	20

Source: Reprinted, with permission, from Yushima et al. (1974).

[a]*cis*-9, *trans*-11-tetradecadienyl acetate/*cis*-9, *trans*-12-tetradecadienyl acetate.

of individual compounds in a multicomponent system to the actual chemical message is poorly understood in most cases. Certainly the synergistic role cannot be minimized. Often no overt change in behavior is elicited by single components and a mixture of the two is essential. In other cases, an inactive component will have a synergistic effect on a component that alone is only slightly active. In some pheromonal blends, it is believed that the various individual components may elicit different behavioral responses from the receiver. For example, major or primary pheromone components may be involved in long-distance detection, whereas minor or secondary components, which are released as part of the pheromone, may function in mediating close-range responses, such as landing or wing fanning (Roelofs, 1981). Minor components might also alter the behaviors associated with primary pheromone components by changing the threshold of receptors to the primary component(s). Finally, there are also data that suggest that certain components of a multicomponent sex attractant pheromone may also function concurrently as an interspecific chemical signal by suppressing the sexual excitement and mating behavior of a sympatric species, thus helping ensure sexual isolation. In such instances, the pheromones are also functioning as allomones.

The functional significance of minor or trace constituents in the complex mixtures that characterize most defensive secretions is also poorly understood. A synergistic role, so important in the multicomponent pheromonal systems of arthropods, has also been demonstrated for some of these minor constituents (Pasteels et al., 1983). Many of the major compounds that serve as tissue irritants are polar, and hence have difficulty spreading over and penetrating the cuticle of arthropod predators without the presence of more lipophilic trace components in the secretion mixture. This function was elucidated by Eisner et al. (1961) in their study of the defensive secretion of the whip scorpion, *Mastigoproctus giganteus*. The major constituent of the defensive secretion of this species is acetic acid (84%). Its effectiveness as a topical irritant is facilitated by the presence of octanoic acid (5%), which serves as a spreading agent and lipid solvent, enabling the two-carbon acid to penetrate into the interior of the body. A similar role has been proposed for the hydrocarbons present in the defensive secretions of formicine ants (Blum, 1981). Hydrocarbons may also help prolong the activity of such defensive secretions by delaying the evaporation of more volatile irritants present in the mixture. The possibility that the complexity of defensive blends might be a way to avoid counteradaptation by predators is intriguing (Pasteels et al., 1983), but thus far there is no evidence to support this hypothesis.

Aquatic Versus Airborne Semiochemicals

What about chemical communication in aquatic environments? Earlier in the chapter a set of "rules" pertaining to expected physicochemical properties of semiochemicals used by terrestrial organisms was discussed. Are the features that characterize airborne messengers the same for chemical signals transmitted in water?

Examination of chemical communication by aquatic organisms indicates that for the most part the physical and chemical prerequisites of waterborne semiochemicals are quite different from those of airborne messengers, especially in the case of long-range semiochemicals. Molecules in air or water are dispersed in the surrounding medium by diffusion (Brownian movement) or passive transport (wind, currents, etc.). To function as airborne semiochemicals, the messenger molecules must necessarily be volatile, which in turn imposes an upper limit to both their molecular weight and number of polar functions (see the section on the physical and chemical features of semiochemicals earlier in this chapter). In aqueous environments, the condition of volatility is replaced by that of solubility; there is little limitation as to the molecular size of the chemical signal (Daloze et al., 1980). Indeed, a survey of the aquatic semiochemicals for which the chemical structure is known shows a range of very small to very large molecules. Many of these are protein or polypeptides with molecular weights from 10,000 to 200,000 (Wilson, 1970). Protein molecules are relatively polar compounds with a moderately high solubility in water despite their large size. It may be possible for quite insoluble substances to function as long-distance chemical signals in aquatic environments if their behavioral threshold concentration is extremely low or if they are propagated as suspensions instead of solutes (Daloze et al., 1980).

A second point of contrast between chemical communication in water and in air involves the diffusion rates of semiochemicals. If the surrounding medium is still, the transport of long-distance semiochemicals will be determined primarily by their diffusion coefficients. For most water-soluble substances with a molecular size similar to that of airborne semiochemicals, the diffusion coefficient is about 10^{-5} cm^2 per sec in water. This rate is about 1000 times less than the diffusivity in air, and results in significant changes in the properties of the active space of the chemical signal. Wilson (1970) provided theoretical data that show that, if the semiochemical is released discontinuously, the maximum radius of the active space is the same in water as in air, but the time required to reach the maximum radius and the interval between release of the chemical and the disappearance of the active space (i.e., "fade-out" time) are approximately 10,000 times greater in water than in air. To compensate, an aquatic species would have to increase the emission rate of the semiochemical one million times or lower the response threshold a millionfold, or alter each of these parameters by some combined equivalent. These requirements become much less demanding if the water is flowing. Organisms that are able to place the semiochemical in natural currents or in currents created by the organism will be able to spread the signal much more quickly.

The solubility and diffusivity problems imposed by water obviously limit the use of lipid-based semiochemicals by aquatic organisms. Nonetheless, lipids or lipid derivatives have been identified as chemical messengers in a

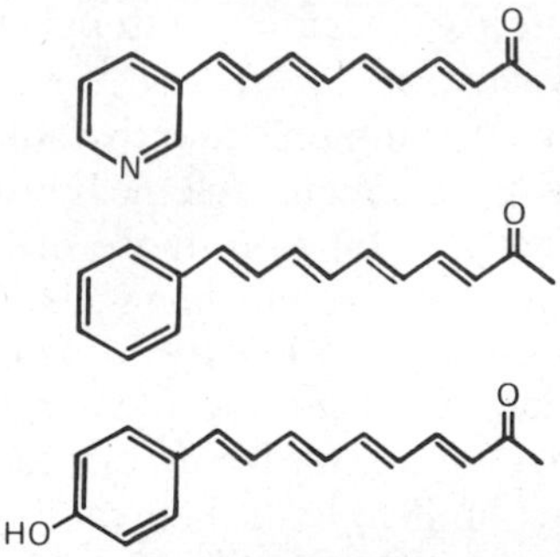

FIGURE 9.4. Active substances secreted by *Navanax*, which induce alarm and avoidance responses in other sea-slugs.

variety of freshwater and marine forms. These compounds are often associated with communication over short distances where the chemical signal needs to diffuse scarcely or not at all outside of the organism that produces it. Some examples here include the sex pheromones of water mold (Fungi), which are oxygenated sesquiterpenes (Machlis et al., 1966), gamete attractants produced by brown algae, which are polyunsaturated linear or cyclic hydrocarbons (Jaenicke, 1977; Müller and Gassmann, 1978), and three methylketones (Figure 9.4) that serve as the active compounds in the alarm substances secreted by the sea-slug *Navanax inermis* (Sleeper and Fenical, 1977). Many higher invertebrates (molluscs, crustacea) also use pheromones to mediate sexual activities. It has been suggested by several investigators that the steroid molting hormone in crustaceans, crustecdysone, may also be released by the female to attract male crabs (*Pachygrapsus crassipes*), although Dunham's (1978) review of the experimental evidence raises some serious doubts. Chemical signals appear to be extremely important in regulating the social behavior of fish (Bardach and Todd, 1970); however, few of these compounds have been chemically characterized and, with the possible exception of estrogen released by females, most appear to be nonlipid compounds.

Chemical Signals and the Regulation of Behavior

The material presented thus far in this chapter is intended to provide background information on the breadth and complexity of chemical communication in organisms, and specifically to identify those properties of lipid compounds that make them well suited to serve as chemical messengers in terrestrial and, to a much more limited extent, aquatic environments. In describing the types of lipids that function as semiochemicals and how physical factors in the environment affect their performance, many examples have already been given on how these chemical substances modify or control the behavior of organisms. The objective in the remaining section on chemical communication is to examine in more detail some of the important behaviors that are directly or indirectly regulated by lipid-based semiochemicals. As mentioned earlier, coverage here necessarily has to be incomplete in view of the enormous literature that exists. The reader interested in additional examples including nonlipid semiochemicals and/or more applied aspects related to chemical communication should consult the comprehensive treatises provided by Shorey (1976), Ritter (1979), Blum (1981), Nordlund et al. (1981), and Harborne (1982).

Chemical Signaling Between Cells

Although the major theme of this chapter concerns semiochemicals and communication between organisms, the use of chemical signals to regulate cellular processes within individuals is basic to all plants and animals, and thus warrants at least brief mention. Chemical signals can operate in a variety of ways, depending on the type of organism and the complexity of its development. Some signaling molecules are secreted into the extracellular fluid and act only on cells in the immediate vicinity. These are termed local mediators; most are taken up or enzymatically destroyed almost as fast as they are released. One important group of lipids that function as local chemical mediators are the prostaglandins, a family of 20-carbon fatty acid derivatives that contain a cyclopentane ring in the middle of the fatty acid chain (Figure 9.5, see also Chapter 1). Members of the prostaglandin family (e.g., PGE, PGF) are determined by differences in the position of oxygen atoms on the cyclopentane ring. Prostaglandins are synthesized from fatty acid moieties cleaved from membrane phospholipids, and are continuously being released to the exterior where they bind to cell-surface receptors (Alberts et al., 1983).

Prostaglandins initiate or influence a wide variety of biological activities. They stimulate uterine contractility, elevate blood pressure by causing arteriolar constriction (PGF_2-alpha), inhibit gastric secretion (PGE group), inhibit transmission of nerve impulses, and are released in trauma and shock (Strand, 1983). The diversity of functions suggests that prostaglandins operate at some elementary level in the cell communication system. One likely

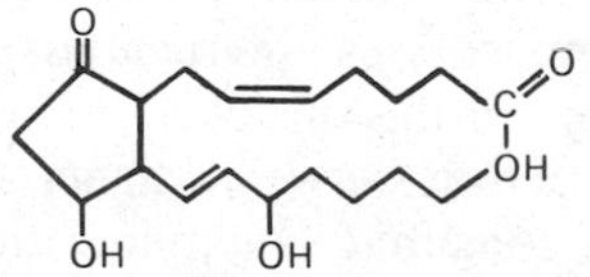

FIGURE 9.5. Structure of the prostaglandin PGE_2. The subscript refers to the two double bonds outside the cyclopentane ring.

possibility is that cyclic AMP, the ubiquitous "second messenger" of cells, serves as the mediator for prostaglandins just as it does for hormones.

A second type of chemical communication at the cellular level that employs lipid messengers is the endocrine system. Steroid hormones secreted by the gonads, placenta, and adrenal cortex of mammals travel via the blood to various target organs located throughout the body. There are three basic types of steroid hormones, each synthesized from a cholesterol precursor: androgens (C_{19} compounds), estrogens (C_{18} compounds), and progesterones and corticosterones (C_{21} compounds). Steroid hormone functions include the regulation of sexual behavior, the development and maintenance of secondary sex characteristics, and, in the case of corticosterones, the regulation of carbohydrate, water, and salt metabolism. More detailed information on these and other functions in mammals, plus additional behavioral and physiological processes influenced by steroids in lower vertebrates, is provided in most basic and comparative physiology texts.

Unlike the majority of hormones, which are peptides or amino acid derivatives, steroid hormones are relatively insoluble in water and must be bound to specific carrier proteins for transport in the blood. The hydrophobic nature and relatively small molecular size of steroid hormones, however, enable them to pass directly through the lipid bilayer of a target-cell plasma membrane and to ultimately bind with specific receptor sites inside the cell. Hydrophilic signaling hormones, in contrast, activate receptor proteins on the surface of the target cell. Steroid hormones also persist in the blood for longer periods than water-soluble hormones or local chemical mediators and, as a result, tend to mediate longer lasting responses.

Hormones also serve as chemical messengers in plants, although the system and chemical composition of these molecules are quite different from that described in animals. Most plant cells have the ability to synthesize hormones, many of which act more like local chemical mediators in that they are often translocated only a short distance within the plant. The only plant hormones that qualify as lipid compounds are the terpenoid-based dormin (= abscisin II), which is a growth inhibitor, and giberellins, a group of some 30 terpenoid compounds that promote stem elongation or cell division in plants (Devlin, 1975). Interestingly, both human male and female sex hormones are present in plants. Their function is open to speculation; they may have a natural role in relation to growth and flowering in plants or they may function as feeding deterents in mammals by disrupting the delicately balanced steroid hormone levels in females (see Harborne, 1982). Two other important lipid compounds, the molting hormone ecdysone (a steroid) and juvenile hormones (sesquiterpenoids), which control the different developmental stages in the life cycle of insects, are also found in plants. Although the interaction between plants and insects produced by these compounds may not constitute communication *per se*, by interfering with the natural development of insects, they provide the plant with a potentially excellent defense mechanism against insect predation.

Orientation and Recognition Pheromones

The use of pheromones to guide members of the same species to a food source, to a sexually receptive partner, to suitable habitats for colonization, or to simply move from one area to another when visual orientation is impaired is widespread among animals. In many cases the pheromone in question is deposited on the substratum in the form of a continuous line or discontinuous streak (Figure 9.6) (Shorey, 1976). The evaporation of the pheromone into the air forms an elongate odor trail that can be followed by individuals who maintain close contact with the surface as they move. Controlled experiments and field observations have shown that animals following such trails typically zigzag back and forth across the trails. This pattern likely reflects the periodic loss of the trail odor and subsequent chemotactic steering reactions necessary to reorient on the trail. It is likely that most vertebrates, and perhaps some invertebrates, supplement their olfactory senses with visual cues to gauge the direction of the trail.

The use of trail pheromones and their chemical nature have been most extensively studied in the social insects. Ants lay scent trails from a newly discovered food source back to the nest. The freshly laid trail contains highly volatile substances that produce a strong recruitment response in addition to guiding the ants to the food source. Other ant species, in searching for food, deposit trails that radiate out from the nest like spokes; these trails are used to help relocate the nest. The most common sources of trail pheromones are the hind-gut contents and a variety of glands, most of which are located in the abdomen (Parry and Morgan, 1979).

Chemically, ant trail pheromones are represented by a diverse number of compounds, which vary depending on the species. Included in this group

(A)

(B)

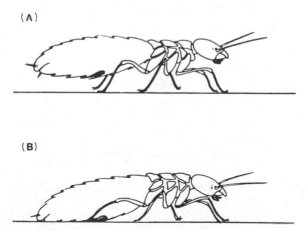

FIGURE 9.6. Schematic representation of a nymph of the termite *Zootermopsis nevadensis* with the sternal gland (shaded area) (A) elevated above the surface during normal activity and (B) contacting the surface during trail laying. Reprinted, by permission, from Stuart (1970).

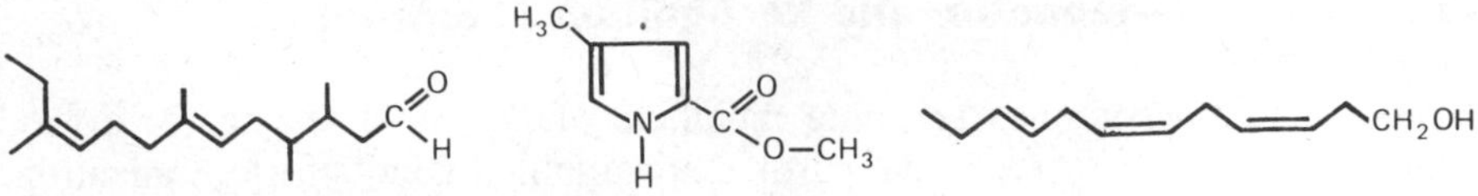

Faranal   4-Methylpyrrole-2-carboxylate   3-cis-6-trans-8-Dodecatriene-1-o1

(a)   (b)   (c)

FIGURE 9.7. Structures of some ant and termite trail pheromones.

are hydrocarbons (C_{15}, C_{16}, C_{17}), fatty acids (6:0–10:0, 12:0), faranal, and several heterocyclic nitrogen compounds. Faranal is a sesquiterpene aldehyde (Figure 9.7a) that bears a striking structural resemblance to juvenile hormone. Among the nitrogen heterocycles is 4-methylpyrrole-2-carboxylate (Figure 9.7b), a major trail pheromone utilized by the leaf-cutting ant *Atta texana* (Tumlinson et al., 1971). Because of the low detection threshold of this highly active substance (ca. 0.8 pg per cm), it has been calculated that 0.33 mg would be sufficient to lay a trail around the world.

Termites also use odor trails to recruit and direct members of the colony to a food source. These terrestrial pheromones are deposited from sternal glands located on the abdomen. One such compound associated with *Reticulitermes flavipes* is 3-cis-6-trans-8-dodecatriene-1-ol (Figure 9.7c). The same or very similar alcohol also occurs in the fungi that attack the wood upon which the termites feed. It is not certain if the termite synthesizes the pheromone *de novo* or simply ingests the fungally derived compound and channels it to the sternal gland before depositing it as a trail substance (Blum and Brand, 1972).

A social insect that does obtain its trail pheromone directly from a plant is the honeybee, *Apis mellifera*. Workers collect the monoterpenoid alcohol geraniol from the flowers that they frequent, concentrate the substance, and then release it to guide other workers to rich sources of food (Harborne, 1982). The pheromone can be released directly into the air, where it forms an aerial trail, or can be deposited as spots on various objects (e.g., leaves, branches), which provides a terrestrial trail to the food source.

Many pheromones produced by vertebrates and invertebrates convey information about the identity or some characteristic of the emitting animal. These pheromones, for example, may be used to recognize: (1) individuals of the same species, (2) the physiological or social status of the species' mates, (3) assemblages that represent colonies or social groups, or (4) an animal's nest, territory, or home range (Shorey, 1976). Mammals have been especially well studied in this regard. Sources of odors in mammals include the urine, feces, and a variety of skin scent glands that occur in different regions of the body (Mykytowycz, 1970). These odors may emanate or be deposited passively as the animal moves about (e.g., mule deer) or be applied directly in gland-to-object contact (e.g., rabbits that rub chin on branch). Chemical analyses of these secretions indicate that they are composed pre-

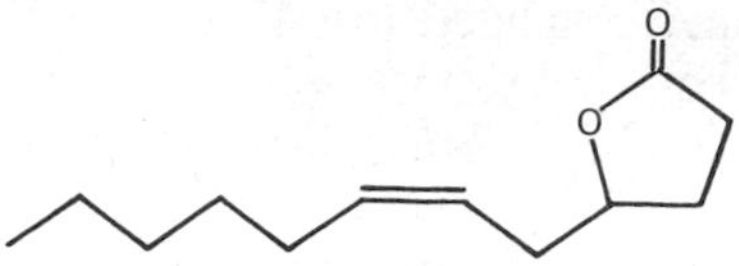

FIGURE 9.8.   Structure of *cis*-4-hydroxydodec-6-enoic acid lactone, the major chemical component in the tarsal gland secretion of male black-tailed deer.

dominantly of lipids or lipids mixed with proteins and sometimes carbohydrates. In the black-tailed deer, a species for which information on sex, age, identity, and mood is provided by odor signals, the principal active ingredient in the tarsal gland secretion is *cis*-4-hydroxydodec-6-enoic acid lactone (Brownlee et al., 1969) (Figure 9.8). The secretion from the nasal gland of the South American marsh deer is a complex lipid mixture containing cholesterol esters, mono- and diester waxes, and cholesterol (Jacob and von Lehmann, 1976). The role of recognition pheromones in nonmammalian vertebrates is no less important, although little is known regarding the chemical basis to the various signals. Oldak (1976) found that it was easy to distinguish the odors produced by the scent gland of snakes belonging to 11 different genera, and that the scent gland secretions of 25 species exhibit an identifiable and reproducible lipid pattern.

Social insects are well known for their ability to recognize conspecifics and to distinguish the caste and sex of individuals with which they interact. There is strong evidence to suggest that pheromones produced by these groups play an integral role in the recognition process. Subterranean termites are an ideal group for study in this regard. They live in a closed, low-volume system of interconnecting galleries distributed between the soil and wood. Semiochemicals operative in such a system must be complex enough to have a high information content but of low enough volatility to minimize sensory habituation (Howard and Blomquist, 1982).

It was initially postulated that cuticular hydrocarbons, which would possess the above physicochemical properties, might be serving as species- and caste-recognition cues in three species of termites: *Reticulitermes flavipes, Zootermopsis angusticollis,* and *Nasuititermes exitiousus* (Howard et al., 1978; Blomquist et al., 1979). Although the hydrocarbon profile for each species is unique, these data did not provide a good test of the hypothesis, since none of the three species is sympatric with one another. Subsequent studies of *R. flavipes* and *R. virginicus,* which are sympatric across much of their range and share similar ecological niches, have shown differences in the cuticular hydrocarbon composition that are probably sufficient to enable termites to distinguish whether or not another termite is of the same species (Howard et al., 1982). Experimental results from a bioassay developed by these authors to test the species-recognition hypothesis are also supportive. Additional evidence that hydrocarbons are serving as species recognition cues comes from the finding that some staphylinid beetles,

Trichopsenius frosti, have managed to integrate themselves into the social life of the *R. flavipes* colony by possessing the same complex mixture of cuticular hydrocarbons as their termite hosts (Howard et al., 1980).

Aggregatory and Sexual Behavior

The pheromones that regulate these behaviors refer to compounds liberated by animals of one sex (usually the female) that have the purpose of attracting a member of the opposite sex from a distance, and also inciting or facilitating the actual mating process. The same compound or blend of compounds may govern both types of behavior, or different substances may be required to both attract an individual from a distance and to stimulate close-range courtship or copulatory behavior. Sex pheromones have been extensively studied, especially in insects, where they have been used to manipulate the behavior of pest species. In the Lepidoptera alone, female pheromones have been recognized in over 200 species and male-released attractants in at least an additional 60 species (Harborne, 1982). In fact, Shorey (1976) estimated that as of the mid-1970s about one-half of the world's pheromone literature related to chemicals released by female moths that stimulate the approach of males prior to mating.

Chemically, sex pheromones are represented by a variety of predominantly lipid compounds. Most of the classes to which these pheromones belong are shown in Figure 9.3. In fact, much· of the discussion on the chemical nature of semiochemicals presented earlier in the chapter pertains to the sex pheromones. In insects, long-chain, unsaturated alcohols, acetates, or carboxylic acids are the most common chemical forms, although aliphatic cyclic compounds and even cyclohexane derivatives serve as sex pheromones in a few species. In moths and butterflies, the group for which the chemical knowledge of these signaling molecules is perhaps most complete, most sex pheromones are straight-chain (12 to 16 carbon atoms), even-numbered molecules with an oxygen function (alcohol, aldehyde, or ester) on one terminal carbon and one or more double bonds usually far removed from the oxygenated carbon (Wolf et al., 1981). Species-specificity is achieved by varying the chain length, functional group, and site and stereochemistry of the double bond, as well as by mixing two or more of these compounds in specific ratios.

The chemical structure of a sex pheromone is usually very specific, and a small change in the molecule normally destroys or greatly diminishes the activity (Harborne, 1982). This fact is especially true of the position and stereochemistry (*cis* or *trans*) of the double bond. The effect has been demonstrated in the female sex pheromone of the cabbage looper, *Trichoplusia ni* (Lepidoptera), by Jacobson et al. (1970). The pheromone produced by the female is a C_{12} acetate ester with a *cis*-7 double bond (Figure 9.9). Of the several synthetic analogues tested, none exhibited as much activity as the natural pheromone. A *trans*-7 isomer showed reduced activity, as did

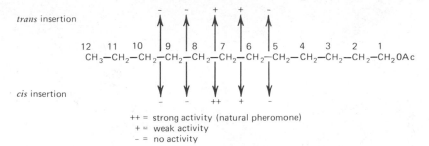

trans insertion

```
         -    -    +    +    -
         ↑    ↑    ↑    ↑    ↑
   12   11   10  │9  │8  │7  │6  │5   4    3    2    1
   CH₃—CH₂—CH₂—CH₂—CH₂—CH₂—CH₂—CH₂—CH₂—CH₂—CH₂—CH₂OAc
                 ↓    ↓    ↓    ↓    ↓
```

cis insertion

```
         -    -   ++   +    -
```

++ = strong activity (natural pheromone)
 + = weak activity
 – = no activity

FIGURE 9.9.   Effect of position of double bond on sexual attraction of C_{12} alcohol acetates to cabbage looper males. With permission from J. B. Harborne, *Introduction to Ecological Biochemistry,* 2nd ed., 1982. Copyright: Academic Press Inc. (London) Ltd.

analogues with the double bond in the six position (*cis* and *trans*). Synthetic compounds with the double bond in the five, eight, or nine position were inactive. Another lepidopteran, the red-banded leaf roller, can distinguish between *cis*-11-tetradecen-1-ol acetate and *trans*-11-tetradecen-1-ol acetate, and in fact is highly sensitive to a precise ratio (Klun et al., 1973).

Pheromones are also intimately involved in regulating the sexual behavior of vertebrates, although with the possible exception of mammals the chemistry of the active components is not well known. There appears to be a link between external and internal communication in many species as often the terpenoid and/or steroid compounds that are employed as sex hormones are also released to the exterior as pheromones (Shorey, 1976). For example, the increased activity and attraction by the male guppy is probably elicited by the release of estrogen from the gonopore of the female guppy (Amouriq, 1965a,b). Garstka and Crews (1981) reported that male garter snakes used their chemosensory vomeronasal system to locate and recognize potential mates by detecting an estrogen-dependent, species-specific pheromone on the skin of females. The pheromone, which is believed to be associated with the circulating yolk lipoprotein vitellogenin, is transported by the female's circulatory system to the dermal vascular bed in the skin. During courtship, hyperventilation by the female moves the adjacent scales apart and stretches the skin in the hinge region, forcing the lipid material to the surface.

In several species of mammals, sexual arousal appears to depend on the release of musk-smelling steroids that are closely related to the male sex hormones androgen and testosterone. The specific compounds have been identified in the boar as 5α-androst-16-en-3-one (see Figure 9.3) and a related 3-ketone. These compounds, which are concentrated in boar urine, saliva, and sweat, induce the female to assume proper position for mating. Females are so sensitive to the compounds that even after male boars are removed from a pen, the remaining odor is sufficient to cause over 80% of the females in estrus to assume a mating stance (Harborne, 1982). The steroids have even been incorporated into an aerosol that is dispensed prior

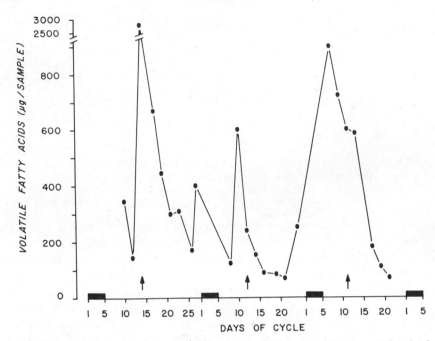

FIGURE 9.10. Volatile fatty acids in samples of vaginal secretions during three menstrual cycles from an individual human female. Consistent declines occur in the acid content during the luteal phases of each cycle. Reprinted, by permission, from Michael et al. (1974). Copyright 1974 by the AAAS.

to artificial insemination to facilitate the injection of sperm (Melrose et al., 1971). The glandular secretions of the civet cat (civetone) and the musk ox (muscone), which also produce a musky odor, have a striking structural resemblance to that of the boar odor compounds.

In addition to steroid and steroid-like pheromones, it has been demonstrated that a series of short-chain, aliphatic acids (acetic, propanoic, methylpropanoic, butanoic, methylbutanoic, and methylpentanoic acid), present in the vaginal secretions of female rhesus monkeys, increase the sexual activity and motivation of male monkeys (Michael et al., 1971). Injection of the steroid estradiol into ovariectomized females increases the concentration of these volatile acids in the vaginal secretions. In a subsequent study, Michael et al. (1974) showed that the same fatty acids are present in vaginal secretions of human females. The proportions of the acids in the rhesus monkey and human female are similar, except for higher levels of acetic and propanoic acids in humans. The amounts of these acids increase near the midpoint of the menstrual cycle in control females (Figure 9.10), whereas in women using contraceptives (birth control pills) the increase is abolished and amounts of the acids are significantly lower. The physiological role of these acids in the sexual behavior of humans has not been established; however, it should be noted that human vaginal secretions possessed male stimulatory properties in cross-taxa experiments with rhesus monkeys.

Alarm and Defense

Many animals release alarm pheromones when a source of potential danger is detected. The response to these alarm pheromones usually involves increased locomotion and often increased aggressiveness, the intensity of which varies from species to species. Specific examples of alarm responses include increased linear and directional (positive and negative) movement, the readying of defensive structures (e.g., mandible, sting), trail laying, recruitment, and attack (Parry and Morgan, 1979). In many cases alarm pheromones are intimately associated with defensive secretions; in fact, in some social and gregarious insects the defensive allomones also function as alarm pheromones. In bees and wasps, alarm pheromones are usually produced in the defensive glands, often in close connection with the main defensive weapons. While the contents from these glands are being discharged during combat onto the aggressor, an alarm is simultaneously communicated to other members of the society by diffusion of pheromonal vapors in the air.

One of the first studies to demonstrate a relationship between alarm and defense function as well as to identify the principal components responsible for invoking the characteristic behavioral responses was performed on the formicine ant *Acanthomyops claviger* (Regnier and Wilson, 1968). When worker ants in the vicinity of the nest are disturbed, they release a mixture of volatile substances that are detected by their nestmates. These individuals respond by simultaneously raising and sweeping their antennae and opening their mandibles, and then moving toward the source of the odor. If the stimulus is sustained, sufficient alarm chemicals are released to mobilize an entire colony.

The chemicals responsible for the alarm response and their origin are shown in Figure 9.11. Two lipid classes are involved: terpenes, which are located in the mandibular gland in the head, and alkanes and ketones produced in the Dufour's gland in the abdomen. All but pentadecane and 2-pentadecanone are effective alarm substances, with undecane (C_{11}) and tridecane (C_{13}) n-alkanes most efficient (see also Figure 9.2). Undecane also functions as a defensive secretion by virtue of its own toxic effects (Gilby and Cox, 1963) and by its function as a spreading agent for formic acid (Figure 9.11), which is the principal venom toxicant in formicine ants. Spreading the venom over the cuticle of an attacker increases the possibility that the toxic components will come in contact with tracheae or cuticular abrasions caused by the ant's biting (Regnier and Wilson, 1968). A similar interaction of formic acid and n-undecane produces alarm, attraction, and aggressive behavior in two other species of formicine ants: *Oecophylla longinoda* (Bradshaw et al., 1975) and *Formica rufa* (Löfqvist, 1976). When *O. longinoda* is attacked by a predator, it releases a mixture of four oxygenated hydrocarbons of differing volatilities that produce a well-orchestrated sequence of alarm behavior, which provides an effective defense of the ant colony (Bradshaw et al., 1979).

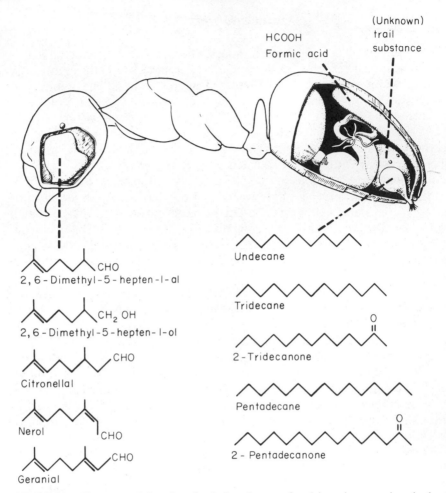

FIGURE 9.11.   The structural formulas of volatile substances found in various exocrine glands of *Acanthomyops claviger*. Reprinted with permission from F. E. Regnier and E. O. Wilson, The alarm-defence system of the ant *Acanthomyops claviger*. *J. Insect Physiol*. **14.** Copyright 1968, Pergamon Press, Ltd.

Whereas various ant species utilize alarm pheromones to alert members of the colony to possible danger, several species of nonsocial arthropods who are likely to encounter predatory ants in nature make use of the same or structurally similar compounds to escape from their formicid adversaries. The use of defensive allomones as cryptic ''alarm pheromones'' has been nicely documented by Blum (1980), whose research has been instrumental in establishing this communicative turnabout. Two groups of arthropods in particular, the opilionids (harvestmen) and mutillids (velvet ants), both of which typically occupy habitats densely populated with ants, manufacture a large variety of compounds identical or closely related to the alarm pheromones of the ants. A major defensive allomone in both groups, 4-methyl-3-heptanone (see Figure 9.3), is also the most widespread alarm pheromone produced by ants (Blum and Hermann, 1978). When opilionids or mutillids are confronted by predatory ants, they discharge this lipid compound, which momentarily disrupts the organized behavior of the ants and provides ample time for the arthropods to escape. Blum (1980) further notes that, because ant alarm pheromones are very nonspecific, it is likely that a broad spectrum of ant species will respond to a compound that is identical or closely related to compounds used by ants as alarm pheromones.

There are a tremendous number and variety of chemicals present in or released by organisms that provide direct defense protection against predation by virtue of their toxic, distasteful, or physical properties. Chemical defense mechanisms have been identified, or at least implicated, in many different plant and animal species, with arthropods as a group again being most thoroughly investigated. Whether the use of chemical substances for defense in the absence of an accompanying alarm reaction constitutes communication is open for debate. The importance of these chemical compounds, especially lipids, for the survival of a species, however, cannot be disputed. The extensive literature reviewed by Eisner (1970) and Pasteels et al. (1983), plus coverage provided in recent books by Bettini (1978) and Blum (1981), suggests that if anything the topic of chemical defense warrants a separate chapter.

The substances used as defensive compounds by organisms are either synthesized *de novo* or obtained from dietary sources. In many cases the chemicals are manufactured and stored in special exocrine glands. Some defense substances are sprayed at a potential enemy, some are ejected with force, and others simply ooze out from glands and spread over the surface of the emitter. Most defensive allomones appear to be effective against a broad spectrum of potential predators.

There appears to be a strong correlation between the chemical nature of defensive compounds and their mode of action. Many defensive compounds act mechanically rather than chemically, that is, they are sticky, slimy or entangling secretions that impair movement or normal function of mouthparts or sensory structures (Pasteels et al., 1983). Chemically, these are often liquid waxes or lipids mixed with proteins or carbohydrates. Examples include the C_{22} through C_{27} hydrocarbons and mucopolysaccharides that comprise the latex of Australian termite soldiers (Moore, 1968), and supercooled waxes that are released from peglike processes on the abdomen of aphids (Edwards, 1966). A second category of quite volatile and rather reactive compounds (e.g., quinones, aliphatic aldehydes, ketones, monoterpenes) exert their action by having an irritating effect on the eyes, other exposed surfaces of the face, and respiratory passages. It is these types of compounds that characterize the elaborate defense systems of over 150 species of tenebrionid beetles (Tschinkel, 1975a). A third group of chemical defense substances contain compounds that are classed as true poisons.

Unlike chemicals in the previous category, which exert a rather immediate, topical effect, true poisons are active in relatively low concentrations and tend to induce a number of systemic effects (i.e., vomiting, irritation of gastrointestinal tract, paralysis), the onset of which may be somewhat delayed (Eisner, 1970). Lipid compounds in this category are typically complex structures that often occur as defensive substances in plants as well. Some examples include the cardiac glycosides in the monarch butterfly and the steroid toxins in dytiscid water beetles (Figure 9.3) and the common toad (*Bufo*).

Of all the elaborate chemical defense systems of animals, perhaps none can match the chemical weaponry and delivery systems present in termites. Prestwich (1983) has recently described how termite soldiers with specially modified heads combine mechanical defense with a variety of lipid secretions

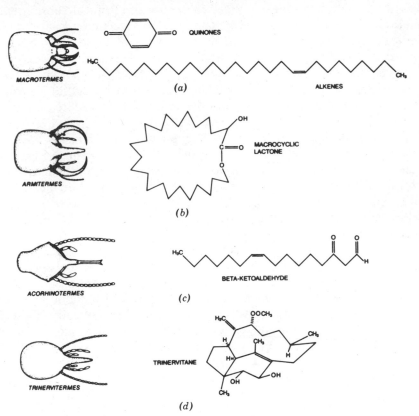

FIGURE 9.12. Modification of the head of termite soldiers and the structure of the corresponding lipid defensive secretion. (*a*) The biting/injecting genus *Macrotermes* with its powerful mandibles. (*b*) The macrolide-secreting genus *Armitermes*. (*c*) The genus *Acorhinotermes* with its enlarged labrum. (*d*) The genus *Trinervitermes* with its snoutlike nasus through which sticky secretions are squirted. Modified from G. D. Prestwich, The chemical defenses of termites. Copyright © 1983 by Scientific American, Inc. All rights reserved.

to subdue potential enemies. Three separate defensive strategies are utilized by termites. Some soldiers, which use enlarged mandibles to bite intruders, simultaneously release an oily substance through a glandular opening in the head and apply it to the punctured cuticle. The substance, which in the genus *Macrotermes* consists of long-chain alkanes, alkenes, or quinones (Figure 9.12*a*), appears to soften the cuticle and to interfere with the natural coagulation of the ant hemolymph. Another biting genus, *Armitermes*, discharges more complex lipids called macrolides; these are modified fatty acids containing 22 to 36 carbon atoms with the two terminal carbons connected by an ester linkage to form a loop (Figure 9.12*b*). A second line of chemical defense involves daubing an intruder with hydrocarbons such as nitroalkenes, vinyl ketone, and highly reactive β-ketoaldehydes. The latter molecule is secreted by soldiers belonging to the genus *Acorhinotermes*. These individuals possess an enlarged labrum, which the termite presses against the intruder to spread the toxin over the cuticle (Figure 9.12*c*). The hydrocarbon portion of the molecule facilitates its penetration through the intruder's cuticle, while the electrophilic group of atoms at one end of the chain causes tissue damage once inside the cuticle. A third method of termite chemical defense involves the production of an irritating, entangling secretion from a snoutlike tube called a nasus. Chemical analysis of this gluelike substance from the genus *Trinervitermes* shows it to be a mixture of diterpenes with a carbon skeleton that does not resemble any other known natural molecular structure (Figure 9.12*d*).

Plant-Animal Interactions

All major categories of chemical messengers described for animals—pheromones, allomones, kairomones, and hormones—have been found in plants. Some of these compounds are identical or very similar to those chemical signals used by animals to govern behavior related to orientation, reproduction, and defense discussed in the previous sections. In fact, many of these compounds are likely obtained by the animal from plants it consumes. Chemicals previously identified as kairomones utilized by parasitoids to locate host species have been found in significant quantities in food plants of the host species. These cues (tricosane, heptanoic acid), which are concentrated and released unaltered by the host insect, attract parasites to areas having high infestations of the host insect and initiate oviposition of the parasite in the host. By serving as the source of these semiochemicals, the plant plays an indirect but important role in determining the ecological balance between insect hosts and their parasites (Hendry et al., 1976).

There are also numerous examples of chemicals present in or released by plants that directly mediate interactions between a plant and an animal, especially plants and insects. Certain chemicals of plant origin may inhibit food intake in phytophagous insects. The chemoreceptive systems of insects detect these plant chemicals, some of which are lipids, and encode this information in the nervous system for future reference (Dethier, 1970). On the other hand, some insects have not only developed resistance against the noxious effects of certain plant defensive compounds, but use them as kairomones to recognize their hosts, stimulate food intake, or stimulate oviposition (Schoonhoven, 1982). Vertebrates, including man, also respond to chemical stimuli present in plant tissues. Apart from nutritional considerations, the selection of plant species as foods by these animals may depend on the presence of an agreeable flavor or odor or the absence of disagreeable toxins.

Although the origin and use of chemicals in this manner is not irrelevant to the concept and definition of chemical communication presented at the beginning of the chapter, I restrict coverage of communicative interactions between plants and animals to situations where a volatile lipid compound or compounds is(are) released by the plant and detected at some distance by the olfactory system of the animal. Two rather classic examples are examined in detail: the chemical interactions between pine trees and bark beetles, and the role of flower scent in insect pollination.

There has been considerable research directed toward understanding the chemical interactions between bark beetles and pine trees because of the damage caused to these commercially important stands by beetle infestations. Many of the semiochemicals produced by both the trees and the beetles have been identified, but not all of their synergistic effects and biological functions are completely understood. The following account pertains to the ponderosa pine, *Pinus ponderosa,* of North America and the Western pine beetle, *Dendroctonus brevicomis,* and is partially based on the lucid summary provided by Harborne (1982). More detailed descriptions of the biological and chemical interactions can be found in papers by Silverstein et al. (1968), Wood and Silverstein (1970), and Wood (1973).

The various stages in the infestation of the pine trees by bark beetles and the structures of the active components produced by the trees and beetles are shown in Figures 9.13 and 9.14, respectively. Female beetles are attracted to the trees by volatile terpenes (myrcene and α-pinene) exuded from the host-tree resins (Stage 1). Soon after arrival, the female bores a hole through the bark and releases a pheromone mixture that attracts males to the site (Stage 2). The mixture consists of three components: myrcene, obtained from the tree and used without structural alteration, and *exo*-brevicomin (*exo*-7-ethyl-5-methyl-6,8-dioxabicyclo [3.2.1] octane) and frontalin (1,5-dimethyl-6,8-dioxabicyclo [3.2.1] octane), the last two compounds apparently synthesized *de novo* by the female. Data from trapping experiments indicate that all three compounds must be present for maximum effectiveness in attracting males. Interchanges between females and arriving males, possibly involving additional pheromones, assure that the two sexes get together in tunnels bored in the bark, mate, and produce a new generation of beetles.

Aggregation pheromones released from newly arrived beetles and by beetles that develop from the eggs continue to attract new individuals to the

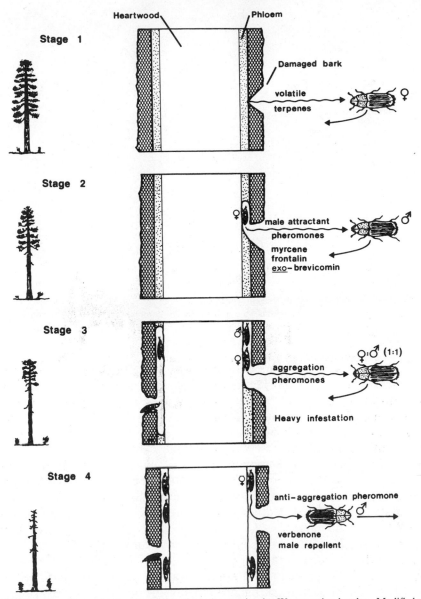

FIGURE 9.13. Stages in the infestation of ponderosa pine by Western pine beetles. Modified with permission from J. B. Harborne. *Introduction to Ecological Biochemistry,* 2nd ed., 1982. Copyright: Academic Press Inc. (London) Ltd.

infestation site (Stage 3). During this time it appears that changing concentrations of pheromones and host-tree volatiles, plus different blends of chemical constituents in the pheromones produced by the females determine the numbers and sex ratios of additional beetles that are recruited. Supporting evidence is based on experiments in which all three of the active chemicals

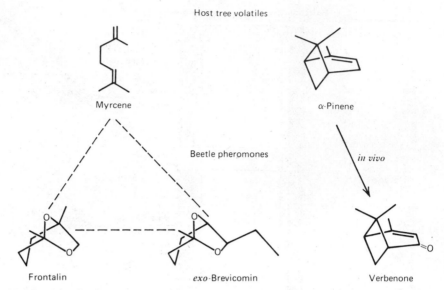

FIGURE 9.14. Semiochemicals involved in pine tree–bark beetle interactions. The pheromone complex released by females to attract males includes the resin-containing compound myrcene. Verbenone, which functions as a male repellent, is synthesized from another host-tree resin compound, α-pinene.

in the male attractant pheromone released by females have been added to the undigested residue deposited at the infestation site. A mixture of myrcene, *exo*-brevicomin, and frontalin attracts male and female bark beetles in a ratio of 1:1; however, when frontalin is omitted from the mixture, the ratio shifts to 2:1 in favor of males.

The beetle population at the infestation site continues to grow until individuals no longer gain reproductively by joining the aggregation. Since additional recruits at this time cannot be sustained by the now limited food supply, they respond to scent cues that indicate that resources in the host tree have been exhausted. Not all of these cues have been firmly identified. It is likely that the amount of terpenes, especially myrcene, present in the resin of the weakened tree is reduced, which in turn diminishes the effectiveness of the aggregating scent. Complementing this action is the switch by females from the production of a male attraction pheromone to the release of a male repellent pheromone called verbenone (Figure 9.14). This ketone, which is synthesized from α-pinene obtained from the host-tree resin, effectively prevents males from reaching the infestation. Thus, in pine bark beetles the natural products of their host trees are used in ways that result in complete colonization of the host without overcrowding.

Another form of chemical communication between plants and animals involves the attraction of pollinators to highly scented flowers. Floral odors are very important in attracting bees, flies, moths, butterflies, and bats to not only brightly colored flowers, but also to flowers that lack color or that are pollinated at night. In fact, floral scents often exhibit temporal patterns in quantity and possibly quality that correspond to the approximate time (i.e., noon, dusk) that pollination occurs in nature. It is possible that non-floral parts of a plant (e.g., leaf surfaces), which also can produce odors, may contribute to the attraction of pollination vectors.

A variety of chemical substances, often in complex mixtures, are responsible for the diverse scents associated with flowers. Frequently identified lipid constituents in these mixtures include *n*-alkanes and *n*-alkenes, monoterpene, sesquiterpene, and phenolic alcohols, aldehydes, ketones, and simple carboxylic acids. The majority of these compounds have molecular weights between 200 and 300, and have low olfactory thresholds; thus, they comply with many of the theoretical constraints postulated for insect pheromones discussed earlier in the chapter. As noted below, the evaporation and transmission characteristics of many of the volatile compounds, which are largely determined by their respective vapor pressures, are quite different. This leads to the formation of spatial and temporal gradients that further compound the complexity of flower-pollinator interactions.

Coevolutionary interactions between flowers and pollinators are extremely complex, and researchers are just beginning to unravel the varied and intricate relationships. One example in which pheromones and flower scents have become interwoven in insect behavior, and for which the chemical composition of the messenger compounds is fairly well known, involves orchids of the genus *Ophrys* and solitary bees (e.g., *Andrena* and *Eucera* sp.). Many of the behavioral and chemical interactions have been worked out by two Swedish biologists, B. Kullenberg and G. Bergström; their findings are summarized in Kullenberg and Bergström (1975) and Bergström (1978).

FIGURE 9.15. Attempted copulation by an *Andrena* sp. male on *Ophrys lutea*. With permission from G. Bergström. Role of volatile chemicals in *Ophrys*-pollinator interactions. In: *Biochemical Aspects of Plant and Animal Coevolution* (J. B. Harborne, Ed.), 1978. Copyright: Academic Press Inc. (London) Ltd.

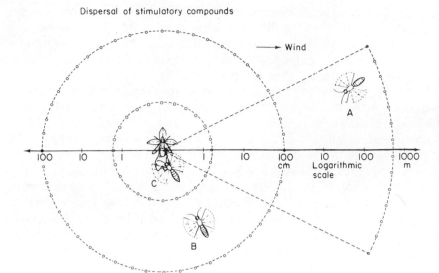

Dispersal of stimulatory compounds

FIGURE 9.16.   Dispersal model for volatile compounds from *Ophrys* orchids. Reprinted, by permission, from Kullenberg and Bergström (1976).

Orchids of the genus *Ophrys* are almost exclusively pollinated by aculeate Hymenoptera. Male bees are attracted to the orchid flower by its shape and color, which resemble the female bee of the species, and a species-specific scent released by the flower. This combination of stimuli attracts the male, who then tries to copulate with the flower labellum (Figure 9.15), pollinating the flower in the process. The volatile chemicals identified in the scent of *Ophrys* include mono-, sesqui-, and diterpenes, mainly hydrocarbons and the derived alcohols, and a variety of fatty acid derivatives—straight-chain hydrocarbons, alcohols, aldehydes, and acetate esters of short- and medium-chain length. The volatile compounds released from the mandibular glands of female *Andrena* bees have also been examined. Many of the same chemicals emitted by the *Ophrys* flowers are also found in the female bees, fueling the hypothesis that the flower scent mimics the sexual odors of the female bee to excite and attract male bees, thus ensuring that the flower is pollinated. Confirmation of this chemical mimicry, however, will require further experimentation.

The multicomponent secretions emanating from *Ophrys* flowers contain substances of high, medium, and low volatility. Each of these compounds will have a different rate of dispersal in air, resulting in different proportions of the chemical blend being present at specific distances from the flower. Kullenberg and Bergström (1976) have used the dispersal characteristics and the resultant concentration gradients in a model to explain how volatility properties of the odor constituents might alter the behavioral phases of approaching male pollinators. Their model includes three regions at different distances from the flower: region A, long distance from the flower, where visual stimulation is not possible and odor compounds are transported by wind; region B, close enough to the flower to permit visual detection, where volatile compounds are transported by a combination of wind and convection; and region C, immediate vicinity of the flower, which can be detected visually and by touch, and where odor can spread by simple diffusion, microcirculation, or chemical contact (Figure 9.16). The authors believe that the highly volatile sesquiterpenes may be active as stimulants in region A, while short-chain fatty acid derivatives and monoterpenes may be active compounds in region B. The least volatile compounds would likely be responsible for releasing the distinct pseudocopulatory behavior at close range. Although there are some field observations under controlled conditions that support the theoretical model, it is difficult to test this potentially interesting relationship between the physicochemical properties of these lipid semiochemicals and the behavioral phases elicited in males because of the rapid flight of these bees.

MECHANICAL COMMUNICATION

In chemical communication, various odors or scents released by plants and animals are perceived by the olfactory senses of animals. These chemical messengers, the majority of which are volatile lipid compounds, provide information in relation to an organism's need for food, reproduction, and protection from predation. They also enable members of the same species to interact with one another in complex social organizations. Similar functions are also provided by mechanical communication, especially in the higher vertebrates including humans. We are concerned here with communication via the production of sounds, specifically echolocation in cetaceans. The role played by lipids in the sonarlike use of auditory signals, although very much reduced in comparison to their all-encompassing function in chemical communication, is still essential for effective signal transmission and receipt in water by these aquatic mammals. Moreover, the mechanisms by which this is achieved depend upon physical properties of lipids quite different from those important in chemical communication and, for that matter, properties that have not been emphasized in previous chapters. Before examining the role of lipids in cetacean communication, let us first review some of the physics and biophysics of sound production and transmission in aquatic environments.

Sound Communication: Some Physical and Biological Considerations

Sound can be defined as a series of self-propagating waves of compression and decompression that have a more-or-less constant and characteristic velocity, depending on the medium (gas, liquid, solid). Because of the greater density of water, sound travels about five times faster in water than it does in air (1,524 m per sec vs. 350 m per sec). Sounds produced in water also do not dissipate as rapidly as in air. As a result, sound waves transmitted underwater can be detected at rather astonishing distances. The bottle-nosed dolphin is reported to communicate with other dolphins over distances of 6 miles, while the finback whale is thought to maintain contact with other whales that may be 500 miles away. In contrast, the maximum transmission of the human voice in air under ideal conditions is limited to less than 1 mile (Lilly, 1978). In addition to traveling great distances at great speeds in water, sound communication can be used at all times and at all depths by these aquatic mammals. Visual communication no doubt complements that of sound, but the visual sense is at best limited because light is restricted to the upper layer of large bodies of water, and even here it is unavailable for approximately one-half of each 24-hour day.

Echolocation: "Seeing" with Sound

Sound communication in several species of dolphins, porpoises, and toothed whales (suborder Odontoceti of the order Cetacea) involves a remarkable form of acoustical function in which the animal emits high-frequency sound pulses that are inaudible to the human ear. These pulses move away from the animal until they encounter an object with a density different from that of water. Upon striking the object, the sounds return in the form of echos, which are received by the emitter through specialized structures and are read and analyzed in higher auditory centers in the brain. This process, known as echolocation or sonar, enables these mammals to determine the distance, direction, speed, shape, texture, density, and even the internal structure of an object. The acoustical ability of cetaceans has been the subject of extensive study and numerous publications. For more detailed information on the experimental recordings and communication potential of this fascinating system, the reader is referred to Norris (1966) and pertinent chapters in Busnel and Fish (1979) and Herman (1980).

Neither the mechanics of sound production nor sound reception are well understood in these animals. Cetaceans have no vocal cords, yet are able to produce complex sounds simultaneously (Ellis, 1982). Several air cavities exist within their bodies that are potential sites for sound production. These include various sacs, plugs, and slits associated with the nasal passage, and the larynx (Figure 9.17). Based on limited experimental evidence, plus conclusions inferred from air passageway morphology, it is believed that the sounds originate from the nasal sac system and that the larynx plays little or no role in this function. As for sound reception, the odontocetes have external ears, but these are quite small and the auditory canal and ear drum of the porpoise (used here interchangeably with dolphin) no longer functions in the hearing process (McCormick et al., 1979). Instead, the porpoise "hears" by detecting reflected sound with its thin, delicate jaw bone. The specific sound receptors are the "acoustic windows" located on each side of the lower mandible (Figure 9.17). Sounds entering the acoustic

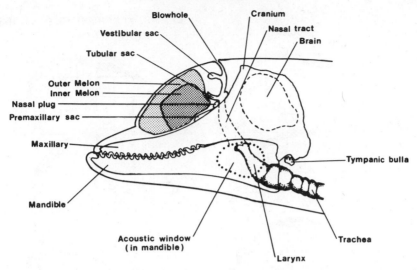

FIGURE 9.17. Schematic drawing of the head of a bottlenose dolphin showing those features associated with sending and receiving sounds. Redrawn, by permission, from Ellis (1982).

window are transmitted to the tympanic bulla, or ear bone, and eventually to the brain (Ellis, 1982).

Lipid Tissues and Sound Processing

Two lipid-rich structures in the head of cetaceans are intimately involved in the transmission and reception of sounds. One of these is the acoustic window found on each side of the rear mandible. The window regions are actually pockets of fat through which reflected external sounds pass to the internal mandibular waveguide. The second of these structures is the melon, a fatty protrusion on the forehead of all odontocetes (Figure 9.17). The melon reaches its greatest degree of specialization in sperm whales (the spermaceti organ), where it is used to regulate the buoyancy of these giant mammals (Chapter 8). Its principal function, however, in most cetaceans, including perhaps the sperm whale, is to give directionality to sound production by acting as an acoustical transducer and sonic lens (Litchfield and Greenberg, 1979). The melon is located directly in front of the sound generator for the echolocation process.

There have been a number of studies of the lipid composition and its distribution in the acoustic tissues of cetaceans (see references in Varanasi et al., 1975; Litchfield et al., 1978). These investigations have shown that the lipid composition differs markedly from that of surrounding tissues (e.g., blubber), and that the types of lipid classes and their molecular composition vary regionally within the melon or mandibular tissue (see Figure 9.18). The two predominant lipid classes are triacylglycerols and wax esters, both rich in short- and medium-chain (C_5 to C_{12}) acids. In the porpoise family Delphinidae, the triacylglycerols in the acoustic window and inner melon

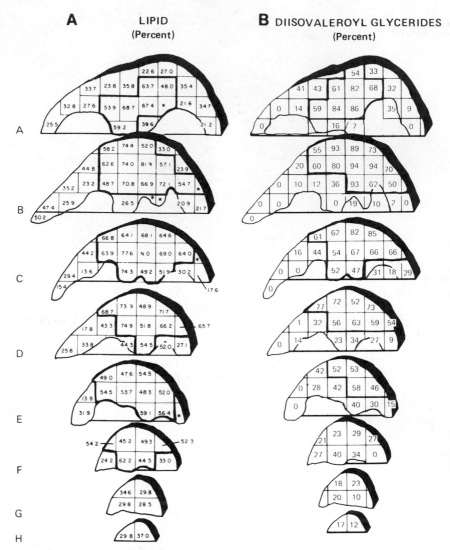

FIGURE 9.18. Compositional topography of lipid content (wt %) and the proportion of diisovaleroylglycerides in serial slices taken from the melon of the porpoise *Delphinus delphis*. The slices A–H are from the blowhole to the rostrum. Areas blocked in by heavy lines represent tissue containing greater than 45% lipid and diisovaleroylglycerides. Reprinted, with permission, from Varanasi et al. (1982).

are composed largely of diisovaleroylglycerides, a triacylglycerol containing two isovaleroyl (= 3-methyl butyric acid) moieties, and a long-chain (C_{10} to C_{22}) acid moiety. In the Amazon River dolphin, *Inia geoffrensis* (Platanistidae), the central inner melon has a much higher wax ester content than the surrounding outer melon, and the molecular weights of both triacylglycerols and wax esters are lower in the inner melon. The melon fat in *Inia*, however, does not contain an appreciable amount of isovaleric acid (Litchfield and Greenberg, 1979).

The discovery that the lipid composition of the acoustic tissues is distinct from adjacent tissues and that certain types and amounts of lipid molecules appear to be preferentially deposited at specific locations within the acoustic tissues strongly suggests that the lipids are intimately associated with cetacean biosonar. This has prompted investigators to examine the physical properties of lipids that are pertinent to echolocation function (e.g., adiabatic compressibility, sound velocity). Varanasi et al. (1975) found that the ultrasonic velocity of porpoise triacylglycerols is much lower (1383 m per sec) than triacylglycerols with three oleic acid moieties (1460 m per sec), and that the velocity decreases as wax esters are added to the former. The authors concluded that the molecular structures of these lipids and their resultant intermolecular associations contributed to the relatively low sound velocity observed in the porpoise melon. In a more recent study, Varanasi et al. (1982) showed that the concentration of isovaleroyl lipids is high in slices taken near the site of sound production (A and B, Figure 9.18), thereby creating a region of low sound velocity and helping channel the sound for efficient emission from regions G and H where the proportion of isovaleroyl lipids is much lower and sound velocity most closely matches that of the sea water. Thus, the three-dimensional lipid matrix acts as a lens to help focus the sound for effective echolocation.

Variations in lipid composition also influence the propagation and refraction of sound rays passing through the melon in the dolphin *I. geoffrensis* (Litchfield and Greenberg, 1979). Analytical tests indicated that the inner melon had the highest levels of lipids, the highest wax ester content, and contained lipids with the lowest average carbon numbers, whereas higher molecular weight lipids were present in the outer melon. These findings have potential functional significance in that for an homologous series of fatty compounds, ultrasonic velocity decreases as the molecular weight becomes smaller (Hustad et al., 1971). Compositional differences between the outer and inner melon should result in sound traveling approximately 5 to 15 m per sec faster in the outer melon. This, in turn, should cause sound rays crossing the inner melon–outer melon boundary to be refracted inward, producing collimation of the sound beam (Litchfield and Greenberg, 1979). Still, the authors feel that in *I. geoffrensis* melon anatomy and forehead shape are probably more important factors in shaping the beam than is melon lipid topography.

CONCLUDING REMARKS

Much of the present chapter has focused on the physical and chemical properties of lipids that make them effective signaling compounds and how organisms utilize lipid messenger molecules to regulate their behavior. The largest body of information here pertains to insect sex pheromones, which are vital to successful mate finding and reproduction. Our knowledge and understanding of insect behavior and physiology has increased greatly during the past two decades. This gain has been accompanied by major advances in chemical technology, which have led to the discovery, identification, and synthesis of many of the semiochemicals of insects. Together, these developments have made possible the practical application of some of these chemical signals in pest-management programs.

A variety of approaches for the use of semiochemicals to modify insect behavior have been either tested or proposed. These include exploiting positive orientation responses of males to the female sex pheromone and then trapping the males to reduce large populations, saturating an environment with an attractant pheromone such that directional orientation is precluded, or releasing more unspecific volatile chemicals, which mask the effect of the pheromone and thus interfere with the signal. In each approach, the strategy is to fool the natural population by creating a false biologic message or condition (Borden, 1977).

There have been at least two instances in which the application of lipid-based semiochemicals has been used successfully to control field populations of insects. Gossyplure, an acetate ester that is the natural sex pheromone of the pink bollworm moth, has been synthesized and used to disrupt the premating communication of adults, resulting in a significant reduction in the numbers of damaging larvae attacking the cotton bolls. Another synthetic material, multilure, a mixture of two female-produced compounds and one compound from the host plant, has been incorporated into sticky traps to help control the spread of the European elm bark beetle in the eastern United States. Research interest in the use of pheromones to monitor and regulate the populations of destructive insects continues to grow. Roelofs (1981) lists 20 species of fruit pests, 15 species of forest pests, 14 species of field pests, and 6 household pests for which attractants are known and field studies for their control have been initiated.

In order for the pheromonal control of insect pests to achieve its maximum potential, there must be continued research directed toward understanding the biology and chemistry of the pheromone-communication system. There must also be integration of this approach with other relatively new management techniques such as the release of sterile males, use of insect growth regulators, production of genetically inferior strains, dissemination of biological control agents (parasites, predators, pathogens), and the implementation of cultural practices that are disadvantageous to insect pests (Koehler et al., 1977). Further experimentation and testing may show that the use of lipid-based semiochemicals may be more important as a means of monitoring the distribution or abundance of pests rather than in directly controlling a pest population. Whatever its eventual role and success, this behavior-modifying tool offers much promise as a very selective and environmentally sound procedure.

REFERENCES

Alberts, B., D. Bray, J. Lewis, M. Raff, K. Roberts, and J. D. Watson (1983). *Molecular Biology of the Cell.* Garland Publishing, New York.

Amouriq, L. (1965a). L'activité et le phénomène social chez *Lebistes reticulatus* (Poeciliidae-Cyprinodontiformes). *Ann. Sci. Nat. Zool. Biol. Animale* 7:151–172.

Amouriq, L. (1965b). Origine de la substance dynamogene emise par *Lebiste reticulatus* femelle (Poisson, Poeciliidae, Cyprinodontiformes). *Compt. Rend.* 260:2344–2335.

Bardach, J. E. and J. H. Todd (1970). Chemical communication in fish. In: *Communication by Chemical Signals* (Johnston, J. W., Jr., D. G. Moulton, and A. Turke, Eds.), pp. 205–240. Appleton-Century-Crofts, New York.

Bedoukian, P. Z. (1970). Purity, identity, and quantification of pheromones. In: *Communication by Chemical Signals* (Johnston, J. W., Jr., D. G. Moulton, and A. Turk, Eds.), pp. 19–34. Appleton-Century-Crofts, New York.

Bergström, G. (1978). Role of volatile chemicals in *Ophrys*-pollinator interactions. In: *Biochemical Aspects of Plant and Animal Coevolution* (Harborne, J. B., Ed.), pp. 207–231. Academic Press, New York.

Bettini, S. (Ed.) (1978). *Arthropod Venoms.* Springer-Verlag, Berlin.

Blomquist, G. J., R. W. Howard, and C. A. McDaniel (1979). Structures of the cuticular hydrocarbons of the termites *Zootermopsis angusticollis* (Hagen). *Insect Biochem.* 9:365–370.

Blum, M. S. (1980). Arthropods and ecomones: Better fitness through ecological chemistry. In: *Animals and Environmental Fitness* (Gilles, R., Ed.), pp. 207–222. Pergamon Press, New York.

Blum, M. S. (1981). *Chemical Defenses of Arthropods.* Academic Press, New York.

Blum, M. S. and J. M. Brand (1972). Social insect pheromones: Their chemistry and function. *Am. Zool.* 12:553–576.

Blum, M. S. and H. R. Hermann (1978). Venoms and venom apparatuses of the Formicidae: Myrmeciinae, Ponerinae, Dorylinae, Pseudomyrmecinae, Myrmicinae, and Formicinae. In: *Arthropod Venoms* (Bettini, S., Ed.), pp. 801–869. Springer-Verlag, Berlin.

Borden, J. H. (1977). Behavioral responses of Coeloptera to pheromones, allomones, and kairomones. In: *Chemical Control of Insect Behavior: Theory and Application* (Shorey, H. H. and J. J. McKelvey, Jr., Eds.), pp. 169–198. Wiley-Interscience, New York.

Bradshaw, J. W. S., R. Baker, and P. E. Howse (1975). Multicomponent alarm pheromones of the weaver ant. *Nature* 258:230–231.

Bradshaw, J. W. S., R. Baker, and P. E. Howse (1979). Multicomponent alarm pheromones in the mandibular glands of major workers of the African weaver ant, *Oecophylla longinoda. Physiol. Entomol.* 4:15–25.

Brand, J. M., M. S. Blum, H. M. Fales, and J. M. Pasteels (1973). The chemistry of the defensive secretion of the beetle *Drusilla canaliculata* (Col. Staphylinidae). *J. Insect Physiol.* 19:369–382.

Brownlee, R. G., R. M. Silverstein, D. Müller-Schwarze, and A. G. Singer (1969). Isolation, identification and function of the chief component of the male tarsal scent in black-tailed deer. *Nature* 221:284–285.

Burghardt, G. M. (1970). Defining "Communication". In: *Communication by Chemical Signals* (Johnston, J. W., Jr., D. G. Moulton, and A. Turk, Eds.), pp. 5–18. Appleton-Century-Crofts, New York.

Busnel, R.-G., and J. F. Fish (Eds.) (1979). *Animal Sonar Systems.* Plenum, New York.

Butenandt, A., R. Beckmann, D. Stamm, and E. Hecker (1959). Über den Sexuallockstoff des Seidenspinners *Bombyx mori* Reindarstellung and Konstitution. *Z. Naturforsch., Teil B.* **14**:283–384.

Croteau, R. (1981). Biosynthesis of monoterpenes. In: *Biosynthesis of Isoprenoid Compounds,* Vol. 1 (Porter, J. W. and S. L. Spurgeon, Eds.), pp. 226–282. Wiley-Interscience, New York.

Daloze, D., J. C. Braekman, and B. Tursch (1980). Chemical communication in the marine environment. In: *Animals and Environmental Fitness* (Gilles, R., Ed.), pp. 243–261. Pergamon Press, New York.

Dethier, V. G. (1970). Chemical interactions between plants and insects. In: *Chemical Ecology* (Sondheimer, E. and J. B. Simeone, Eds.), pp. 83–102. Academic Press, New York.

Dethier, V. G., L. Barton-Browne, and C. N. Smith (1960). The designation of chemicals in terms of the responses they elicit from insects. *J. Econ. Entomol.* **53**:134–136.

Devlin, R. M. (1975). *Plant Physiology,* 3rd ed. Van Nostrand, New York.

Dunham, P. J. (1978). Sex pheromones in Crustacea. *Biol. Rev.* **53**:555–583.

Edwards, J. S. (1966). Defence by smear: Supercooling in the cornicle wax of aphids. *Nature* **211**:73–74.

Eisner, T. (1970). Chemical defense against predation in arthropods. In: *Chemical Ecology* (Sondheimer, E. and J. B. Simeone, Eds.), pp. 157–217. Academic Press, New York.

Eisner, T., J. Meinwald, A. Morro, and R. Ghent (1961). Defence mechanisms of arthropods. I. The composition and function of the spray of the whipscorpion, *Mastigoproctus giganteus* (Lucas) (Arachnida, Pedipalpida). *J. Insect Physiol.* **6**:272–298.

Ellis, R. (1982). *Dolphins and Porpoises.* Knopf, New York.

Frings, H. and M. Frings (1977). Animal Communication, 2nd ed. Univ. Oklahoma Press, Norman.

Garstka, W. R. and D. Crews (1981). Female sex pheromone in the skin and circulation of a garter snake. *Science* **214**:281–683.

Gary, N. E. (1962). Chemical mating attractants in the queen honey bee. *Science* **136**:773–774.

Gilby, A. R. and M. E. Cox (1963). The cuticular lipids of the cockroach, *Periplaneta americana* (L.). *J. Insect Physiol.* **9**:671–681.

Harborne, J. B. (1982). *Introduction to Ecological Biochemistry,* 2nd ed. Academic Press, New York.

Hendry, L. B., J. K. Wichmann, D. M. Hindenlang, K. M. Weaver, and S. H. Korzeniowski (1976). Plants—the origin of kairomones utilized by parasitoids of phytophagous insects? *J. Chem. Ecol.* **2**:271–283.

Herman, L. M. (Ed.) (1980). *Cetacean Behavior: Mechanisms and Functions.* Wiley, New York.

Howard, R. W. and G. J. Blomquist (1982). Chemical ecology and biochemistry of insect hydrocarbons. *Ann. Rev. Entomol.* **27**:149–172.

Howard, R. W., C. A. McDaniel, and G. J. Blomquist (1978). Cuticular hydrocarbons of the eastern subterranean termite, *Reticulitermes flavipes* (Kollar) (Isoptera: Rhinotermitidae). *J. Chem. Ecol.* **4**:233–245.

Howard, R. W., C. A. McDaniel, and G. J. Blomquist (1980). Chemical mimicry as an integrating mechanism: Cuticular hydrocarbons of a termitophile and its host. *Science* **210**:431–433.

Howard, R. W., C. A. McDaniel, D. R. Nelson, G. J. Blomquist, L. T. Gelbaum, and L. H. Zalkow (1982). Cuticular hydrocarbons of *Reticulitermes virginicus* (Banks) and their role as potential species- and caste-recognition cues. *J. Chem. Ecol.* **8**:1227–1239.

Hustad, G. O., T. Richardson, W. C. Winder, and M. P. Dean (1971). Acoustic properties of some lipids. *Chem. Phys. Lipids* 7:61–74.

Jacob, J. and E. von Lehmann (1976). Chemical composition of the nasal gland secretion from the marsh deer *Odocoileus (Dorcelaphus) dichotomus* (Illiger). *Z. Naturforsch.* 31:496–498.

Jacobson, M., N. Green, D. Warthen, C. Harding, and H. H. Toba (1970). Sex pheromones of the Lepidoptera. Structure-activity relationships. In: *Chemicals Controlling Insect Behaviour* (Beroza, M., Ed.), pp. 3–20. Academic Press, New York.

Jaenicke, L. (1977). Sex hormones of brown algae. *Naturwissenschaften* 64:69–75.

Klun, J. A., O. L. Chapman, K. C. Mattes, P. W. Wojtkowski, M. Beroza, and P. E. Sonnet (1973). Insect sex attractants: Minor amounts of opposite geometric isomer critical to attraction. *Science* 181:661.

Koehler, C. S., J. J. McKelvey, Jr., W. L. Roelofs, H. H. Shorey, R. M. Silverstein, and D. L. Wood (1977). Advancing toward operational behavior-modifying chemicals. In: *Chemical Control of Insect Behavior: Theory and Application* (Shorey, H. H. and J. J. McKelvey, Jr., Eds.), pp. 395–400. Wiley-Interscience, New York.

Kullenberg, B. and G. Bergström (1975). Chemical communication between living organisms. *Endeavour* 34:59–66.

Kullenberg, B. and G. Bergström (1976). The pollination of *Ophrys* orchids. *Bot. Notiser* 129:11–19.

Law, J. H. and F. E. Regnier (1971). Pheromones. *Ann. Rev. Biochem.* 40:533–548.

Lilly, J. C. (1978). *Communication Between Man and Dolphin: The Possibility of Talking with Other Species.* Crown, New York.

Litchfield, C. and A. J. Greenberg (1979). Compositional topography of melon lipids in the Amazon River dolphin, *Inia geoffrensis:* Implications for echolocation. *Comp. Biochem. Physiol.* 63A:183–187.

Litchfield, C., A. J. Greenberg, R. G. Ackman, and C. A. Eaton (1978). Distinctive medium chain wax esters, triglycerides, and diacyl glyceryl ethers in the head fats of the Pacific beaked whale, *Berardius bairdi. Lipids* 13:860–866.

Löfqvist, J. (1976). Formic acid and saturated hydrocarbons as alarm pheromones for the ant *Formica rufa. J. Insect Physiol.* 22:1331–1346.

Machlis, L., W. H. Nutting, M. W. Williams, and H. Rapoport (1966). Production, isolation, and characterization of sirenin. *Biochemistry* 5:2147–2159.

McCormick, J. G., E. G. Wever, S. H. Ridgway, and J. Palin (1979). Sound reception in the porpoise as it is related to echolocation. In: *Animal Sonar Systems* (Busnel, R.-G. and J. G. Fish, Eds.), pp. 449–467. Plenum Press, New York.

Melrose, D. R., H. C. B. Reed, and R. L. S. Patterson (1971). Androgen steroids associated with boar odour as an aid to the detection of oestrus in pig artificial insemination. *Br. Vet. J.* 127:497–501.

Michael, R. P., E. B. Keverne, and R. W. Bonsall (1971). Pheromones: Isolation of male sex attractants from a female primate. *Science* 172:964–966.

Michael, R. P., R. W. Bonsall, and P. Warner (1974). Human vaginal secretions: Volatile fatty acid content. *Science* 186:1217–1219.

Moore, B. P. (1968). Studies on the chemical composition and function of the cephalic gland secretion in Australian termites. *J. Insect Physiol.* 14:33–39.

Müller, D. G. and G. Gassmann (1978). Identification of the sex attractant in the marine brown alga *Fucus vesiculosus. Naturwissenschaften.* 65:389.

Mykytowycz, R. (1970). The role of skin glands in mammalian communication. In: *Communication by Chemical Signals* (Johnston, J. W., Jr., D. G. Moulton, and A. Turk, Eds.), pp. 327–360. Appleton-Century-Crofts, New York.

Needham, A. E. (1965). *The Uniqueness of Biological Materials.* Pergamon Press, London.

Nordlund, D. A. (1981). Semiochemicals: A review of the terminology. In: *Semiochemicals: Their Role in Pest Control* (Nordlund, D. A., R. L. Jones, and W. J. Lewis, Eds.), pp. 13–28. Wiley-Interscience, New York.

Nordlund, D. A., R. L. Jones, and W. J. Lewis (Eds.) (1981). *Semiochemicals: Their Role in Pest Control.* Wiley-Interscience, New York.

Norris, K. S. (1966). *Whales, Dolphins, and Porpoises.* Univ. Calif. Press, Berkeley.

Oldak, P. D. (1976). Comparison of the scent gland secretion lipids of twenty-five snakes: Implications for biochemical systematics. *Copeia* **1976**:320–326.

Parry, K. and E. D. Morgan (1979). Pheromones of ants: A review. *Physiol. Entomol.* **4**:161–189.

Pasteels, J. M., J.-C. Gregoire, and M. Rowell-Rahier (1983). The chemical ecology of defense in arthropods. *Ann. Rev. Entomol.* **28**:263–289.

Prestwich, G. D. (1983). The chemical defenses of termites. *Sci. Am.* **249**:78–87.

Regnier, F. E. and E. O. Wilson (1968). The alarm-defence system of the ant *Acanthomyops claviger. J. Insect Physiol.* **14**:955–970.

Ritter, F. J. (Ed.) (1979). *Chemical Ecology: Odour Communication in Animals.* Elsevier, Amsterdam.

Robinson, T. (1980). *The Organic Constituents of Higher Plants,* 4th ed. Cordus Press, North Amherst, Massachusetts.

Roelofs, W. L. (1981). Attractive and aggregating pheromones. In: *Semiochemicals: Their Role in Pest Control* (Nordlund, D. A., R. L. Jones, and W. J. Lewis, Eds.), pp. 215–235. Wiley-Interscience, New York.

Schoonhoven, L. M. (1982). Biological aspects of antifeedants. *Ent. Exp. Appl.* **31**:57–69.

Shorey, H. H. (1976). *Animal Communication by Pheromones.* Academic Press, New York.

Silverstein, R. M. (1977). Complexity, diversity, and specificity of behavior-modifying chemicals: Examples mainly from Coleoptera and Hymenoptera. In: *Chemical Control of Insect Behavior* (Shorey, H. H. and J. J. McKelvey, Jr., Eds.), pp. 231–251. Wiley-Interscience, New York.

Silverstein, R. M., J. O. Rodin, and D. L. Wood (1966). Sex attractants in frass from bark beetles. *Science* **156**:105.

Silverstein, R. M., J. O. Rodin, W. E. Burkholder, and J. E. Gorman (1967). Sex attractant of the black carpet beetle. *Science* **157**:85–86.

Silverstein, R. M., R. G. Brownlee, T. E. Bellas, D. L. Wood, and L. E. Browne (1968). Brevicomin: Principal sex attractant in the frass of the female Western pine beetle. *Science* **159**:889–890.

Sleeper, H. L. and W. Fenical (1977). Navenones A-C: Trail-breaking alarm pheromones from the marine opisthobranch *Navanax inermis. J. Am. Chem. Soc.* **99**:2367–2368.

Strand, F. L. (1983). *Physiology: A Regulatory Systems Approach,* 2nd ed. Macmillan, New York.

Stuart, A. M. (1970). The role of chemicals in termite communication. In: *Communication by Chemical Signals* (Johnston, J. W., Jr., D. G. Moulton, and A. Turk, Eds.), pp. 79–106. Appleton-Century-Crofts, New York.

Tamaki, Y. (1977). Complexity, diversity, and specificity of behavior-modifying chemicals in Lepidoptera and Diptera. In: *Chemical Control of Insect Behavior* (Shorey, H. H. and J. J. McKelvey, Jr., Eds.), pp. 253–285. Wiley-Interscience, New York.

Tschinkel, W. R. (1975a). A comparative study of the chemical defensive system of tenebrionid beetles: Chemistry of the secretions. *J. Insect Physiol.* **21**:753–783.

Tschinkel, W. R. (1975b). Unusual occurrence of aldehydes and ketones in the defensive secretion of the tenebrionid beetle, *Eleodes beameri. J. Insect Physiol.* **21**:659–671.

Tumlinson, J. H., R. M. Silverstein, J. C. Moser, R. G. Brownlee, and J. M. Ruth (1971). Identification of the trail pheromone of a leaf-cutting ant, *Atta texana. Nature* **234**:348–349.

Varanasi, U., H. R. Feldman, and D. C. Malins (1975). Molecular basis for formation of lipid sound lens in echolocating cetaceans. *Nature* **255**:340–343.

Varanasi, U., D. Markey, and D. C. Malins (1982). Role of isovaleroyl lipids in channeling of sound in the porpoise melon. *Chem. Phys. Lipids.* **31**:237–244.

Wilson, E. O. (1963). Pheromones. *Sci. Am.* **208**:100–114.

Wilson, E. O. (1970). Chemical communication within animal species. In: *Chemical Ecology* (Sondheimer, E. and J. B. Simeone, Eds.), pp. 133–155. Academic Press, New York.

Wilson, E. O. and W. H. Bossert (1963). Chemical communication among animals. *Recent Prog. Horm. Res.* **19**:673–716.

Wolf, W. A., L. B. Bjostad, and W. L. Roelofs (1981). Correlation of fatty acid and pheromone component structures in sex pheromone glands of ten lepidopteran species. *Environ. Entomol.* **10**:943–946.

Wood, D. L. (1973). Selection and colonization of ponderosa pine by bark beetles. In: *Insect-Plant Relationships* (Van Emden, H., Ed.), pp. 101–118. Blackwells, Oxford.

Wood, D. L. and R. M. Silverstein (1970). Bark beetle pheromones. *Nature* **225**:557–558.

Yushima, T., Y. Tamaki, S. Kamano, and M. Oyama (1974). Field evaluation of a synthetic sex pheromone, "litlure," as an attractant for males of *Spodoptera litura* (F.) (Lepidoptera: Noctuidae). *Appl. Entomol. Zool.* **9**:147–152.

10

LUBRICATION, DETERGENCY, AND OTHER INTERFACIAL PHENOMENA

Interfaces may be defined as organized molecular layers that are found at the junction between two media (Needham, 1965). In biological systems, lipid interfaces are often present at air-water junctions as well as at boundaries between aqueous solutions and immiscible organic liquids. Such lipids are typically organized into a monolayer in which the molecules are parallel and oriented normal to the surface. Moreover, these lipids are amphiphilic molecules; that is, they have both hydrophilic (water-soluble) and hydrophobic (water-insoluble) regions. At a gas-liquid interface (e.g., air-water), the lipid is oriented so that the polar or hydrophilic portion of the molecule is in the liquid phase and the nonpolar or hydrophobic portion is in the gas phase (see Chapter 1, Figure 1.1). At an interface between fat globules and the surrounding aqueous phase, the hydrophobic portion of the molecule is dissolved in the fat and the hydrophilic portion dissolved in water. As shown in Chapter 3, the lipid component of biological membranes exhibits a similar organization, except that the lipid moiety in biomembranes occurs in bilayers rather than as a monolayer.

LUBRICATION

An important property of lipids relevant to the formation and function of organized layers is lubrication. Naturally occurring fats and oils have a long history of use as lubricants. The large stones used to build the pyramids were moved on a film of olive oil. In addition, deposits of fats have been found on the axle of a chariot dating back to 1400 B.C. (Friedrich, 1979). The hydrophilic portions of fats and oils form firm bonds with an object, while the hydrophobic hydrocarbon chains slide readily on one another. Solid lipid compounds such as paraffins and long-chain wax esters are also good lubricants even though these are not amphiphilic compounds, and thus do not typically form organized monolayers. Their effectiveness in reducing friction between two moving surfaces is due largely to the fact that layers formed by these compounds shear readily on each other because the molecules are held together only by weak van der Waals forces.

The functional significance of lipid-based lubricants in biological systems is less well documented. Although the presence of lipid constituents in synovial fluid may contribute to the ease with which one bone moves in relation to another at a joint, the principal lubricant in this fluid is hyaluronic acid. Similarly, spherical droplets formed during the emulsification of fats in digestion should enhance the movement of materials along the digestive tract; however, the compound primarily responsible for lubricating the tract is mucin, a protein secreted by gland cells that mixes with water to form mucus. There is stronger evidence (to be discussed) that the lipid film present on the inside of alveolar walls facilitates the sliding of alveoli past one another during expansion and compression of the lungs.

DETERGENCY

A closely related phenomenon to lubrication is that of detergency. Detergents are compounds that tend to accumulate in excess at surfaces and interfaces when mixed in solutions. Here they act as bridges between dissimilar substances (e.g., oil-water, air-water), wetting and dispersing oily substances and stabilizing emulsions and foams (Clements, 1962b). Classic examples of detergents are of course soaps, which have a long nonpolar hydrocarbon chain and a polar salt group at one end (see Chapter 1). Molecules of this sort are able to remove grease or dirt from a surface because the hydrocarbon chain associates with the nonpolar oily material, forcing it to the center of the aggregate, while the polar salt group at the other end associates with the aqueous solvent on the periphery of the aggregate. Once surfaces are coated with detergent, water can flow easily between them and rinse off the soil and dirt particles.

All detergents are amphiphiles. When placed in an aqueous solution, they respond to the hydrophobic effect by migrating to the surface, where they form oriented monolayers that exhibit a weak intermolecular cohesion characteristic of hydrocarbon chains (Tanford, 1980). Because these detergent molecules are attracted less strongly to the molecules of the liquid than are the liquid molecules attracted to one another, they dilute the molecules of the liquid at the surface. This in turn results in a dramatic lowering of the surface tension. Such substances are called surface-active substances, or surfactants. Although detergents and surfactants are often used interchangeably in describing this surface tension lowering behavior, detergents do differ from surfactants in that they are entirely soluble in the liquid phase, and thus do not change the surface tension as surface area changes (Slonim and Hamilton, 1981).

Lipid surface-active agents have widespread and often critical roles in the normal physiological functioning of organisms. Two examples have been selected for detailed examination. The first is a lipid-protein complex referred to as the pulmonary surfactant, which forms an insoluble film at the air-liquid interface in mammalian lungs. The primary function of this interfacial material is to lower the surface tension and thereby reduce the muscular effort necessary to ventilate the lungs and keep them aerated. The second example involves surface-active bile salts, which accumulate at the interface between fat globules and the surrounding aqueous phase. This accumulation reduces the surface tension of the fat globules, enabling them to be broken into smaller globules. The emulsification process greatly increases the area of fatty surface on which pancreatic lipase can operate.

PULMONARY SURFACTANT

Surface Effects in the Lung

The pressure-volume behavior of lungs is strongly influenced by the surface tension on the inner surface of alveoli. Surface tension is the force (dynes per cm) acting across the surface of a liquid. In the lungs it arises because the attraction between the molecules of the fluid lining the alveoli is stronger than forces between the fluid and the air inside the alveoli. Since the molecules of the fluid are more strongly attracted to their neighbors below the surface than they are to the sparser population of molecules in the air above the surface, the net pull is downward, and the surface tends to contract to the smallest possible surface area (West, 1979). Surface tension can be measured with a Maxwell frame or with a surface balance (Figure 10.1A,B). In the former, a film of liquid stretched out on a U-shaped frame tends to pull the cross wire to the bottom of the U. The force necessary to resist this pull is a measure of surface tension (Clements, 1962b). With a surface balance, the material to be tested is placed in a shallow tray. A platinum strip is then suspended in the fluid and attached to a sensitive strain gauge or force transducer. The pull on the thin strip provides a measure of surface tension. A movable barrier enables the investigator to alternately expand and compress the surface area of the film while continuously measuring surface tension.

That surface tension is an important factor in producing the elastic recoil of the lung was first demonstrated by von Neergaard (1929), who showed

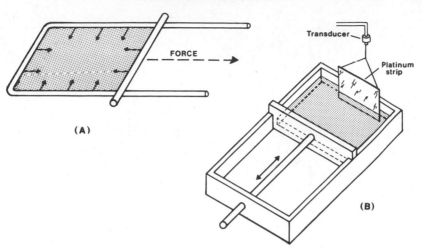

FIGURE 10.1. Instruments for measuring surface tension. (*A*) Maxwell frame. (*B*) Surface balance. Principles of operation are described in the text.

that a lung completely filled with and immersed in water had an elastance (= pressure required to change lung volume) that was less than the normal value obtained when the lung was filled with air. Since saline abolished the surface tension forces at the alveolar air/fluid interface, but presumably did not affect the tissue forces of the lung, he concluded that surface tension contributed a large part of the static recoil force of the lungs. Many years passed, however, before Pattle (1955), who studied the tiny air bubbles present in edema foam coming from the lungs of animals exposed to lethal gas, and Clements (1957), who studied saline extracts from normal lungs, proposed that a highly surface-active material present in the alveolar tissue is responsible for the lower than predicted surface tension values observed for the lung.

To test and quantify the effects of the pulmonary surfactant on surface tension, researchers have isolated this material from lung washings and spread it on a surface balance tray containing saline. By using the movable barrier to alternately increase and decrease the surface area of the fluid, the investigators are able to mimic the expansion and compression of the alveolar lining that occurs during inspiration and expiration, respectively. A typical tracing is shown in Figure 10.2 for the surface tension/surface area relationship for pulmonary surfactant as well as surface tension values recorded for water and a nonspecific detergent. When the film is expanded, surface tension values rise to 40 to 50 dynes per cm, but they drop nearly to zero when the film is compressed. In comparison, the surface tension of pure water exerts a pull of about 72 dynes per cm. When a detergent is added to the water, the surface tension falls to about 30 dynes per cm, but it does not change in response to changing the surface area of the fluid.

These experiments clearly demonstrate that the pulmonary surfactant is

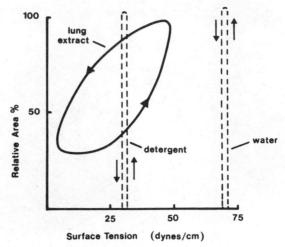

FIGURE 10.2. Plots of surface tension and area obtained with a surface balance. Redrawn, by permission, from Clements (1962a).

capable of lowering surface tension of aqueous solutions and that tension varies with size of the surface. The hysteresis loop recorded for the area-surface tension relationship exhibited by the alveolar extract (Figure 10.2) is presumably caused by the change in orientation of surfactant molecules as their concentration changes (Mingins and Taylor, 1973); however, the physiological significance of this is not fully understood. In some cases the observed hysteresis in force-area curves may be an artifact due to contamination of the surfactant film with spreading solvent, loss of surfactant from the film by leakage or chemical denaturation, or changes in the contact angle of the platinum strip of the surface balance. Barrow and Hills (1983) found little if any hysteresis when individual lipid components of the surfactant were studied on a surface balance under simulated physiological conditions (e.g., 37°C, 100% RH), but that hysteresis reappears when mixtures of these components are tested.

Origin and Chemical Composition of Surfactant

The exchange of gases in the lungs takes place between alveolar air and venous blood flowing through lung capillaries. The alveolar sacs are the last generation of air passages in the lungs. Structurally, they appear as numerous blind pouches that project off each alveolar duct. Approximately 300 million alveoli are estimated to be present in the human lungs (Nunn, 1977). The air within the alveoli is separated from the blood by a very thin respiratory membrane formed by the alveolar epithelium and its basement membrane, and the endothelium of the capillary and its basement membrane (Figures 10.3 and 10.4). The pulmonary surfactant covers the alveolar ep-

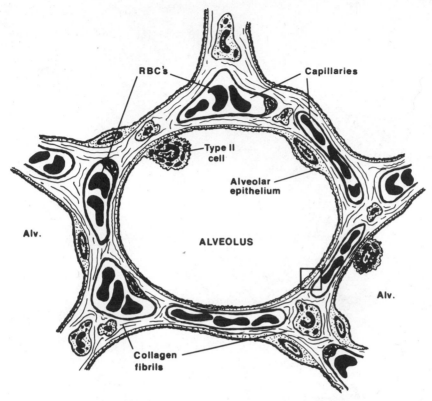

RBC's

Capillaries

Type II cell

Alveolar epithelium

Alv.

ALVEOLUS

Alv.

Collagen fibrils

FIGURE 10.3. Schematic drawing of alveolar septa of inflated dog lung as seen with the electron microscope. The capillary network is interwoven in a continuous network of collagen fibrils. Higher magnification micrograph of portion of respiratory membrane outlined in rectangular box appears in Figure 10.4.

ithelium, forming a thin film interface with the alveolar air (Figure 10.4).

The pulmonary surfactant appears to be synthesized in Type II alveolar epithelial cells, which are typically located at the junction of alveolar septa. These cells have large nuclei and numerous microvilli on the surface. Type II cells are also characterized by the presence of intracellular membrane-bound vesicles, called inclusion bodies, which contain lamellar osmiophilic material (see Figure 10.3). These vesicles are thought to be the subcellular organelles involved in the metabolism and/or secretion of the pulmonary surfactant (King, 1974). Presumably these cells fuse with the plasma membrane of the Type II cell, and the material contained in the lamellar bodies is extruded into the alveolar space, where it spreads on the surface as an organized monolayer. The evidence in support of these events is based on studies with the electron microscope showing material being discharged from the inclusion bodies, plus the fact that early in the life of the fetus the lungs have no alveolar cells and no extractable surfactant; when one appears so does the other (Comroe, 1974).

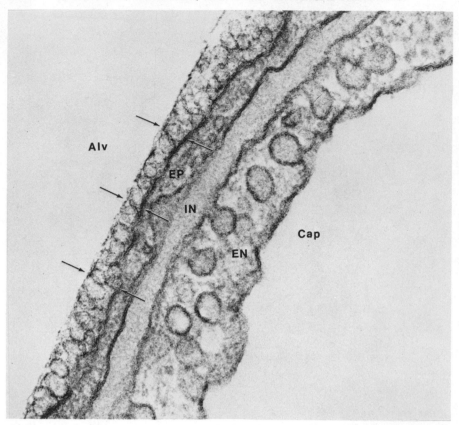

FIGURE 10.4. Electron micrograph of respiratory membrane showing surfactant (between arrows) lining inner surface of alveolus. Alv, alveolus; Cap, capillary; EN, endothelium; EP, epithelium; IN, interstitial space. × 120,000. Reprinted, by permission, from Weibel (1973).

The composition of pulmonary surfactant has been determined for several mammalian species. Although the exact chemical nature of this material is not known, analyses invariably show that the substance is a complex mixture of lipid and protein, with the predominant component being dipalmitoyl phosphatidylcholine (50 to 60% by weight). This phospholipid molecule, also known as lecithin (see Chapter 1), consists of palmitic acid (16:0) esterified to both positions one and two. Other constituents include monoenoic phosphatidylcholine (PC) (20 to 25%), phosphatidylglycerol (5 to 10%), cholesterol (5 to 10%), and protein (5 to 6%) (King and MacBeth, 1981). An early study claimed the presence of polysaccharides, which on hydrolysis yielded glucose, galactose, galactosamine, and other carbohydrate residues (Scarpelli et al., 1967). Carbohydrates, however, are not typically recognized as significant surfactant components and, if present, are not equated with the surface-active properties of this fluid.

The preponderance of evidence indicates that it is the high concentration of dipalmitoyl PC that is responsible for the unique surface-active forces exhibited in the lung. Of the constituents found in this material, only dipalmitoyl PC is capable of lowering the surface tension of an air-liquid interface to less than 10 dynes per cm. The saturated fatty acid moieties present in the molecule can be compressed to molecular areas that are nearly twice those calculated for the molecular surface area of each individual fatty acid chain (King, 1974). This close proximity of hydrocarbon units enhances the stability of the monolayer film by increasing the inter- and intramolecular attractive forces. Electrostatic interactions between positive and negative groups associated with the polar portion of the molecule, which is anchored in the aqueous phase, also contribute to the stability of the fluid lining. In addition to reducing surface tension, dipalmitoyl PC also meets the prerequisite of being able to maintain the observed low tensions for periods sufficient to prevent alveolar collapse (King, 1974).

Experimental data based on the results of dietary studies also support the contention that dipalmitoyl PC is primarily responsible for lowering alveolar surface tension. Burnell et al. (1978) fed male rats fat-free diets for 14 weeks to induce essential fatty acid (EFA) deficiency. The EFA-deficient rats, in addition to having reduced amounts of linoleic and arachidonic acids, contained significantly less palmitic acid (61.4 vs. 77.4%) than controls; there was no change in the total amount of phospholipids in lung tissue and lung lavage fluid, nor was there a reduction in the quantity of lecithins as a proportion of total phospholipids. When the lung surfactant from experimental rats was tested with a surface balance, it exhibited a significant increase in the minimal surface tension as compared to surfactant obtained from control rats. Within seven days of adding safflower oil to the diet of the experimental rats, the fatty acid composition and the surface activity of the lung lavage lipids approached normal values. The authors concluded that the impairment of surface activity observed in the EFA-deficient rats is due to the reduced content of palmitic acid in the lecithins present in their pulmonary surfactant.

The functional roles for the other surfactant components are not as well defined, nor have they been as extensively investigated. The presence of monoenoic PC, in which one of the fatty acid moieties contains a double bond, and cholesterol may facilitate the adsorption of the surfactant to the alveolar surface (King and Clements, 1972). This function is probably linked to the physical state of the surfactant. At physiological temperatures (37°C), dipalmitoyl PC is in a solid gel state. The addition of monoenoic PC and cholesterol lowers the phase transition temperature so that the surfactant is now in a liquid-crystalline state. This more fluid state may be necessary if the surfactant is to adsorb rapidly to the interface. The physiological functions of phosphatidylglycerol, which occurs in relatively high proportions compared with that found in other mammalian lipoprotein systems, and specific proteins associated with the surfactant have not been established. King and MacBeth (1981) suggest that both the protein and phosphatidylglycerol may be essential to the organization and binding of this complex interfacial material.

Physiological Importance of Pulmonary Surfactant

We have seen that the surface tension of the alveolar lining fluid, which might otherwise be comparable to values for the surface tension of water (ca. 72 dynes per cm), is reduced to a mean tension of 20 to 25 dynes per cm by the interruption of the fluid by the surfactant molecules. As the alveolar surface area becomes smaller during expiration, the number of surfactant molecules per unit area increases. This causes a decrease in the surface tension within the alveoli. With inflation and an increased alveolar surface area, the surfactant concentration is reduced and the surface tension of the alveoli is correspondingly increased. Thus, surface tension is least at the beginning of inflation and greatest at the beginning of deflation. The combined effect is one of increased lung compliance, that is, an increase in the ease by which the lungs are distended. The low surface tension at the beginning of inflation minimizes the amount of energy that must be expended to inflate the lungs, making the work associated with breathing easier. Sanderson et al. (1976), who have likened the surfactant's function to that of an "anti-glue" or abhesive, comment that it is the presence of low surface tension that makes the evolution of alveoli having fragile walls possible.

The presence of surfactant also helps stabilize the alveoli. Laplace's law states that pressure within a bubble (i.e., an alveolus) is directly proportional to the surface tension and inversely proportional to the radius. This relationship is described by the equation:

$$P = \frac{2T}{r}$$

where P is pressure (dynes per cm^2), T is surface tension (dynes per cm), and r is radius (cm) (Slonim and Hamilton, 1981). As surface tension increases or the radius decreases, more pressure will be required to maintain the volume of the bubble. Therefore, if surface tensions are the same in all alveoli, regardless of their size, there will be a tendency for smaller alveoli, with their smaller radii and hence greater pressure, to empty into interconnecting larger alveoli. This would lead to hyperinflation of the larger alveoli, while at the same time causing collapse of the smaller alveoli. The presence of pulmonary surfactant, however, causes the alveolar surface tension to decrease more rapidly than the alveolar radius. This enables small alveoli to remain inflated and to enlarge using less pressure than is required by large alveoli. As a result, a stable balance is reached between small and large alveoli, and the collapse of smaller alveoli is prevented.

The presence of sufficient pulmonary surfactant is also necessary to prevent fluid accumulation in the alveoli. Just as surface tension forces tend to collapse alveoli, they also tend to draw fluid from the pulmonary capillaries and interstitial spaces into the alveoli. Normally this fluid-drawing effect is countered by the significantly higher colloidal osmotic pressure exerted by the plasma proteins. However, if pulmonary surfactant is absent or its activity is diminished, the surface tension in the alveoli could increase to the point that the pressure gradient across the respiratory membrane is reversed and transudation into the alveoli occurs. Indeed, alveoli filled with transudate is one of the pathophysiological features associated with surfactant deficiency.

Surfactant Deficiency and Respiratory Distress Syndrome

A number of clinical disorders are attributed to either an absence or insufficient quantity of pulmonary surfactant. One of these disorders, the respiratory distress syndrome (RDS), or hyaline membrane disease, is responsible for more infant deaths than any other disease. In view of its medical importance, it is not surprising that much of our knowledge of the chemical and biophysical characteristics of pulmonary surfactant has been generated by researchers attempting to understand and prevent this disease.

Before birth, the alveoli and the air-conducting pathways are filled with liquid, part of which is amniotic fluid inhaled during respiratory movements *in utero*. At birth the respiratory passages must become filled with air and the collapsed alveoli must expand to function in gas exchange. The success of this transition depends largely on surface forces generated by pulmonary surfactant. The first breath requires considerable respiratory effort in order to overcome the viscous resistance created by the fluid. This effort is essentially the same for newborn babies with and without adequate surfactant. Because of the higher surface tension in the alveoli of surfactant-deficient babies, however, the alveoli will collapse to almost their original uninflated state during expiration (Tortora and Anagnostakos, 1981). Moreover, the high surface tension of the alveolar lining fluid increases the forces that favor the movement of fluid from the capillaries into the alveoli. As the transudate fills the alveoli, the radius of curvature of the gas-liquid interface shrinks and surface tension becomes even greater, drawing more fluid into the alveoli (Slonim and Hamilton, 1981). Affected infants continue to exhibit difficult and labored breathing and, unless oxygen-rich air is provided to the baby to keep the alveoli open, death often occurs in a few hours. Microscopic examination of lung tissues during autopsy typically shows a glassy, pale blue membrane lining the alveoli. It is this protein-rich material that gives rise to the older terminology "hyaline membrane disease."

It is essential that pulmonary surfactant be synthesized and "readied" for function in advance of the initial ventilatory movements of a newborn. Analysis of fetal lung tissue and fluids shows that this surface-active material

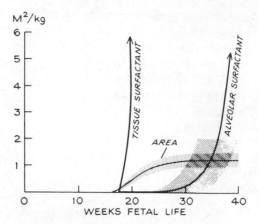

FIGURE 10.5. Time course of surfactant appearance in fetal lung. Large amounts of surfactant can be obtained from minced lung tissue before it can be detected on the alveolar surface. The area band shows the potential area of the lungs if maximally expanded. When the other band showing the amount of alveolar surfactant exceeds this, there is enough surfactant to cover all alveolar surfaces. Width of bands indicates variability about mean values. Courtesy of Dr. John Clements. Modified with permission from J. H. Comroe: *Physiology of Respiration,* 2nd ed. Copyright © 1974 by Year Book Medical Publishers, Inc., Chicago.

is present in the lungs as early as the 18th to 20th week of gestation, but that the surfactant does not move onto alveolar surfaces until about the 30th week (Comroe, 1974; Figure 10.5). These observations account for the high frequency of respiratory distress syndrome exhibited by infants born prematurely. The factors in fetal physiology that control the timing of surfactant synthesis and its appearance are not completely understood. Certainly the development of Type II alveolar cells is critical, as is an adequate flow of pulmonary blood during fetal development. There is some evidence that cell maturation and surfactant production may be controlled by steroid hormones, especially those produced by the adrenal cortex (Hitchcock, 1980). Because some of the fetal lung fluid ultimately diffuses into the amniotic cavity, it is possible to test for the presence of surfactant in the amniotic fluid near the time of birth. If no surfactant, or perhaps too little, is detected, the physician may seek ways to delay parturition, or at least anticipate respiratory difficulties upon delivery.

Pulmonary Surfactant in Nonmammalian Vertebrates

Information on the occurrence and function of surfactant in the lungs of vertebrates other than mammals is presently limited to a few scattered taxa. There have been some studies of either lung ultrastructure or its pulmonary surface properties in a given species, but seldom have investigators addressed both aspects in the same species. There is a tendency for the lungs of endothermic vertebrates to have a well-developed, stable surfactant that is capable of lowering surface tension well below that of ordinary liquids, whereas in lower vertebrates the surface-active properties of this material are less well developed and the surfactant is present in smaller quantities. Chemically, the composition of the surfactant appears to be similar through-out the vertebrate groups, except that the lung linings of birds and mammals appears to contain more fully saturated lecithin, a feature that is consonant with the surfactant's greater capacity to lower surface tension at air-liquid interfaces in these groups (Pattle, 1976). There are also differences in the ultrastructure of the lamellated osmiophilic bodies, which are believed to be the site of surfactant synthesis and storage, but the functional significance of these differences is presently unknown.

A closer examination of individual vertebrate classes reveals that the surfactant layer may serve several functions whose relative importance may vary in different groups, and that physical factors in the environment may strongly influence the behavior of the surfactant, and hence its functional effectiveness. The air capillaries in the lungs of birds are strengthened by connective tissue, creating a more rigid structure than is characteristic of the mammalian lung. With the added support, there would appear to be little need for a surface-active lining in the bird lung; however, a lining similar in location and composition to that of mammals is apparently present in the chicken, goose, and turkey (see references in Hylka and Doneen, 1982). It is believed that the pulmonary surfactant in the avian lung promotes the distension of narrow air capillaries, and that its presence may be crucial during the hatching process. Amphibians, with their large diameter alveolar units, also would not appear to require the surface-active forces provided by a surfactant, but ultrastructural studies plus chemical analyses of lung washes indicate the presence of an alveolar film lining in the frog *Rana pipiens* (Hughes and Vergara, 1978; Vergara and Hughes, 1980). These authors suggest that the surfactant may be important during the early stages of lung inflation, especially in small frogs, and that it may also function as an "anti-glue" to facilitate the sliding of alveoli over one another as they open and close. In the tortoise *Testudo hermanni* (Reptilia), the physical properties of the surfactant are dependent on temperature. Meban (1980) found that a decrease in temperature from 34 to 19°C produced a twofold increase in the static pressure and a twentyfold increase in the viscosity of the surfactant. If the same responses occur *in vivo,* the decreased efficiency of the film at higher temperatures would necessitate a greater expenditure of energy in breathing, for it would be more difficult for the tortoise to expand its terminal air passages. These findings would seem to have po-tentially important implications for lizards and snakes active in hot, dry desert environments where ambient temperatures approach 50°C.

The presence of a surfactant in the swimbladder of fish and in the lungs of Dipnoi (lungfish) and certain primitive bony fish is of special interest because of the supposed evolutionary relationship between these structures and the lungs of higher vertebrates. There is fairly strong evidence for the role of surfactant in maintaining alveolar patency and stability in air-breathing fish who use their lungs in gas exchange, although some disagreement exists as to the surpellic qualities of this material (Pattle, 1976; Phleger and Saunders, 1978). A surfactant with relatively poor surface-active properties has been identified in the swimbladder of several fish, but its principal function in these species is probably to dissolve oxygen gas to help maintain neutral buoyancy (Chapter 8). Other suggested functions for the surfactant-like material include protection against oxygen poisoning because of the hyperbaric oxygen pressures found at ocean depths and to prevent the leakage of water into the swimbladder. While neither of these latter functions can be discounted, there is also no evidence presently available to support their validity.

BILE SALTS

Chemistry and Physical Properties

Bile is a complex mixture of organic and inorganic components. The most abundant bile constituents are the bile acids. These are steroid compounds that contain a branched side chain of five carbon atoms that ends in a carboxyl group (Figure 10.6). In most mammals, the two primary bile acids are cholic and chenodeoxycholic acid. Before they are secreted by the liver they are conjugated with the nitrogenous bases taurine or glycine. Conjugation makes the bile acids more water-soluble by adding a negative charge. The conjugates have much lower pKs than bile acids and exist as dissociated salts at pHs found in the biliary tract and duodenum (Weisbrodt, 1977). Only primary bile acids are actually synthesized in the liver from cholesterol. The chief secondary bile acids, deoxycholic and lithocholic acids, are products of bacterial metabolism of primary bile acids in the digestive tract. These secondary acids are absorbed in the terminal ileum along with the primary bile acids, removed from the blood by liver hepatocytes, conjugated with taurine or glycine, and secreted in the bile along with newly synthesized and recycled primary acids. In fishes and amphibians, bile sterols occur as sulphate esters. Haslewood (1965) regards these evolutionarily as "primitive" types of bile salts.

Structurally, bile salts and their analogues in lower vertebrates, the alcohol sulfates, have several unique features that account for their behavior in solution. Like lecithin, the major lipid in pulmonary surfactant, bile salts are surface-active, amphiphilic compounds, and hence have a tendency to orient in relation to water molecules. The bulk of the bile salt molecule consists of the hydrophobic sterane nucleus and methyl groupings; the hydroxyl and dissociated carboxyl (or sulfonate) groups, which represent the hydrophilic moieties of the molecule, are located on the same side of the molecule. The interaction of the polar groups with water is strong enough

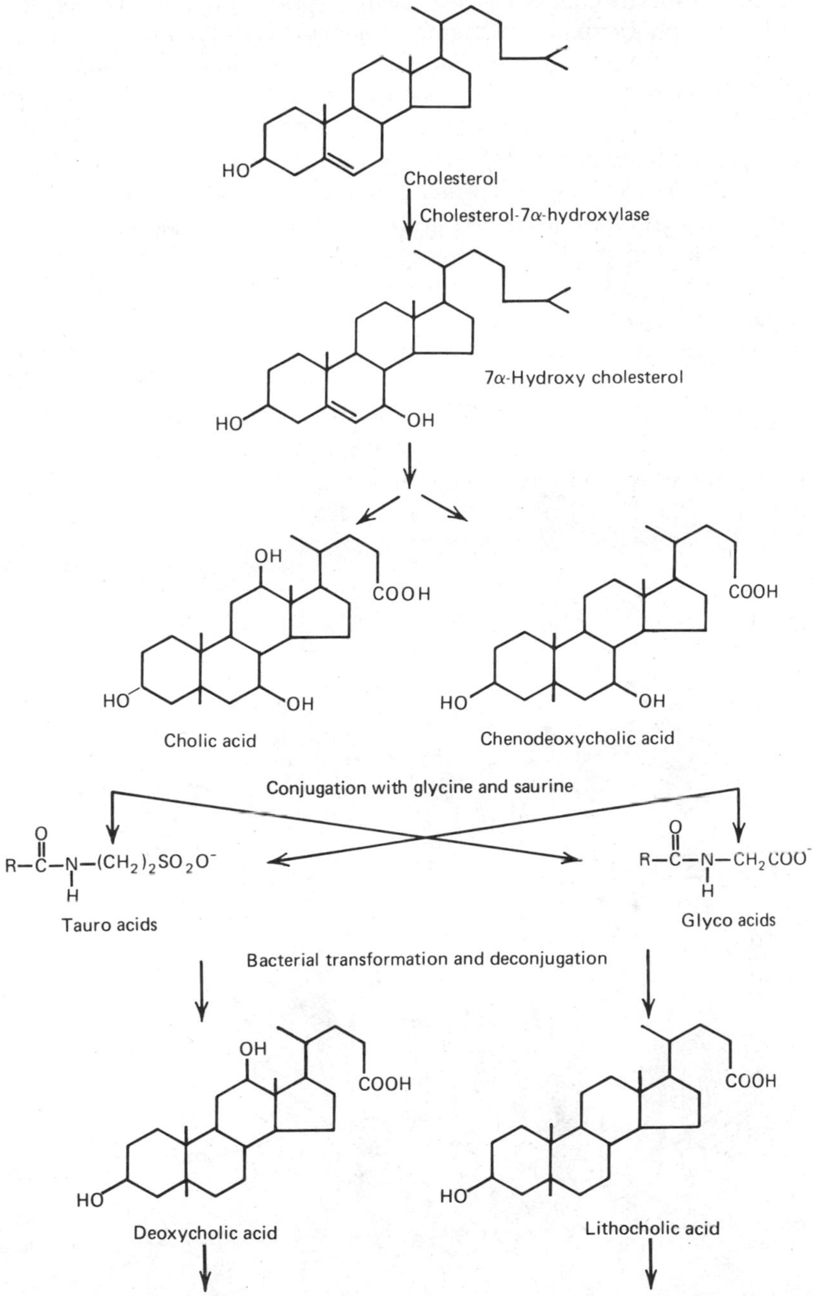

FIGURE 10.6. Transformation of cholesterol to bile acids. Modified by permission, from Simons and Gibson (1980).

303

relative to the hydrophobic portion to render the entire molecule soluble in water. Lecithin, in contrast, with its hydrophilic phosphorylcholine head but two hydrophobic paraffin chains, is not sufficiently polar to exist as free molecules in solution in more than trace concentrations (Simmonds, 1974).

The behavior of bile salts in solution depends on their concentration. In dilute solution, they exist as unassociated molecules, forming oriented monolayers with whatever interfaces are present. For example, bile salts will interact with a triacylglycerol-water mixture by coating the oil-water interface with a hydrophilic layer. At higher concentrations, however, bile salts spontaneously form spherical aggregates, usually of a single size, called micelles. These particles are organized so that the polar groups of the bile salts face out toward the water and the nonpolar portion faces the interior. Adjacent bile salt molecules in a micelle also appear to orient with the nonpolar sides of the sterol nuclei back-to-back (Simmonds, 1974) (Figure 10.7). The point at which micelles are formed is termed the critical micellar concentration. The transformation from an ordinary (monomolecular) solution to a micellar solution also requires that the solution be above a certain temperature, the Krafft point. Below the Krafft point, bile salts, despite a

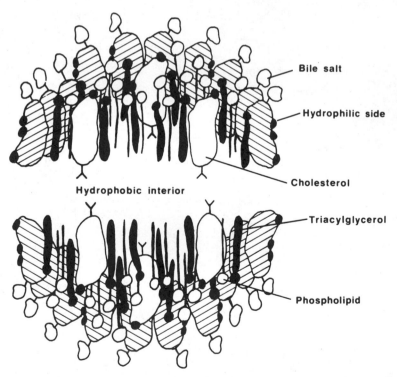

FIGURE 10.7. Diagrammatic representation of proposed organization of a mixed bile salt micelle. The solubilization of nonpolar lipids occurs in the hydrophobic interior of the aggregate. Modified from Castro (1977).

high concentration, will exist as a suspension of crystals. Heating the solution beyond the Krafft point produces a clear micellar solution (Hargreaves, 1968).

Micelles comprised solely of conjugated bile salts are termed simple micelles. In the intestine of mammals, it is more common for some of the products of lipid digestion (e.g., monoacylglycerols, lysolecithin) and other major organic constituents of bile (e.g., phospholipids) to be incorporated into the shell to form a mixed micelle (Figure 10.7). Monoacylglycerols and phospholipids such as lecithin are amphiphilic molecules; they are insoluble in water but form liquid crystals that swell in solution, hence the category "swelling amphiphiles." In the presence of bile salts, these crystals are broken up and dissolved as a component of the micelles. A mixed micelle may contain two moles of phospholipid for each mole of bile salts (Weisbrodt, 1977). As we shall see in the next section, mixed micelles are superior to simple micelles in their emulsifying activity. Furthermore, the incorporation of swelling amphiphiles within the bile salt aggregate greatly increases the solubilization of nonswelling amphiphiles such as cholesterol, itself a bile component, and nonionized long-chain fatty acids during the digestive process. Nonpolar lipids or bulky, nonswelling amphiphiles such as triacylglycerol are also solubilized to some extent in mixed micelles.

Lipid Digestion: Emulsification and Solubilization

The basic processes in the digestion and absorption of lipids were briefly outlined in Chapter 2. Triacylglycerols, which represent the largest fraction of ingested lipid, along with some phospholipids, cholesterol, and plant sterols are delivered essentially unchanged to the duodenum of the small intestine. Here they are prepared for enzymatic hydrolysis by a complex of pancreatic lipases. The predominant pancreatic enzyme, lipase B, converts triacylglycerols into free fatty acids, 2-monoacylglycerols, and some glycerol. A second pancreatic lipase acts on cholesterol esters, converting these to cholesterol, while pancreatic phospholipase hydrolyzes the two position of lecithin, yielding free fatty acid and lysolecithin. Both the cholesterol esterase and phospholipase require bile salts for activation.

Because these dietary lipids are primarily insoluble triacylglycerols, they occur as globules floating in the watery chyme of the intestinal lumen. The pancreatic lipases are secreted in large excess by the pancreas, so the activity of these enzymes depends on the substrate concentration or, in this case, the size and resulting surface area of the globules. The dispersion of large fat globules to smaller uniformly distributed particles, that is, emulsification, is one of the major functions of bile salts. By coating the oil-water interface with a hydrophilic layer, bile salts lower the interfacial tension, permitting shearing forces to form smaller droplets. Before emulsification, fat globules have an average diameter of about 1000 Å; they are reduced to only about 50 Å after bile salt emulsification (Sernka and Jacobson, 1983). The droplets are also negatively charged due to the bile acid anions. This creates a repulsion between droplets and delays their coalescence into larger droplets (Simmonds, 1974).

A second important function of bile salts that complements events occurring during emulsification and that is essential for good absorption of lipid from the intestine is solubilization. This is the process by which endproducts of lipid digestion are incorporated and "dissolved" in the central hydrophobic core of the bile salt micelle. As noted earlier, the chemical composition of the micelle is altered during digestion. Before digestion begins, the micelle contains bile salts plus some endogenous biliary phospholipid and cholesterol. The mixed micelle solubilizes the polar lipids released from dietary triacylglycerols and phospholipids during digestion. With the addition of these components to the micelle particle, the micelle can now more readily dissolve dietary cholesterol and fat-soluble vitamins. The redistribution of lipids brought about by solubilization helps ensure that lipolysis continues at an optimal rate. Without solubilization, the more polar lipid digestive products would tend to displace triacylglycerols from the interface, thereby reducing the availability of substrate and slowing the hydrolytic activity of the lipase enzymes. It is also possible that the accumulation of polar lipids might reduce lipase activity toward the remaining substrate by making the surface of the fat globule more hydrophilic (Simmonds, 1974). By the cooperative action of micellar bile salts and swelling amphiphiles, a mechanism is provided for guaranteeing the rapid removal of digestive end-products from the interface, thus maintaining both a large and hydrophilic surface for maximum enzyme absorption.

Lipid Absorption

In addition to their functions as emulsifying and solubilizing agents, micellar bile salts also play an important role in the absorption of lipolytic products. Current evidence strongly supports the entry of lipids into absorptive cells of the small intestine by passive (lipid-soluble) transport. The micelle is important here because it serves as a vehicle for the diffusion of the primary hydrolytic products—monoacylglycerols, free fatty acids, lysolecithin—to the site of absorption. The region through which the micelle must diffuse is the unstirred water layer overlying the glycocalyx (Figure 10.8). This relatively undisturbed layer between the lumen and the brush border (microvilli) of the intestinal walls persists regardless of how vigorously the bulk of liquid in the digestive tract may be mixed. Because the micelle is a stable, charged particle in the aqueous medium, it can slowly diffuse through this unstirred layer. In contrast, lipid contents of the micelle, although having smaller molecular sizes, are virtually insoluble and unstable in the unstirred layer (Sernka and Jacobson, 1983).

Micellar transport across the unstirred layer is only one of the ways that micelle particles promote lipid absorption. The solubilization of hydrolytic

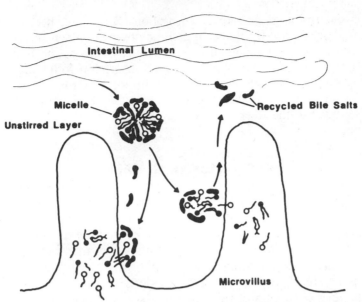

FIGURE 10.8. Schematic drawing showing micellar diffusion in the unstirred layer and the ultimate absorption of solubilized lipolytic products into a microvillus. Redrawn, by permission, from Sernka and Jacobson (1983). © 1983, the Williams & Wilkins Co., Baltimore.

products by bile salt micelles is equally important, for this increases the lipid concentration in the aqueous phase, which in turn provides the driving force for uptake. This relationship has been experimentally demonstrated for long-chain fatty acids, which are absorbed more rapidly into everted sacs of intestine when micellar bile salts are present than when bile salts are absent (Dawson and Isselbacher, 1960; Johnston and Borgström, 1964). Such fatty acids have a low concentration in monomolecular solution, but are readily solubilizcd by mixed bile salt micelles. Simmonds (1974) estimated that micellar solubilization of an 18-carbon fatty acid would increase the driving force for diffusion about 1000 times, while resistance due to the greater sizc of the micellar aggregates compared with single molecules in solution increases only about threefold.

Evidence from ultrastructural studies, plus data on the transport of radiolabeled lipids strongly suggest that micelles are not absorbed across the cell membranes of the microvilli as intact aggregates. Instead, it appears the lipid molecules contained in the micelle are released very near the microvillus membrane, and that the molecules then cross the membrane passively. The rate at which they enter the intestinal epithelial cells is determined in part by their solubility in the luminal fluid. Monoacylglycerols are more soluble than long-chain fatty acids, which in turn are more soluble than cholesterol. Consequently, the free fatty acids and cholesterol remain in the lumen longer than do monoacylglycerols. The permeability of the cell membrane to selective lipid molecules is of course another important factor in determining the rate of absorption. Bile salts, because of their electrical charges, are insoluble in cell membranes. When the contents of the micelle have been absorbed, the bile salts return to the lumen where they re-form new micelles. When all lipid absorption following a meal is complete, the bile salts are then actively transported across the ileum (most absorption of lipolytic products occurs across the duodenum and jejunum) and travel back to the liver via the enterohepatic circulation.

CONCLUDING REMARKS

Although their role as surface agents is perhaps not as spectacular or as well popularized as some of the other functions discussed for lipids, physiologically it is of critical importance and illustrates well the functional diversity of these organic compounds. Lipid participation in interfacial phenomena also highlights the relationship between lipid structure and function, which, like adaptive significance, has been an underlying theme in each chapter.

At the very beginning of this book it was noted that the insolubility of lipids in water was central to a simple operational definition of lipids. The chapters that have followed show that the low degree of polarity inherent in the definition of lipids is indeed the key to their biological function. We have seen how hydrocarbons, wax esters, and triacylglycerols, which have little if any affinity for water, are particularly effective in situations where an interaction with water is either unnecessary or undesirable (e.g., water-proofing, fuel storage, buoyancy). A strict adherence to nonpolarity would, however, severely limit lipid's potential as a biological material, especially in view of the abundance and importance of water in living organisms. The bridge between the two extremes of the polarity spectrum is provided by amphiphilic lipids such as phospholipid, glycolipids, and cholesterol derivatives. Although these compounds are still insoluble in water, they are capable of interacting with hydrophilic substances because of polar moieties in their chemical structures. Thus, we find amphiphilic molecules serving as biomembrane constituents and as surface active agents at air-water boundaries and at the interface between nonpolar lipids and water. This complementary relationship between the amphiphilic compounds and the more hydrophobic members of the class greatly broadens the functions of lipids in biological systems and is a major contributing factor to their uniqueness as a biological material.

REFERENCES

Barrow, R. E. and B. A. Hills (1983). Properties of four lung surfactants and their mixtures under physiological conditions. *Resp. Physiol.* **51**:79–93.

Burnell, J. M., E. C. Kyriakides, R. H. Edmonds, and J. A. Balint (1978). The relationship of fatty acid composition and surface activity of lung extracts. *Resp. Physiol.* **32**:195–206.

Castro, G. A. (1977). Digestion and absorption of specific nutrients. In: *Gastrointestinal Physiology* (Johnson, L. R., Ed.), pp. 122–138. C. V. Mosby, St. Louis.

Clements, J. A. (1957). Surface tension of lung extracts. *Proc. Soc. Exp. Biol. Med.* **95**:170–172.

Clements, J. A. (1962a). Surface phenomena in relation to pulmonary function. *Physiologist* **5**:12–28.

Clements, J. A. (1962b). Surface tension in the lungs. *Sci. Am.* **207**:120–130.

Comroe, J. H. (1974). *Physiology of Respiration,* 2nd ed. Year Book Medical Publishers, Chicago.

Dawson, A. M. and K. J. Isselbacher (1960). Studies on lipid metabolism in the small intestine with observations on the role of bile salts. *J. Clin. Invest.* **39**:730–740.

Friedrich, J. P. (1979). Lubricants. In: *Fatty Acids* (Pryde, E. H., Ed.), pp. 591–607. Amer. Oil Chem. Soc., Champaign, Illinois.

Hargreaves, T. (1968). *The Liver and Bile Metabolism.* Appleton-Century-Crofts, New York.

Haslewood, G. A. D. (1965). Comparative biochemistry of bile salts. In: *The Biliary System* (Taylor, W., Ed.), pp. 107–115. F. A. Davis Co., Philadelphia.

Hitchcock, K. R. (1980). Lung development and the pulmonary surfactant system: Hormonal influences. *Anat. Rec.* **198**:13–14.

Hughes, G. M. and G. A. Vergara (1978). Static pressure-volume curves for the lung of the frog (*Rana pipiens*). *J. Exp. Biol.* **76**:149–165.

Hylka, V. W. and B. A. Doneen (1982). Lung phospholipids in the embryonic and immature chicken: Changes in lipid composition and biosynthesis during maturation of the surfactant system. *J. Exp. Zool.* **220**:71–80.

Johnston, J. M. and B. Borgström (1964). The intestinal absorption and metabolism of micellar solutions of lipids. *Biochim. Biophys. Acta* **84**:412–423.

King, R. J. (1974). The surfactant system of the lung. *Fed. Proc.* **33**:2238–2247.

King, R. J. and J. A. Clements (1972). Surface active materials from dog lung. II. Composition and physiological correlations. *Am. J. Physiol.* **223**:715–726.

King, R. J. and M. C. MacBeth (1981). Interaction of the lipid and protein components of pulmonary surfactant. Role of phosphatidylglycerol and calcium. *Biochim. Biophys. Acta* **647**:159–168.

Meban, C. (1980). Physical properties of surfactant from the lungs of the tortoise *Testudo hermanni. Comp. Biochem. Physiol.* **67A**:253–257.

Mingins, J. and J. A. G. Taylor (1973). Physicochemical properties of phospholipid monomolecular layers. *Proc. Roy. Soc. Med.* **66**:383–385.

Needham, A. E. (1965). *The Uniqueness of Biological Materials.* Pergamon Press, New York.

Nunn, J. F. (1977). *Applied Respiratory Physiology,* 2nd ed. Butterworths, London.

Pattle, R. E. (1955). Properties, function and origin of the alveolar lining layer. *Nature* **175**:1125–1126.

Pattle, R. E. (1976). The lung surfactant in the evolutionary tree. In: *Respiration of Amphibious Vertebrates* (Hughes, G. M., Ed.), pp. 233–255. Academic Press, New York.

Phleger, C. F. and B. S. Saunders (1978). Swim-bladder surfactants of Amazon air-breathing fishes. *Can. J. Zool.* **56**:946–952.

Sanderson, R. J., G. W. Paul, A. E. Vatter, and G. F. Filley (1976). Morphological and physical basis for lung surfactant action. *Resp. Physiol.* **27**:379–392.

Scarpelli, E. M., B. C. Clutario, and F. A. Taylor (1967). Preliminary identification of the lung surfactant system. *J. Appl. Physiol.* **23**:880–886.

Sernka, T. J. and E. D. Jacobson (1983). *Gastrointestinal Physiology: The Essentials,* 2nd ed. Williams and Wilkins, Baltimore.

Simmonds, W. J. (1974). Absorption of lipids. In: *Gastrointestinal Physiology* (Jacobson, E. D. and L. L. Shanbour, Eds.), pp. 343–376. Univ. Park Press, Baltimore.

Simons, L. A. and J. C. Gibson (1980). *Lipids: A Clinician's Guide.* Univ. Park Press, Baltimore.

Slonim, N. B. and L. H. Hamilton (1981). *Respiratory Physiology,* 4th ed. C. V. Mosby Co., New York.

Tanford, C. (1980). *The Hydrophobic Effect: Formation of Micelles and Biological Membranes,* 2nd ed. Wiley-Interscience, New York.

Tortora, G. J. and N. P. Anagnostakos (1981). *Principles of Anatomy and Physiology,* 3rd ed. Canfield Press, San Francisco.

Vergara, G. A. and G. M. Hughes (1980). Phospholipids in washings from the lungs of the frog (*Rana pipiens*). *J. Comp. Physiol.* **139**:117–120.

von Neergaard, K. (1929). Neue Auffassungen über einen Grundbefriff der Atemmechanik. Die Retraktionskraft der Lunge, abhängig von der Oberflächenspannung in den Alveolen. *Z. Gesamte Exp. Med.* **66**:373–394.

Weibel, E. R. (1973). Morphological basis of alveolar-capillary gas exchange. *Physiol Rev.* **53**:419–495.

Weisbrodt, N. W. (1977). Bile production, secretion, and storage. In: *Gastrointestinal Physiology* (Johnson, L. R., Ed.), pp. 84–94. C. V. Mosby, St. Louis.

West, J. B. (1979). *Respiratory Physiology—The Essentials,* 2nd ed. Williams and Wilkens, Baltimore.

INDEX